Seaweeds and their Uses

V. J. Chapman
O.B.E., M.A., Ph.D. (Cantab.), F.L.S.
Emeritus Professor of Botany, Auckland University

with chapters by
D. J. Chapman
B. Sc., Ph. D. (Cal.), F.L.S.
Professor of Biology, University of California, Los Angeles

THIRD EDITION

1980
London   New York
Chapman and Hall
150th Anniversary

First published 1950
Second edition 1970
by Methuen and Co. Ltd.
Third edition 1980
by Chapman and Hall Ltd.,
11 New Fetter Lane, London EC4P 4EE

Published in the USA by
Chapman and Hall
in association with Methuen, Inc.
733 Third Avenue, New York, NY 10017

© *1980 V. J. Chapman and D. J. Chapman*

Typeset in Great Britain by
C. Josée Utteridge-Faivre

Printed in the United States of America

ISBN 0 412 15740 3

All rights reserved. No part of
this book may be reprinted, or reproduced
or utilized in any form or by any electronic,
mechanical or other means, now known or hereafter
invented, including photocopying and recording,
or in any information storage and retrieval
system, without permission in writing
from the publisher.

British Library Cataloguing in Publication Data
Chapman, Valentine Jackson
   Seaweeds and their uses. – 3rd ed.
   1. Marine algae – Economic aspects
   I. Title II. Chapman, D.J.
   589'.39'2   QH390   79-42819

   ISBN 0-412-15740-3

Contents

	Preface to the First Edition	page vii
	Preface to the Third Edition	x
1	Occurrence, Distribution and Historical Perspective by D.J. Chapman	1
2	Seaweed as Animal Fodder, Manure and for Energy	30
3	Sea Vegetables (Algae as Food for Man)	62
4	Laver or 'Nori' Industry and Carragheen or Irish Moss	98
5	Agar-agar	148
6	Algin and Alginates	194
7	Minor Uses of Algae and their Products	226
8	Mariculture of Seaweeds by D.J. Chapman	241
9	Looking for Seaweeds — the World's Supplies	253
	Bibliography	279
	Author Index	313
	Plant Index	320
	Subject Index	327

SEA WRACK

by Moira O'Neill

The wrack was dark an' shiny where it floated in the sea,
There was no one in the brown boat but only him an' me;
Him to cut the sea wrack, me to mind the boat,
An' not a word between us the hours we were afloat.

> *The wet wrack,*
> *The sea wrack,*
> *The wrack was strong to cut.*

We laid it on the grey rocks to wither in the sun,
An' what should call my lad then, to sail from Cushendeen?
With a low moon, a full tide, a swell upon the deep,
Him to sail the old boat, me to fall asleep.

> *The dry wrack,*
> *The sea wrack,*
> *The wrack was dead so soon.*

There' a fire low upon the rocks to burn the wrack to kelp,
There' a boat gone down upon the Moyle, an' sorra' one to help!
Him beneath the salt sea, me upon the shore,
By sunlight or moonlight we'll lift the wrack no more.

> *The dark wrack,*
> *The sea wrack,*
> *The wrack may drift ashore.*

Preface to the First Edition

The 1939–45 war forced the Allied countries to seek alternative sources of raw materials and, as in the First World War, attention was paid by all belligerents to the marine algae or seaweeds. These occur in considerable quantities in various parts of the world, and attempts to make use of this cheap and readily accessible, though not so readily harvestable, raw material have been made almost from time immemorial. Much of the work on the economic utilization of seaweeds has been published only in scientific journals and has never been collected within the compass of a single book. Tressler's work on *The Marine Products of Commerce* contains three useful chapters on this subject, whilst Sauvageau's book, *Les utilisations des Algues Marines,* is a mine of valuable information, especially as regards the use of seaweeds in France. Both these volumes are, however, somewhat out of date, Tressler's being published in 1923 and Sauvageau's in 1920. Furthermore there is no book wholly on this subject in the English language, and so the present volume has been undertaken in order to fill this gap. The opportunity has also been taken to incorporate the results of researches carried out since 1920. In certain aspects of the subject it will be found that considerable advances have been made, and in the present volume particular reference to such advances will be found in the chapters on agar and alginic acid.

This book has been written not only for those with technical knowledge but also for the general reader. It is hoped that sufficient information has been introduced to make it of value to the former class of person, whilst not burdening it unduly to make it unreadable to the second category. For the benefit of those who want to study any aspect in more detail the bibliography has been made as complete as possible.

The subject matter refers solely to the algae, and therefore the economic uses of the marine phanerogams have not been considered. The reason for this decision is that, compared with the algae, the marine phanerogams do not occupy such an important position.

When consideration has been given to all the different methods throughout the centuries whereby mankind has endeavoured to make use of the seaweeds, and a survey has been made of the sorry sequence of failures and abortive efforts, the question 'Why have the industries failed?' comes foremost. The supply of raw material is vast and inexhaustible, and yet there is really no flourishing industry other than in Japan. There are, I think, various reasons that provide an answer to this question, and in answering it, one may also tentatively suggest the requisite conditions for the establishment of an economically successful industry. The recent formation of the Scottish Seaweed Research Association (Anon., 1944) may be regarded as one step in the right direction.

The following causes appear to have contributed generally towards the failure of past enterprises in Europe and America. The primary cause has been the restriction of the enterprise to one aspect only, e.g. production of iodine, production of a manure or of a cattle feed, or production of alginic acid compounds. Alternatively, attempts to make use of by-products were unsuccessful because the chemical processes employed led to uneconomic results. In Europe, another important general contributory cause has been the failure, so far, to evolve a cheap means of harvesting. It is the cheapness of the labour in Japan that has enabled the industry in that country to flourish. Cheap labour will never be obtainable in Europe and therefore the development of efficient mechanical harvesting is a first essential.

It is perhaps pertinent to attempt some analysis of the individual industries. Thus the kelp industry in Europe failed because it could not compete with iodine and sodium salts manufactured from other sources. The answer is comparatively simple: the methods of production were inefficient and the chemical processes produced by-products that were not pure enough. Far too often a seaweed industry has been erected upon the success of laboratory experiments and trials, and then subsequently it has been found that the laboratory methods, when applied on an industrial scale, do not yield products of the necessary commercial purity. Such failure has spelt the death of the company before it has become established.

The Pacific coast kelp industry failed because it was conceived as a war-time industry when finance was immaterial. Under peace conditions the methods of production were far too costly. The agar industry gravitated naturally to Japan because the type of weed they used and their method of preparation produced such an excellent and cheap commodity that there seemed little reason for other countries to spend money on finding alternative supplies, which would probably only have yielded a more expensive material. Industries associated with alginic acid have never been wholly successful from the very start, on account of undue optimism, uneconomic methods of harvesting and a failure of the chemical processes on a commercial scale to produce a sufficiently pure product.

Seaweeds as food for cattle and beasts have never achieved popularity, partly perhaps due to lack of advertising and partly on account of inadequate experiments to determine their food values. So far as the land is concerned there is no doubt that the brown algae are a good potassic and nitrogen manure, but so far no adequate method of dealing with their enormous wet bulk has been devised. Their use is therefore restricted to areas close to the region of supply. The spread of civilization with its greater range of foods is slowly eliminating the use of seaweeds as human foods, even in such strongholds as Japan and Hawaii. One must therefore accept the fact that the present known types of edible algae will never again achieve a wide use.

The somewhat unhappy story related above should not, however, be a deterrent to future generations. If the difficulties can be overcome there is no reason why a flourishing industry should not arise. One may venture to prophesy that success is more likely to be achieved if more than one product is manufactured. Thus it should be possible to establish a successful industry based upon the production of alginic acid, manures and animal foods. If one product temporarily goes out of favour the others will help to tide over the depression. In the early stages of an industry it is probable that a wider range of products might avert a crisis until the industry has become established.

Summing up the present position it is doubtful whether an improvement can be made on de Launay's comment, even though it was written in 1902: 'Mais, en industrie comme en science, les recommencements sont fréquents; plus d'une méthode ou d'une idée que l'on avait abandonées comme ayant fait leur temps, reparaissent un beau jour, un peu transformées, avec des airs de merveilleuse nouveauté.' Another comment equally to the point was made by the author of the *Report on Home Industries in the Highlands and Islands.* This writer said: 'No doubt the utilization of marine algae will bring many problems, many of which will be economic, others chemical and mechanical, while some will be social, but with courage and foresight these should be capable of a satisfactory solution, to be attempted each in its turn as a particular question arises.'

In preparing this volume I have been much aided by valuable criticism and advice from Dr E. M. Delf; Professor J. B. Speakman also very kindly allowed me to see the manuscript of two papers on alginic acid derivatives before publication. Considerable assistance in the preparation of the manuscript, index and figures has been afforded me by my wife and Mrs. B. O. Parks, and to both of them I am very grateful.

Auckland, New Zealand   *V. J. Chapman*
   1946

Delays caused by factors outside the author's and publisher's control have resulted in some time elapsing between the writing of the book and publication. The results of subsequent work have been incorporated as far as possible in the text or by addenda.

Auckland, New Zealand   *V. J. Chapman*
   1949

Preface to the Third Edition

Since the last edition of this book was published there has been increased activity in the study of marine algae in relation to compounds that have or could have commercial applications. There has also been greatly increased emphasis upon algal mariculture. The increasing world population will inevitably impose greater pressures upon the available land resources and it is therefore inevitable that more and more attention will need to be given to the oceans and what they can provide as a food resource and also as a resource for industry. The proposed kelp farm is a particular example of this type of pressure. For this reason less emphasis has been placed in this edition upon the historical aspects and more emphasis has been placed upon new developments. I have invited Professor D. J. Chapman to join with me in the preparation of this new edition, revising and combining two chapters and writing a new one.

In preparing this new edition numerous persons have responded to my requests for information and I am much indebted to them. In particular I would like to express my thanks to the following: Arramara Teo Ltd , Atlantic Mariculture Ltd , Cellulose Products of India, Canadian Benthic Ltd, FAO, Kelco Ltd, Kyowa Hakko Kogyo Co., Kelp Industry Pty Ltd, Marine Colloids Ltd, Stauffer Chemical Co., Drs M. Doty, A. Jensen, D. Luxton, M. Neushel, Wheeler North, J. Pringle, W. Shurtleff and H. Wilcox.

Auckland, New Zealand   *V. J. Chapman*
    1979

1
Occurrence, Distribution and Historical Perspective

The plant life in the sea is extremely rich and some exploitation of these resources has taken place over hundreds of years. At the present time, when man is increasingly turning his attention to the ocean as a major source of food and industrial chemicals the plant life, both attached and floating, is becoming of great importance. Where seaweed has been used, both in the past and at present, it has been as freshly gathered plants, but there are a few industries that can use seaweed cast up on the shore as drift, provided it is soon collected. The great amount of attached seaweed existing in the world is probably not fully realized: if it were, it is possible that greater efforts would have been made in the past to find means of collecting and using all this raw material. The minute floating plants of the sea, the phytoplankton, form the basic foodstuff for small animals and fish, and do not in themselves have a direct commercial use, although recent work (see Chapter 8) indicates that 'artificial food chains' in mariculture systems utilizing phytoplankton as the basic trophic level are feasible.

1.1 CLASSIFICATION

Seaweeds belong to a rather ill-defined assemblage of plants known as the algae. The term 'seaweed' itself does not have any taxonomic value, but is rather a popular term used to describe the common large attached (benthic) marine algae found in the groups Chlorophyceae, Rhodophyceae, Phaeophyceae or green, red and brown algae respectively.

The algae differ from the higher plants in that they do not possess true roots, stems or leaves. However, some of the larger species, upon which the industries are primarily based, possess attachment organs, or hold-fasts, that have the appearance of roots, and there may also be a stem-like portion called a stipe, which flattens out into a broad leaf-like portion or lamina (e.g. *Laminaria*, Fig. 1.4). Some species consist simply of a flat plate of tissue (e.g. *Ulva*), whilst in others the plant body, or thallus, is composed of a narrow, compressed or tubular axis with similar branches arising from it (e.g. *Gelidium*, Fig. 1.10a). The smaller species differ from those described above in that they are mainly filamentous.

In the primary classification the algae are defined into 15 classes, excluding the Cyanophyceae (blue-green algae) which are true prokaryotes.

Of these 15 classes, three are represented by macroscopic forms, present in sufficient quantities in nature to have direct commercial importance. These are the Chlorophyceae, Rhodophyceae, Phaeophyceae. The other classes are principally planktonic (unicellular or colonial), and with the exception of the 'food chain groups', the diatoms (Bacillariophyceae), haptophytes (Haptophyceae) dinoflagellates (Dinophyceae) and planktonic Chlorophyceae, have no commercial importance.

It is obvious that a knowledge of the means of reproduction of algae is extremely important if they are to be cropped annually for economic purposes or cultivated for maximum yield. However, this is not the place to describe these features in full, and further details about the life-histories of the various algae that are of economic use can be found in Fritsch (1935, 1945), Smith (1938), Dawson (1966), Boney (1966), Chapman and Chapman (1973) and Bold and Wynne (1977).

Very few, if any, of the large benthic seaweeds are cosmopolitan in distribution. Geographical areas are characterized by their own distinctive algal flora and this in turn determines the type of seaweed industry present.

1.2 DISTRIBUTION

1.2.1 Europe and North Atlantic

On the rocky coasts of the northern hemisphere, the most conspicuous plants are usually the different kinds of brown seaweed, although towards low-water mark there may be a moss-like carpet of red algae. Around the coasts of Europe each one of the principal kinds of brown seaweed, and also some of the Rhodophyceae, occupy much the same relative position on the shore. Observations of the distribution from high water mark to below the low tide level reveal a characteristic zonation of algae. *Pelvetia canaliculata* will always be found near high-water mark (Fig. 1.1). Though they are small the plants may live for 4—5 years and they only start reproducing in their third year (Subrahmanyan, 1960). Below *Pelvetia* there will commonly be a larger species, *Fucus spiralis* (Fig. 1.2), or alternatively a closely allied form *F. platycarpus* (= *F. spiralis* var. *platycarpus*). Both of these have a broader branching system or thallus, with a well-marked stipe at the base arising from a disc-like hold-fast. Further down, and about the middle of the shore, two different brown algae are to be found; these are often collectively referred to as the bladder wracks because of their vesicles. The first, which often grows to quite considerable lengths, is known as *Ascophyllum nodosum* (Fig. 1.2).

The true bladder wrack, which occurs in a zone either above or below the *Ascophyllum* or else mixed with it, is *Fucus vesiculosus* (Fig. 1.2). The vesicles, like

Fig. 1.1 *Pelvetia canaliculata.*

those of *Ascophyllum*, are normally full of gas, and so enable the plant to float near the surface when the tide comes in.

Near low-water mark there is another species, readily recognizable by the serrated edge of the thallus, *Fucus serratus*, which does not have any vesicles (Fig. 1.3a).

At lower levels, the nature of the vegetation changes. Thus, around low-tide mark of ordinary spring tides, the first of the big oarweeds is to be found. This has a basal attachment portion, then a stem-like stipe which expands into a broad divided blade from which it derives its name of *Laminaria digitata* (Fig. 1.4a). Although several varieties of this species have been recognized, in particular vars. *stenophylla* and *flexicaulis,* commercial users of the alga do not usually trouble to distinguish these varieties, but it may be important that they should, because analyses (see p. 93) show that the varieties possibly differ from the parent species in their chemical composition.

In many places where rocks are replaced by stones or shingle another type of oarweed is commonly found (though it can also grow alongside *L. digitata*) called *Laminaria saccharina* (Fig. 1.5a) or 'sugar wrack'. It is so-called because it is sweet to the taste, owing to the presence of mannite, a sugar alcohol (c.f. p. 229). *L. digitata* normally grows from low-water mark down to about two or three fathoms, where it is replaced by a related species, *Laminaria hyperborea*, which normally lives between four and twenty fathoms (Fig. 1.4b). In Long Island Sound (USA), where there is a strong current, oarweeds are said to have been pulled up from much greater depths.

Fig. 1.2(a) *Fucus vesiculosus*;   (b) *F. spiralis*;   (c) *Ascophyllum nodosum*.

One may sometimes find, also growing in sandy or rocky places below low-water mark, two other members of the Laminariaceae, *Saccorhiza polyschides* (Fig. 1.5b) and *Alaria esculenta*. The oarweeds (*Laminaria* and species of allied genera) are widely distributed in the colder waters of the northern hemisphere. They are not able to grow in warm waters, because the microscopic sexual generation is unable to reproduce when the temperature of the sea is more than $12-16°C$, and they do not grow higher up on the shore because they cannot tolerate exposure for any length of time: for nearly all their life they must be covered with water. They will only occur at higher elevations if there are deep rock-pools in which they can grow submerged.

Apart from the principal brown algae a carpet of moss-like red seaweeds will often be found growing in the lower half of the zone exposed at low tide: the plants not only form an open carpet but they may also grow under the brown seaweeds. There are two or three different types of these algae which are often collectively known as Irish Moss. The true Irish Moss is *Chondrus crispus* (Fig. 1.6); another plant very

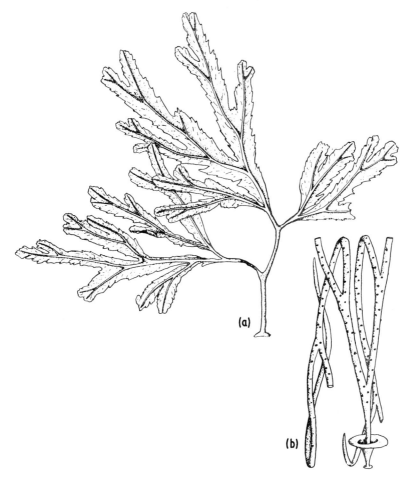

Fig. 1.3 (a) *Fucus serratus*; (b) *Himanthalia lorea*.

like *Chondrus*, is *Gigartina stellata*. Towards high-water mark quite a different kind of red seaweed, *Porphyra umbilicalis*, can frequently be found, especially in spring and summer.

There is one more red alga of the North Atlantic which is used commercially. This is *Palmaria palmata* (Fig. 1.7): it grows on rocks near low-water, but it is often found in abundance on the stems of *Laminaria hyperborea*. Considerable quantities of this species frequently occur in the drift cast up on the shore.

1.2.2 America

On the Pacific coast of America there are a number of very large brown algae,

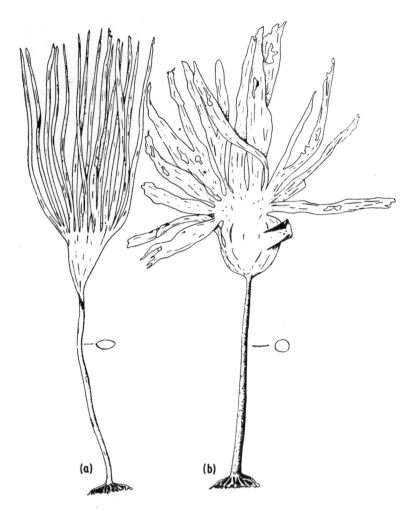

Fig. 1.4 (a) *Laminaria digitata*;   (b) *L. hyperborea* (= *L. cloustoni*).

closely allied to the British oarweeds, which have at different times assumed considerable importance. These seaweeds, which are generally known as kelps, grow together in quantity to form a real submarine forest.

One of the largest of these brown algae is *Macrocystis* (Fig. 1.8) which may grow up to 150 ft (45·7 m) long. It appears to have a life of about five years, though individual fronds have only an average age of six months or so (North, 1961), which means that the growth rate is very high; the average rate of elongation of frond apices has been measured at 7·1 cm ± 4·3 cm per day (Sargent and Lantrip, 1952). At a depth of 60 ft. (18·3m) whole fronds can grow as much as 45 cm per day, and

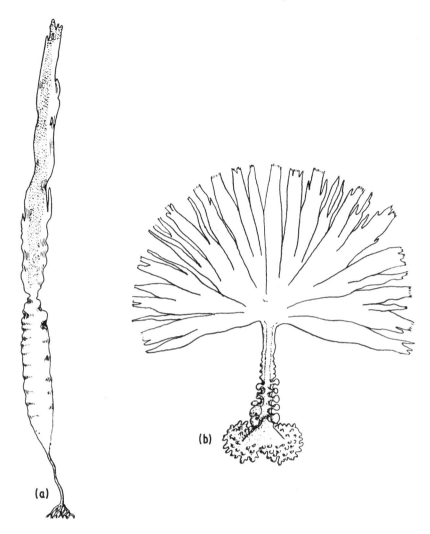

Fig. 1.5 (a) *Laminaria saccharina*; (b) *Saccorhiza polyschides*. (Figs. 1.2–1.5 by courtesy of the Ministry of Supply.)

at this rate they represent the most rapid plant growth known. By means of bladders, which are gas-filled and located at the base of the laminae, the fronds are kept floating at the surface of the sea. In view of their great size these seaweeds can have their rooting portion down at depths of 10–15 fathoms (18·25–27·4m), though they achieve their best growth in depths of about 8 fathoms (14·6m).

Often growing with *Macrocystis* is another giant alga, *Nereocystis luetkeana*, which is commonly known as 'bull kelp'. This seaweed also grows in deep waters, at

Fig. 1.6 Two growth forms of Irish Moss (*Chondrus crispus*).

depths of 15–75 ft (4·57–22·8 m), where a single plant may reach a length of 120 ft (36·6 m). As this plant appears to be an annual, the rate of growth must be considerable, especially in view of the length it may attain. Specimens have been measured with a leaf area of 754 ft^2 (70 m^2).

Another species very similar to the bull kelp, though somewhat larger, is the 'elk kelp' or *Pelagophycus*. Individual plants of this species frequently attain 120 ft (36·6 m) in length, but this species, which has a restricted distribution, does not grow in great abundance, and occurs as single plants or in small patches associated with beds of *Macrocystis*. Yet another Pacific coast seaweed that is of importance, though neither so large nor so heavy, is *Alaria fistulosa*, which is closely allied to the European *Alaria esculenta*.

Macrocystis pyrifera has a wide distribution, and in the north extends from southern California up to Alaska, its southern limit being set approximately by the 20°C water isotherm of the warmest month. The genus is unable to spread into warmer waters because at temperatures higher than 18–20°C the gametophytes do not form any reproductive bodies. *Nereocystis* does not grow so far south and is first encountered around central California, but it spreads farther north in Alaska. *Pelagophycus*, on the other hand, is a restricted southern form, and grows only on the coasts of southern California. *Alaria fistulosa* is even more northerly than *Nereocystis*, and is principally confined to Alaska and British Columbia.

Additional seaweeds present in abundance, but often of a restricted distribution are *Pelvetia*, *Hesperophycus* and species of *Fucus*. A plant that could form an

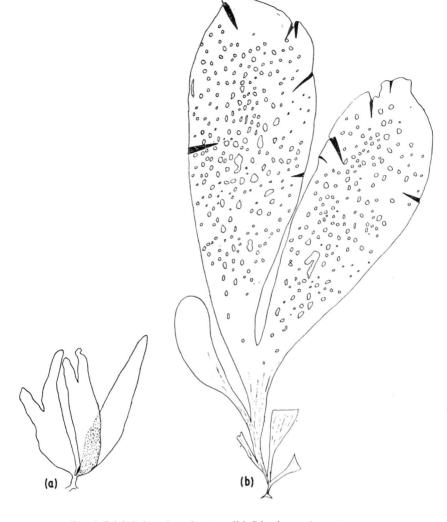

Fig. 1.7 (a) *Palmaria palmata*; (b) *Rhodymenia pertusa*.

important raw material is *Egregia menziesii* or *E. laevigata* (Fig. 1.9), which, unlike other Laminariaceae, are able to tolerate some exposure and grow in the lower intertidal. There are also a number of species of *Laminaria* and other oarweeds belonging to genera such as *Thalassiophyllum*, *Cymathere* and *Costaria*. These, however, are not likely to assume any commercial importance so long as extensive beds of the larger kelps remain to be exploited. The same is probably true of the fucoid *Sargassum muticum*, a recent 'migrant' to Pacific North America and the English south coast from Japan. This species grows abundantly in localized areas and may

Fig. 1.8 The big kelp of California, *Macrocystis pyrifera.*

yet prove to be commercially harvestable.

In addition to the large members of the Phaeophyceae are several red algal species of the *Gelidium–Pterocladia* complex, which occur in sufficient abundance, especially in Baja California, to be of commercial importance. Another genus, *Eucheuma*, has acquired commercial importance in recent years. This pantropical genus is found in Florida and the Caribbean, the Gulf of California and the Central Pacific. All of these algae are found principally in the sublittoral and lower intertidal.

1.2.3 Central Pacific and Hawaii

Dense growths of red and green seaweeds can be observed growing here in the pools amidst the corals, or else forming a close carpet in the lagoons. A very large number of these seaweeds have been used for hundreds of years, and indeed still are used, as significant articles of food. Because of their small size they are principally collected by hand and have to be picked out from other, useless forms. By far the greater quantity of species used are red seaweeds, but a certain number of green and brown ones are also collected. The genus *Eucheuma* is the most important commercial genus of this area.

1.2.4 Japan

In Japan many of the red seaweeds that are to be found growing around low-water mark, in tidal pools or in the shallow sublittoral, are of commercial importance. Extensive use is made of *Porphyra* and so much is required that the seaweed is cultivated (c.f. p. 99). The important agar industry is concerned largely with

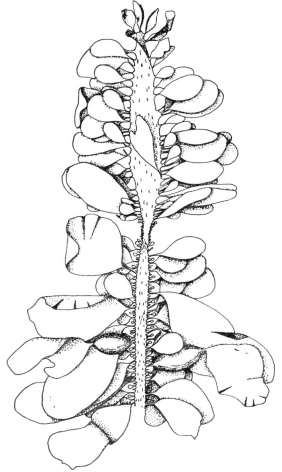

Fig. 1.9 Portion of plant of *Egregia laevigata*.

species of the genus *Gelidium*, especially *G. amansii* (Fig. 1.10a), whilst another industry makes use of the red alga *Gloiopeltis furcata* (see p. 145).

Kelps also grow abundantly in the waters around Japan, especially in the north, and a large number of species are known. The Japanese utilize these algae in considerable quantity, and several important industries are associated with them. These kelps all grow in the deeper waters off-shore where they form very extensive beds. The principal species involved are *Ecklonia cava, Eisenia bicyclis,* (Fig. 3.6), *Laminaria* spp., *Undaria pinnatifida* (Fig. 3.5) and a few species of *Sargassum*.

1.2.5 Indonesia

Farther south, in the Malayan Archipelago, there is much use of the smaller red

Fig. 1.10 (a) *Gelidium amansii*; (b) *Grateloupia filicina*.

seaweeds, which grow totally submerged; but, as in Hawaii, there are no kelps that can be employed. A number of species are gathered, some of the commoner ones belonging to the Rhodophycean genera *Eucheuma* and *Gracilaria* (see p. 167).

1.2.6 New Zealand

The giant kelp *Macrocystis pyrifera* grows in the waters around New Zealand in sufficient quantity to be of economic value, and the major beds have been mapped (Rapson *et al.*, 1943). There are other brown algae belonging to the genus *Ecklonia* which grow just below low water mark, and these too might usefully be employed. Near high-water mark there is *Porphyra columbina*, whilst near low-water mark occur species of *Pterocladia* that can be used for agar production (see p. 176). Species of *Gigartina* also show promise as a source of carrageen-like materials. Also near low-water

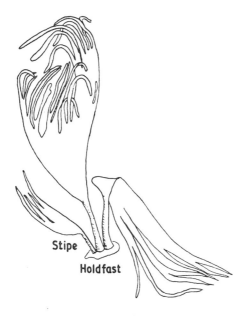

Fig. 1.11 *Durvillea antarctica.*

mark on exposed coasts, one may find a zone occupied by a remarkable brown alga, the 'bull kelp' or *Durvillea* (Fig. 1.11). The common species is *D. antarctica,* and its fronds when slit open are used by mutton-bird catchers as bags.

1.2.7 Australia

The giant *Macrocystis* grows along the shores of eastern and southern Australia and Tasmania where it is now being harvested as a source of alginates. Certain red seaweeds, especially *Gracilaria verrucosa*, growing near low-water mark or down to several fathoms (see p. 175), are also collected for agar manufacture.

1.2.8 South America

Beds of *Macrocystis* are widely distributed down the coast from Peru to the Straits of Magellan. There is very little information, other than an early account by Skottsberg (1921), about the extent of these beds, or whether the alga is present in sufficient quantity to be of economic importance. *Durvillea antarctica* and *D. harveyi* occur near low-water mark in sufficient abundance to form well-marked zones. Red seaweeds, which no doubt could be used in the same way as in other parts of the world, grow on these shores and below low-water mark. *Porphyra umbilicalis,* for example, occupies a zone at the mid-littoral, whilst there are two species of *Rhodymenia* that

Fig. 1.12 *Lessonia fucescens.*

form deep-water communities. *Lessonia* (Fig. 1.12) in addition to *Macrocystis* is found in the Falklands, Chile and around Cape Horn.

1.2.9 Caribbean

Some small use is made of marine algae in the Caribbean: *Eucheuma* in Florida, Antigua and Barbados; *Ulva* in Jamaica and Trinidad and in the last named island *Gracilaria* sp. The most important alga is 'sea moss' which is probably a species of *Codium*. This is dried and exported from Barbados. The carrageenan producer, *Hypnea*, may also prove to be commercially harvestable in the Caribbean.

1.2.10 South Africa

The most important economic members which grow from mid-tide to below low-tide mark are species of *Gelidium, Hypnea, Gracilaria* and *Suhria. Porphyra capensis* occurs in quantity on the west coast, but so far no use has been found for it. South Africa, however, has far greater potentialities because there are fairly

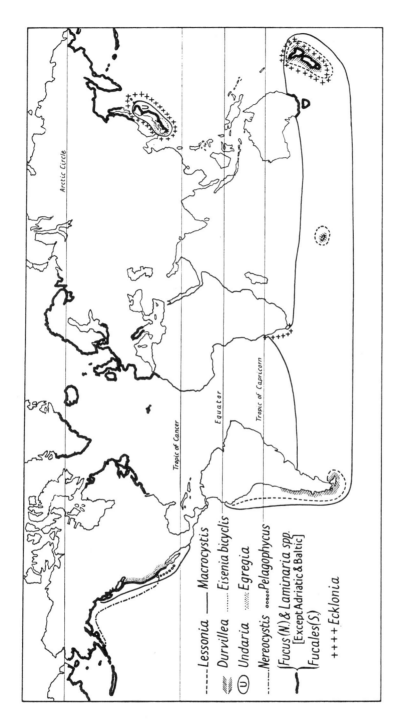

Fig. 1.13 Distribution of some of the more important brown seaweeds.

extensive beds of *Macrocystis* on the west coast just north of Cape Town.

Associated with *Macrocystis* in South Africa is the genus *Laminaria* (*L. pallida*), and *Ecklonia maxima*. Both the latter occur in some quantity and could be harvested; the *Ecklonia* evidently reaches a considerable size, plants of nearly 33 ft (10·06 m) long having been measured. *Laminaria pallida* dominates in the colder waters and *Ecklonia* in the warmer, and both are usually more abundant as subtidal beds on sheltered bays rather than on exposed coasts.

The seaweeds of the world may be regarded as forming three great zones so far as their usefulness is concerned. There is the north temperate and cold-water zone in which large brown seaweeds occur in great quantity, though one can also find small quantities of useful red seaweeds. There is the central warm-water zone in which the principal valuable seaweeds are Rhodophyceae, and then there is the south temperate and cold-water zone where again big areas of brown seaweed are to be found, though there are also quite considerable quantities of red algae. In Fig. 1.14 the distribution of the major algal industries has been set out in relation to the occurrence of the principal species involved, and this map should be compared with the brown seaweed distribution map of Fig. 1.13.

1.3 PLANKTONIC MARINE ALGAE

Although a number of marine planktonic algae, e.g. *Dunaliella,* diatoms, can be grown in bulk culture either directly or in sewage effluent or nutrient enriched ponds (Eddy, 1956; Dunstan and Tenore, 1972; Goldman and Stanley, 1974), the direct commercial advantage is probably minimal, because of the cost involved in harvesting and recovery. However, such mariculture procedures can be used, and in fact are being used (Chapter 8) to provide algae populations as food for commercially harvestable marine organisms such as molluscs, shrimps and other herbivores.

1.4 FRESH-WATER ALGAE

So far only seaweeds have been considered, but there are algae that grow in fresh water or in the soil, but commercially these are of little value. There is a small green filamentous alga (*Chaetomorpha* sp.) that grows in sufficient quantity in Jamaica so that it can be collected and used.

Fresh-water algae serve a valuable purpose in rice-growing where certain blue-green algae found in the rice fields are capable of fixing atmospheric nitrogen. Inoculation of the soil by some species has been proposed as a means of increasing yield. The blue-green alga *Nostoc commune* is used as food in China (p. 69) and recently *Spirulina platensis* has been reported (*I.F.P.,* 1967; Léonard and Compère 1967; Anon. 1968) as a food used in Central Africa. The bulk cultivation of uni-cellular green algae, such as *Chlorella* and *Chlamydomonas,* has been proposed as a

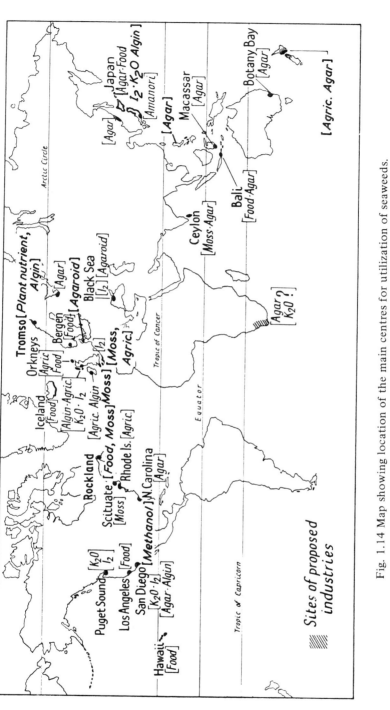

Fig. 1.14 Map showing location of the main centres for utilization of seaweeds.

means of producing highly nutritious food in arid areas, e.g. Israel, and where there is a dense population, e.g. Japan, East Pakistan (Burlew, 1953; Kok and Oorschot, 1954; Reisner and Thompson, 1956; Hua and Li, 1962). There is no reason why such freshwater species as *Chaetomorpha* or *Cladophora* could not be grown in commercial quantities. Freshwater aquaculture is a field badly in need of further exploration.

1.5 HISTORICAL

1.5.1 European and Japanese industry

The history of the European industry falls into three distinct periods. The Laminarians were first collected for the soda they contained; later they were used for the manufacture of iodine; finally they became the source for the alginate industry (see p. 194). The fucoids formed the principal source of supply for soda because they contain a higher percentage of this substance than the oarweeds. The kelp soda was a substitute for an expensive substance derived from certain coastal salt-rich plants in Spain. The kelp was not so good in respect of yield or in quality, but it served quite well in the manufacture of ordinary glass. Iodine is present in sea-water in very small quantities (0.01–0.07 ppm), but certain seaweeds, especially the oarweeds, have the capacity to accumulate it, hence their value as a source of the element. Even a fucoid like *Ascophyllum* can concentrate the iodine up to 220 times that in sea water (Baily and Kelly, 1955).

1.5.2 Species employed

In Europe quite a number of brown seaweeds have been employed in the production of burnt ash of seaweeds or kelp. The term 'kelp' here refers only to the ash, in contrast to the American usage to refer to the algae themselves. First and foremost there were the species of *Laminaria* and *Saccorhiza polyschides*, though their usage was largely confined to areas where enormous quantities were regularly thrown up by storms, e.g. western Ireland and Scotland. Next in importance were the common rockweed species, *Fucus vesiculosus, F. serratus* and *Ascophyllum*. Other species that have been used in the kelp industry are *Himanthalia lorea, Chorda filum, Halidrys* and occasionally *Alaria esculenta.*

Although the production of soda ash was carried out by French peasants in the 17th century, the production of kelp (ash) in Great Britain did not commence until about 1720. The industry between 1726 and 1735 became extended to the outer Hebrides and eventually some 60 000 persons in Scotland, the Orkneys and Hebrides, derived a livelihood from it. By 1790, 3000 tons (3048 tonnes) were being contributed to the overall Scottish production.

The importation of soda derived from salt pans, after 1810, reduced the need for kelp, and by 1845 the industry in Great Britain was almost dead. The discovery of iodine in kelp ash by Courtois in 1811, and its subsequent medical use revived the industry. In 1846, for example, there were twenty manufacturers in Glasgow alone. With the discovery of Chilean mineral deposits the kelp industry in Scotland entered its second decline. The third revival took place mainly during the Second World War with the rise of the alginate industry (see p. 194).

Japan entered late into the iodine trade as compared with the western countries, but soon outstripped them and was even producing considerable quantities as late as 1929. This rapid development can only be accounted for by the cheapness of labour. Production in Japan started in 1900. By 1911, iodine production was 60 tons (60·96 tonnes), reaching a maximum of 294 tons in 1916 (298 tonnes), declining to 115 tons (116·8 tonnes) in 1929.

The enormous amount of labour involved in this industry can be gauged from the fact that it required between one and two million tons of wet seaweed to produce 115 tons (116·8 tonnes) of iodine, and mechanical harvesters were not employed.

The chief places for the manufacture of iodine were in the island of Hokkaido and the prefectures of Chiba, Kanagawa, Yamaguchi and Schizuoka. In Hokkaido, only species of *Laminaria* were used, but in the other places *Ecklonia cava, Eisenia bicyclis* and *Sargassum* spp. were commonly collected.

The original procedures (cf. Chapman, 1970) for the processing of harvested algae were essentially those of burning and recovery of the ash. There is no doubt that this was wasteful. Despite unaccounted for waste, there was certainly considerable contamination of the crude product.

1.5.3 New processes

The destructive distillation, or char process, was suggested in 1862 by the chemist Stanford. This was designed not only to overcome the wastage of iodine but also to utilize the winter cast of old senile plants, which differs from the 'may' or 'leaf' cast in the very high proportion of stems that it contains.

The efficiency of the char process over the kelp process can be gauged from Table 1.1. Data from another process, also invented by Stanford, and known as the wet or 'lixiviation' process is included, but this is described in detail later (cf. p. 208) as it is of more importance in the production of algin. The figures in all three cases are based on an initial quantity of 100 tons (101·6 tonnes) of wet weed.

Stanford illustrated the advantages of the char over the old process by showing the additional materials that he estimated could be extracted from one ton of kelp derived from different seaweeds (Table 1.2). It is possible that failure of the char process has done much to prevent the brown algae being utilized on a large scale.

1.5.4 Iodine from seaweeds in Russia

In all countries other than the USSR it is fucoids and laminarians that provide the

Table 1.1 Yields and waste using different treatment processes

Items	Original burning process	Char process	Wet process
Dry weed obtained and utilized (tons*)	18	36	70
Crude ash (tons)	18	36	33
Salts extracted (tons)	9	15	20
Iodine extracted (lb†)	270	600	600
Kelp wasted (tons)	18	–	–
Charcoal (tons)	–	36	–
Tar, ammonia (tons)	–	some	–
Algin (tons)	–	–	20
Cellulose	–	–	15

* 1 ton = 1·016 tonnes.   † 1 lb = 0·454 kg.

natural source of iodine. In Russia, however, a red seaweed, *Phyllophora nervosa*, was collected in the Black Sea and Sea of Azov (see p. 141) and treated industrially because of its richness in iodine. (A number of other red algae, mainly small forms of no economic value, contain considerable quantities of iodine in special iodine cells', 'blasenzellen', or gland cells. In these species the iodine is present as a compound and not in the free state. Special nets that collect 600–1000 kg of weed are used for the harvesting. The ash of *Phyllophora* may contain from 1·3–3·8% of iodine (0·27–0·58% dry weight; Vinogradova, 1953) and it is therefore as valuable as the species of *Laminaria*. Related species of *Phyllophora* occur on the coasts of England and France, but not in sufficient quantity to enable them to be exploited. The use of *Phyllophora* only commenced in 1927 when the Black Sea factory produced, according to Pentegow (1930), 2·2 tons (2·23 tonnes) of iodine per month. Prior to that date the Russians had operated factories during 1914–17 on the White Sea, Black Sea and at Vladivostock. These had all employed brown algae and, with the exception of the one on the White Sea, were closed down after the First World War.

1.5.5 Composition of eastern kelps

Analyses of the algae used commercially have been published by a number of investigators, but the results are by no means uniform. This is only to be expected, because the iodine content of the fresh weeds varies at different seasons of the year and also with locality. The analytical results also depend upon the relative proportions of frond and stipe in the sample and whether the plant is fruiting or sterile. It is likely also that the presence of epiphytes, e.g. Bryozoa and Rhodophyceae, may affect the results. Similarly the proportion of iodine obtained from the kelp ash varied depending on the species of weed used, the time of year and the amount of care devoted to its manufacture. The Scottish kelp, for example, was regarded as poor

Table 1.2

	Laminaria digitata		L. saccharina		Fucus vesiculosus		Fucus serratus		Ascophyllum	
	cwt*	lb†	cwt	lb	cwt	lb	cwt	lb	cwt	lb
Old Process:										
Potash	6	56	8	20	2	72	4	96	4	46
Soda	6	45	5	53	6	78	5	24	7	41
Ash	3	89	3	103	4	63	6	38	5	53
Iodine		12.5	—	—	—	—	—	—	—	—
Additional with Char Process:	gal‡		gal		gal		gal		gal	
Volatile oil	4 3/4		4 1/2		9 2/3		6 1/2		—	
Paraffin oil	4 3/4		5 1/6		13 1/3		7 1/3		14 1/3	
Naphtha	3	2	2 3/4		5 2/3		2 1/2		—	
	cwt	lb	cwt	lb	cwt	lb	cwt	lb	cwt	lb
Ammonium sulphate	1	46	2	17	2	34	2	38	3	108
Calcium acetate		17 1/2		21		75		36 3/4		—
Colouring matter		2 3/4		5 3/4		11 1/2		6		—
Pure charcoal	8	35	8	59	15	10	14	41	20	49
Gas	3615 ft³ §		2771 ft³		4313 ft³		3811 ft³		8272 ft³	
Iodine	19.4 lb		—		—		—		—	

* 1 cwt = 50·802 kg. † 1 lb = 0·454 kg. ‡ 1 gal = 4·546 l. § ft = 0·3048 m

Table 1.3

	% Iodine in Wet weight alga	% Iodine in Ash of alga	% Ash in alga
Europe			
Fucus vesiculosus	0·003–0·103	0·04–0·2	6·4
Fucus serratus	0·007–0·067	0·2	5·6
Ascophyllum	0·0014–0·23	0·23–0·4	6·2
Laminaria digitata			
Leaf	0·09–0·25	1·36–1·7	5·3
Stem	0·12–0·44	1·04–1·65	6·1
Japan			
Ecklonia maxima	0·25	0·5	47·2
Eisenia	0·3	0·5	50·9
Sargassum	0·05	0·1	52·0
Laminaria angustata	0·23	0·99	23·9
Laminaria japonica	0·26	0·92	26·8
Laminaria religiosa	0·16	0·68	24·2

whilst the Irish kelp was richer because it was made at a lower temperature so that less iodine was lost.

Tables 1.3 and 1.4 give some average representative values of iodine composition that illustrate well the variability in composition.

Work (Annual Report Scottish Seaweed Research Association, 1946) on the chemical composition of both littoral (*Ascophyllum*) and bottom weeds (*Laminaria*) has shown that in the former the ash, nitrogen and iodine contents are at a maximum during January and February. The ash and iodine content do not exhibit any marked variation, the former ranging between 14 and 24% and the latter from 0·03 to 0·15% of dry weight. The iodine content of 'loch' weed is lower than that of 'open sea' weed, a feature which is probably due to the lowered salinity of loch water.

In the case of the sublittoral weeds the fluctuations are considerably greater. The curves for the ash and protein contents (Fig. 1.15) run more or less parallel and show a maximum in the spring (*L. digitata* and *L. saccharina*) or in the early summer (*L. hyperborea*), whilst minimum values are reached in summer (*L. digitata, L. saccharina*) when photosynthesis is at its maximum, or in winter (*L. hyperborea*) when the metabolic processes are at their minimum. The variations are therefore not uniform and they suggest that the physiological processes of the species may vary. The iodine content behaves somewhat similarly.

In Europe, Cauer (1938) published data which show that the respective yield from the parent species and one of its varieties may be sufficiently divergent to be of economic significance (Table 1.5). The failure of many of the earlier workers to distinguish the various varieties may therefore render their analyses of little value.

Table 1.4

	% Iodine of the dry weight					
L. hyperborea	January (Roscoff)	February (Roscoff)		February (Portrieux)		March (Portrieux)
Stipe	0.81	0.55		0.66		0.66
Frond	0.74	0.49–0.54		0.88		0.7–0.75
L. saccharina	March	July		August		September
	0.6	0.71		0.76		0.46
Ecklonia	March	April	May	June	July	August	September
Young stalk	0.13	0.14	–	–	–	–	0.39
Old stalk	0.25	0.26	0.3	0.5	0.5	0.35	0.59
Young leaf	0.13	0.13	0.19	–	–	–	0.29
Old leaf	0.21	0.26	0.17	0.6	0.72	0.26	0.53

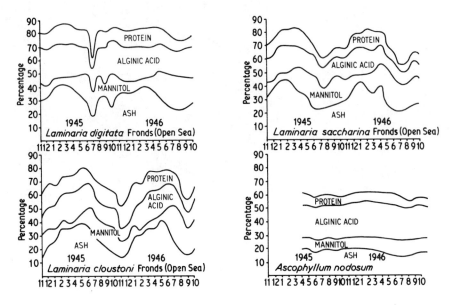

Fig. 1.15 Seasonal variation of the ash and some organic constituents of the fronds of *Laminaria* and *Ascophyllum*.

Table 1.5

	% Iodine in ash
Laminaria digitata, leaf of young plant	1·45
Laminaria digitata, stem of young plant	1·65
L. digitata var. *flexicaulis*, leaf of old plant	0·54–0·61
L. digitata var. *Flexicaulis*, stem of old plant	5·5

In any event it is clear that more attention should be paid to the naming of the specimens, their age, locality, condition and season of collection.

1.5.6 North American industry

The first reference to the giant Pacific brown seaweeds appears to have been to *Macrocystis* in the sixteenth century. *Pelagophycus,* the elk kelp of southern California, has been known since the days of the early Spanish navigators who used it as an unfailing sign of proximity to land. The other two genera of commercial importance are *Nereocystis* and *Alaria*.

However, in about 1910, when a controversy arose between merchants in America and the Kali Syndicate of Germany, who at that time more or less possessed a monopoly of potash, America began to pay attention to these natural resources. There are certain mineral resources in America that could be used for the manufacture of potash, e.g. deposits of the mineral alunite and also the salt deposits of old lakes and inland seas, but it was pointed out that the kelps formed one of the richest natural sources of potash. Two very important documents give details of the resources. The first is Senate Document 90 of the 62nd Congress and the second is Report 100 of the US Department of Agriculture, Bureau of Soils.

As a result of these developments the production of fertilizers and potash salts from seaweeds was commenced on a commercial scale in 1912. The three kelps which occur in sufficient economic quantity are *Macrocystis, Nereocystis* and *Alaria fistulosa. Pelagophycus* contains a higher percentage potash than the other three, but it is not sufficiently abundant to be worked by itself.

These algae often form what foresters would call 'pure stands', that is areas of weed in which there is only one kind of dominant seaweed. South of Point Sur in mid-California *Macrocystis* is the principal species, though there is often a fringe of *Pelagophycus* on the outside, because the latter favours rather deeper water. North of Point Sur *Macrocystis* and *Nereocystis* grow either in separate communities or else in intermingled patches. In Alaska there are considerable beds of *Alaria,* often with *Nereocystis* growing outside. *Macrocystis* grows best where there is a continuous swell, whereas *Nereocystis* and *Alaria* thrive primarily in rapid tideways. *Macrocystis* is regarded as a particularly valuable plant because when the fronds are cut and removed new stipes arise from the base whilst the severed ones die away. Actually, the rate at which regeneration in *Macrocystis* takes place is largely dependent upon the temperature, and there is only one region where even two crops a year would be economically possible (see p. 271).

Since *Nereocystis* is an annual and the large leafy part of *Alaria* is similar, both these plants have to be harvested after sporing has taken place, otherwise there might be no young plants to provide an abundant supply for the following season. North of Point Sur to Vancouver Island, July is suggested as the earliest month for reaping, whilst farther north in Alaska, August would be the earliest because development is somewhat later on account of the shorter growing season.

At one time it was suggested that for these two annuals a 'closed season' would have to be instituted by law in order to ensure that the beds were not cut too soon. However, since no extensive cutting has ever been carried out in areas where these two species are most abundant, the necessity for such legislation has not yet arisen. In the case of *Macrocystis*, a heavy mat of weed is to be found floating at the surface towards mid winter. At this season growth is rapid and more than compensates for the wearing away by wave action. Later the rate of growth decreases and the mat becomes thinner. A new crop of leaves is then formed between April and June, but as the summer becomes hotter the kelp tends to decay and the beds become thinner again. There are thus two periods in the year when the beds are thick, and two periods when they are thin. The effect of warm water on the rate of decay of *Macrocystis* is striking, and if the temperature of the water rises much above $20°C$ whole beds may disappear entirely. Several beds did, in fact, disappear completely as a result of this in the warm summer of 1917 and again later. Destruction appears to be caused by bacterial action at these higher temperatures (see p. 273). In other places the kelp beds exist near sites where wastes are discharged into the sea. Historical records shows that two beds have deteriorated since 1945 because of the local discharge. In addition to the wastes, turbidity in such areas can also be a significant factor as well as sea urchin grazing.

1.5.7 Composition of western kelps

It will be seen from the range of figures for potash and iodine in Table 1.6 that *Alaria* is the least valuable of the four species.

Table 1.6

Species		% dry wt Potash	% dry wt Iodine
Nereocystis	(av. content)	19·6	0·19
	(range)	6·6–31·6	0·13–0·3
Macrocystis	(av. content)	15·6	0·23
	(range)	3·2–27·65	0·14–0·27
Alaria	(range)	2·9–13·1	trace
Pelagophycus	(av.)	19·9	0·4

1.5.8 Harvesting of kelps

At first the plants were gathered by men operating from flat-bottomed scows and using sickles with which to cut the plants below the water. The cut plants were then either hauled aboard the scows or allowed to drift ashore. The Coronado Kelp Company improved on this somewhat primitive method by using a barge with a rotating knife affixed to the front end. No attempt, however, was made to collect the weed into the boat and it was allowed to drift ashore. Both these methods were wasteful, because, unless the bed was close to the land, much of the cut weed would come ashore at places where it could not be gathered, and some must have floated out to sea. An improved method of harvesting was worked out by the Pacific Kelp Mulch Company, and, with further improvements, was adopted by the Hercules Powder Company. The cutter consisted of a dumb barge or scow fitted with an endless belt at the front. This belt passed down into the water to a depth of about four feet, but could be pulled up for inspection or for travelling at speed. At the bottom, and in front of the belt under the water, was the cutting device; this consisted of a number of knives with a sideways movement working over fixed blades. As the weed was cut the forward motion of the boat floated the weed on to the endless belt and thus it was conveyed out of the water on to the scow. In spite of every precaution some weed was inevitably lost in the first part of its passage up the conveyor belt. At the top of the belt the weed fell into a chopper. As the weed came out of this chopper it passed on to another conveyor belt which carried it to a transport barge lying alongside. This particular system required a crew of four and could account for 25 tons (25·4 tonnes) of fresh *Macrocystis* per hour. The collecting ships of the Hercules Powder Company and the Lorned Manufacturing Company were larger and more efficient: they employed six to nine men, and could collect up to 50 tons (50·8 tonnes) per hour whilst operating at a speed of nine knots. Hoffmann (1939) has calculated that one of these large boats must have collected about 700 tons (711·2 tonnes) per day. At the present time the Kelco Company, which harvests kelps for alginate manufacture, uses a self propelled barge very similar to that employed by the Hercules Company.

By arranging the cutting knives at 4–6 ft (1·2–1·8 m) below the surface, it was calculated that about 50% by weight of each *Macrocystis* plant was cut. In the case of *Nereocystis* it is probable that rather more than this proportion would be cut. If the knives were lower still the increase in yield was not raised to any great extent whereas the costs were increased considerably. The expense of the harvesting operation is one of the major items in estimating cost of production, especially when it is realized that about ten tons (10·2 tonnes) of wet kelp are required in order to produce one ton (1·02 tonnes) of dried weed.

When, at the end of the 1914–18 war, the results of the cutting operations were analysed, it was realized that the policy adopted with regard to the actual cutting operations was of considerable importance. There are two possible ways of harvesting. The beds can either be cut completely, or else only the better parts can be cut. The latter is a wasteful process, and naturally leads to intense competition, if there is more than one firm, for the best beds. When the good areas have been cut the harvesting of the remaining thin beds is a wasteful process. If the cutting is systematically carried out over a whole bed, however, the loss in value and time due to the thin portions is compensated for by the better areas. The Hercules Powder Company probably evolved the best technique. They employed three boats, steaming one behind the other; in this way not only was the bed cut systematically but the second and third boats picked up some of the debris from the boat in front.

Today harvesting is carried out under regulations of the California Fish and Game Commission. This has improved the beds very greatly and has also lessened the amount of drift material that formerly floated on to the shores.

By 1915 the majority of workers had concluded that direct use of the dried algae as fertilizer was the most economical. The earliest attempts to provide potash and iodine involved incineration similar to the European procedures. The drying of the algae prior to incineration required substantial fuel oil for heating, such that incineration was basically uneconomical except when potash prices were high.

It was believed that destructive distillation, provided all the by-products were used, might prove an economical process. Between 1917 and 1921 the US Bureau of Soils operated an experimental plant, designed to treat 100 tons (101·6 tonnes) of wet kelp daily.

In the destructive distillation process described, diagrammatically illustrated below (see Chapman, 1970), 100 tons (101·6 tonnes) of *Macrocystis* would yield on an average 12 tons (12·2 tonnes) of dried kelp, which on further treatment gave 2·3 tons (2·33 tonnes) of gas, 3·3 tons (3·34 tonnes) of ammonia, 2·1 tons (2·1 tonnes) of tar, 3 tons (3·04 tonnes) of potash salts, 1·2 tons (1·22 tonnes) of kelpchar and 20 lb (9·07 kg) of iodine. According to the average analyses for *Macrocystis* given earlier (p. 26), the yield of potash was high, but there should have been nearly three times as much iodine. The disadvantage of this process was the high cost of obtaining the by-products, which could not then be sold at competitive prices, and as it could only be profitable if all the by-products were obtained and utilized, it was not particularly successful.

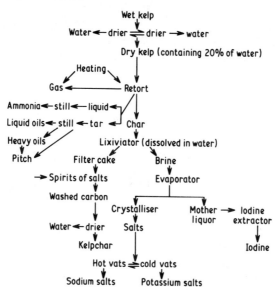

In the 1914–18 War there was a serious lack of acetone, and so the Hercules Powder Company of Los Angeles set out to obtain acetone from kelp with potash as a by-product by a fermentation process; this was a large undertaking and as the acetone was produced for war aims regardless of cost, it was closed down at the end of the war because it could not run profitably in peace-time. The plant, which was erected at San Diego, utilized 24 000 tons (24 384 tonnes) of wet kelp per month, or 1200 tons (1219 tonnes) per working day. The daily intake yielded 13 tons (13·2 tonnes) of 95% potassium chloride as against an expected 16 tons (16·3 tonnes) and 1575 litres of acetone (Crossman, 1918). A flow sheet of the process is given overleaf.

There is a vast supply of weed in the Pacific, and it would appear that the best methods were not employed in the early stages of the industry, and perhaps also because some of the work was carried out during abnormal socio-economic conditions.

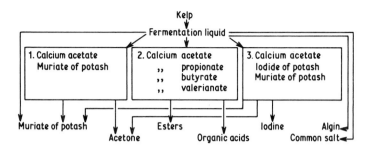

While the potash and iodine industry will probably never be resuscitated, the kelps have proved invaluable as a source of alginates (p. 204), and it is in this field that the kelp industry now operates.

2
Seaweed as Animal Fodder, Manure and for Energy

Historical records show that the use of seaweeds in agriculture is a very old and widespread practice wherever there are rich supplies. This involves their use not only as food for animals, but also as manure for the soil.

2.1 SEAWEED AS ANIMAL FOOD

Today in a number of countries, animals still regularly feed in certain regions upon fresh seaweed or are given a prepared seaweed food. In Iceland fresh seaweeds are commonly employed as a food for sheep, cattle and horses; the animals are encouraged to stay browsing on the shore during the whole of the winter and in some places during the summer as well.

The Icelanders lay in a store of seaweed for a winter supply by washing the plants and then packing them in trenches where they are compressed with heavy oak planks and stones. The compressed mass is broken up as required and fed to the animals without the smell or taste of milk from cows being affected. Sometimes *Alaria*, after washing, is air-dried and then stored in layers in barns, each layer alternating with a layer of hay. At least one seaweed meal factory has existed since 1960 and some weed is extracted to provide a liquid plant nutrient. In Iceland, geothermal heat provides a possible cheap means of drying the weed (Hallsson, 1964). In Finland, both *Laminaria* and *Alaria* are used as fodder for cattle.

In certain coastal areas of Norway, sheep are fed regularly on seaweed and after several generations it has been found that they digest it far better than inland animals. This accords with experimental evidence on the utilization of seaweed foods (see p. 38).

In Scotland, sheep and cattle wander down on to the foreshore and eat various algae. In the small, most northerly island of the Orkneys, North Ronaldsay, there is a local race of small black sheep which feed entirely on seaweed. The whole island is surrounded by a wall which keeps the animals out on the shore. Several hundred (up to 2000) sheep browse on the seaweeds and are allowed to enter a pasture only when in lamb or just before slaughter. They suffer from very few diseases (Stephenson, 1974). Opinion differs as to whether the meat tastes fishy or not, but the wool is regarded as being of a superior quality. On the west coast of Scotland, around Loch Feochan, *Pelvetia* is fed to pigs when they are being fattened for market, the weed being given raw or boiled up and mixed with oatmeal, when it is also fed to calves.

In France, Sauvageau (1920), and Deschiens (1926) have recorded how in Normandy and Brittany dulse (*Palmaria*) is readily eaten by cattle (goémon à vache or goémon à bestiaux), but the animals are rarely stall-fed with seaweed.

Outside of Europe in the Commander Islands in the Behring Sea, polar foxes are fed seaweed as part of their normal diet (Kardakova-Prejeutzoffa, 1938). *Alaria fistulosa, Laminaria bongardiana, f. elliptica* and *Fucus evanescens* are collected, mixed with meal and fed to pigs, especially after farrowing (Kirby, 1953). In Cuba, experiments have been carried out with *Ulva* spp. as an additive to poultry meal with 10% addition giving optimal results (Diaz-Piferrer, Campa and Losa, 1961). In Canada a few tons of rockweed are used annually in Ontario as mink feed (Sharp, personal communication). In New Zealand, meal prepared from imported Norwegian *Ascophyllum* is sold to farmers as a feed additive.

The only reference to seaweed used for animal food in the tropics comes from Hong Kong where species of *Sargassum* are dried and used as pigfeed (Kirby, 1953).

2.2 CATTLE FEED FACTORIES

In earlier times it is obvious that use of seaweed as feed was restricted to coastal areas near the source of supply. Modern practice is to dry the seaweed and grind it into a meal. Factories have therefore been built in various parts of the world in order to manufacture cattle food from the large brown algae. In the old fermentation process of the Pacific Coast industry, some of the weed left after fermentation was dried and made into a form suitable for cattle, whilst a factory was established in Los Angeles devoted to the production of cattle, poultry and pig feeds from dried seaweeds. At the present time there are proposals for 100 000 acre (404·7 km^2) kelp farms in the Pacific Ocean primarily for energy production (see p. 6), but cattle food is regarded as a significant by-product. The initial quarter acre farm was established in 1978 (Chapman, 1978). On the Atlantic Coast the Bonda Company of Yarmouth used to manufacture dried *Ascophyllum* meal (Macfarlane, 1964) but it is now out of production (Sharp, personal communication) and the only company operating is R. and K. Murphy Enterprises Ltd of Wedgeport, Nova Scotia.

Factories have also been built in Denmark, Norway, Eire and Scotland for the preparation of cattle, pig or poultry meal. The growth of the Norwegian seaweed meal industry is shown in Table 2.1.

Since 1965 the *Ascophyllum* harvests have remained fairly constant at around 15 000 tons (15 240 tonnes) of dry meal (= 50 000 tons (51 110 tonnes) wet weed)

Table 2.1 Norwegian seaweed meal production (after Jensen, 1966)

Tons*		Tons	
1950	160	1958	12 000
1951	840	1959	13 000
1952	1 630	1960	14 000
1953	3 350	1961	15 000
1954	5 450	1962	14 000
1955	5 180	1963	12 000
1956	8 300	1964	12 000
1957	10 000	1965	15 000

* 1 ton = 1·016 tonnes.

and a similar quantity of *Laminaria,* mainly *L. hyperborea,* is also collected annually, the latter mainly for alginate production (Jensen, personal communication).

The products of Orcadian and Norwegian factories are largely composed of ground-up rockweeds (*Fucus* spp. and *Ascophyllum*). Pigs are said to thrive on a meal prepared from species of *Laminaria,* although its digestibility is lower for them than for sheep. Another meal prepared from *Fucus vesiculosus, f. serratus* and *Ascophyllum* is said to be lower in digestibility than the *Laminaria* meal. All these meals have some value because of their strong laxative effect.

Two different methods of preparing meal are currently used in Norway, one involving natural drying and the other drum drying, the latter producing a product of quite constant quality (Jensen *et al.,* 1968). 12–15 % moisture is optimal and colour and carotene content must exceed 40–50 mg kg^{-1}. The carotene content of the meal is affected by increased heat, which occurs when naturally dried material is ground, and also by increased moisture content above 15%. A Danish factory cooks the crude weed with superheated steam, drains it and presses it into cakes. These cakes are then dried in a vacuum and group up, after which the meal is said to be more digestible.

It has been found that provided the seaweed can be rotary drum-dried at temperatures decreasing from 800 to 80°C any original microbial population (10^3 organisms per gramme dry weight) does not increase and no mould develops. Mould and bacteria increase in the presence of warm air or with naturally dried weed (Sieburth and Jensen, 1967).

2.3 FOOD VALUE

It is evident that during the First World War there was considerable exaggeration of the nutritional value of seaweed meals, though claims have since been made that their use increased the fertility and birth-rate of animals. Stephenson (1974) suggests that this may be due to the presence of tocopherol, the anti-sterility Vitamin E. The use of seaweed meals because of the fucoxanthin they contain improves the yolk colour of eggs (Jensen, 1966) and their iodine content. In the case of cows they need iodine in proportion to the amount of butterfat they produce (2 mg per gallon of milk with 4% butterfat). The extent to which addition of seaweed increases butterfat still needs further study as experimental results so far are conflicting. In all cases however, it has to be remembered that one cannot expect a quick return after adding seaweed meal to animal rations; continued use is essential. It has been calculated that if each animal in Norway were fed on 500 g of seaweed meal per day, they would consume 445 000 tons (452 100 tonnes) dry seaweed per year, which is equivalent to about 20% of the annual Norwegian hay harvest (Jensen, loc. cit.).

In the case of poultry it appears that with well balanced rations the addition of seaweed meal has no effect on chicks or hens. Use of the weed is beneficial if the ordinary ration is deficient in Vitamin A or B_2 (Jensen, 1972). It is however, clear that one can add up to 7% meal for chicks and up to 15% for laying hens without harm (Hand, 1953). Higher quantities are unfavourable (Black, 1955; Hoie and Sannay, 1960). In the USA 6000 hens were fed 1·25% meal in their normal ration and this reduced the proportion of thin-shelled eggs from 3 to 1·9%; when, after three months, the seaweed addition was stopped the proportion of thin-shelled eggs increased again to 3%. A further advantage is that the trace elements are present in organic form which makes them more readily assimilated (SSRA Report, 1967). Experiments have shown that the effectiveness of the meal varies with the algal species (*Alaria esculenta* meal being better than *Laminaria* or *Ascophyllum*), the locality, time of harvest, method of preparation and storage time.

The meal is intended to be used with ordinary food stuffs, since it is low in carbohydrates, in order to provide a balanced ration (Black, 1955). If normal rations are fully balanced then addition of seaweed meal has no effect (Jensen, 1972). Its principal value is regarded as being in the iodine and mineral content (especially trace elements) because the various elements are naturally dispensed. The amounts used never approach values that could be dangerous. A further advantage is that the trace elements are present in organic form which makes them more readily assimilated. It is, therefore, valuable on mineral-deficient pasture land. It is also claimed that both animals and humans can absorb iodine from eating seaweed when they cannot absorb it from other substances. After two years of trials Saeter and Jensen (1957) found that addition of seaweed meal had no effect on breeding or sterility of sheep. All breeds showed an increased winter wool production. The effects of the meal were more pronounced after a dry summer when hay quality would be reduced.

A first test of the relative nutritional value of these meals can be made by comparing analyses of the raw weeds and their commercial products with other regular fodder (Tables 2.2 and 2.3). This, however, does not lead to direct conclusions, because there is still the problem of relative digestibility.

Average good quality Norwegian meal prepared from *Ascophyllum* has a

Table 2.2

	% Water	% Raw protein	% Fat	% Ash	% Fibre	% (Carbohydrate) N-free extract
Fucus vesiculosus *						
F. serratus	12·4	4·95	1·95	13·1	5·5	62·0
Fucus serratus *						
F. balticus	12·3	4·4	0·8	16·0	5·65	68·85
Ascophyllum nodosum	9–15	5–12	2·6–4·0	15–30	4–7	42–64
Laminaria hyperborea †	12·4	5·86	0·77–1·67	13·67	3·6	63·68
Laminaria saccharina †	14·6	6·37	0·7	16·64	3·28	59·4
Laminaria digitata † (rich in laminarin)	18·6	5·8	0·6	11·3	26·7	36·6
Laminaria digitata † (poor in laminarin)	18·6	9·0	0·6	11·3	26·6	47·0
Californian kelp meal	9·1	5·6	0·4	38·5	5·8	40·6
Seaweed meal A	10·0	6·2	3·8	15·1	3·5	61·4
Seaweed meal B	9·1	5·6	0·4	38·5	5·8	40·6
Seaweed meal C	14·3	7·75	0·5	10·65	10·6	39·5
Norwegian (av.) ‡	12–15	5–10	2–5	18–30	4–5	42–64

* Stephenson (1974).
† Augier and Santimone (1978).
‡ Algea Produkter.

(continued on the next page)

Table 2.2 (continued)

	% Water	% Raw protein	% Fat	% Ash	% Fibre	% (Carbohydrate) N-free extract
Seaweed meal E (Denmark)	5·0	13·1	1·07	5·93	9·0	66·75
Seaweed meal F (Norwegian) (Beharrel, 1942)	13·58	6·9	4·40	16·10	5·07	35·95
Seaweed meal G (Scottish) (Beharrel, 1942)	15·50	10·9	1·5	27·5	9·30	35·3
Seaweed meal (Neptune's Bounty) *	13	6–7·0	1·5	21·5	6·0	51·0
Good hay	14·3	9·7	2·5	–	26·3	41·4
Oats	13·3	710·3	4·8	–	10·3	58·2
Potato tops	–	7·27	0·37	5·12	2·75	84·49
Algit meal	7–9	7–12	1–3	16–19	2·5–8·5	55–62
Goemar †	–	4·25	–	–	–	–

* Stephenson (1974).
† Augier and Santimone (1978).
‡ Algea Produkter.

Table 2.3 Vitamin content of seaweed meals (parts per million)

	Ascophyllum* meal	Algit ‡ (Ascophy.)	Norwegian meal (Ascophy.)	Laminaria* meal	Grass*
Fucoxanthin	90–258	—	—	469	—
B Carotene	16–25	40–65	30–60	11–14	100–450
B_1 (Thiamine)	1·4–5·4	6–8	1–5	1·3–7·2	2–3·3
B_2 (Riboflavin)	5–10	6·0	5–10	2·4	8·6–12
Pantothenic acid	0·2	3·0	—	0·28	8·6–15·5
Niacin	10–30	70·0	10–30	19·4	29·7–40·7
B_{12}	0·004–0·08	0·004	0·004	0·6–0·12	—
C	500–2674	200–400	50–200	2094	—
D_3	0·01	4	—	nil	0·01–0·03
E	156–298	70	150–300	24·4–29·9	111
K	10–14·2	—	10	nil	16·7

* After Black and Woodward (1957).
† After Jensen et al., (1968).
‡ From Mineral Additives leaflet.

composition equal to that of good hay and oats. The figures also show that *Ascophyllum* and potato tops have comparable protein (nitrogen) contents, though less than that of hay. Considerable differences are to be found in the carbohydrate content.

Table 2.3 shows that both *Ascophyllum* and *Laminaria* meals are amply rich in vitamins. Table 2.4 further shows that *Ascophyllum* meal has a wide range of trace elements which are of value to the animals consuming it and that, except for manganese, the amounts are greater than those to be found in grass.

In all these figures (Tables 2.2–2.4) it has to be remembered that analyses of raw plants and manufactured products vary from locality to locality, and are also dependent upon the season of the year and the relative proportions of the different parts of the plants in the samples. It is known, for example, that the carbohydrate content of the algae is highest in the autumn. In addition the composition of the commercial product depends upon the method of preparation (see also p. 32).

Beckmann (1915, 1916) was able to demonstrate that pigs, cows, ducks and sheep can eat a seaweed meal for many months as an additional food and that they thrive as well as control animals fed on a normal diet. In the case of pullets, if the normal diet is somewhat deficient, addition (8%) of *Ascophyllum* or *Laminaria* meal increased egg production by 20–30% (Stephenson, 1974). Chicks given seaweed meal increased in weight more than did controls (Hoie and Sandvik, 1955) unless yeast or grass meal was included in the normal ration. A study by Kim (1972) on the effects of seaweed meal upon the incidence of *Salmonella gallinarum* infection in chicks showed that the use of *Sargassum natans* reduced it by 50% whereas *Ascophyllum* meal had no effect.

A recent development has been the use of seaweed meal as a ration for minks being bred for their fur. The animals are given 0·5 oz (14·2 g) seaweed meal per day and that results in better litters, greater survival rate and more docile animals (Stephenson, 1974).

Although suitable food materials are obviously present in algae, the extent to which they are digestible still remains a debatable problem.

Black (1955) showed that the digestibility coefficient of *Ascophyllum* meal dry matter is 29·7 for sheep and 26·2 for pigs, whilst with *Laminaria* meal it is 66·2 and 71·0 respectively. There is also high digestibility, especially with *Laminaria* meal, of nitrogen free fat. *Laminaria* meal is therefore much more suitable for sheep and pigs.

Experiments with pigs in Norway have shown that 3% of seaweed meal fortified with Ca, P and Vitamin D can supply all extra minerals and vitamins required by the animals. The addition of up to 3–5% seaweed meal has no adverse effects (Jensen, 1971). The same meal, derived from *Ascophyllum* and fortified was given to a set of identical twin cows, one of each pair of twins acting as the control. The seaweed proved more effective than the equivalent standard mineral mixture (Nebb and Jensen, 1966), and it also increased milk production by around 6% in lactating cows (Nebb, 1967; Jensen, Nebb and Saeter, 1968).

Table 2.4 Trace elements in seaweed meal

Element	Grass	*Ascophyllum* meal mg kg^{-1}	Algit mg kg^{-1}	*Ascophyllum** powder
Na	–	30 000–40 000	16 400	19 400†
Mg	–	5000–9000	2500	10 600
Ca	–	10 000–30 000	12 300	11 900
Bo	10	40–100	–10	50
Co	0·14	1–10	3·3	2·0
Cu	4·6	1–10	4	5
Fe	56	150–1100	220	600
Mo	0·82	0·3–1·0	0·1	0·6
Ni	–	2–5	10	5
Ba	–	15–50	–	–
V	0·006	(0·6)–1·5–3	1·1	0·7
Zn	56	50–200	43	33
Gr	–	–	1·0	0·4
Pb	–	–	1·0	0·4
Mn	95	10–50	38·0	24
I	0·5	500	535	1960
Al	–	–	405	200

* Algea Produkter pamphlet.
† This seems low in relation to other analyses.

Experiments have been carried out on the digestion of the various components in seaweed meal and these show that much of the nitrogenous material is undigested. The most important digestible component would appear to be laminaran (see p. 226), which forms 3–6% of *Ascophyllum* meal. Healthy animals appear to digest about 20%, whereas those in poor condition will take up to 50%. Experiments in Ireland using pigs indicated that the chief value of adding seaweed meal was to improve the amount of basic ration that could be diggested (Sheehy *et al.*, 1942, 1947).

Experiments using 3500 sheep showed that an addition of 35 g per day of seaweed meal gave a 3·3% increase in winter wool which was increased a further 17% if the sheep had no mineral supplement. In the case of cows, use of seaweed meal increased butterfat content by 6·8% over a seven year experimental period and also reduced the incidence of mastitis (Jensen, 1972).

In conclusion, so far as animals are concerned, it would seem that for cattle, horses and poultry one can add seaweed meal up to 10% to the basic ration and obtain beneficial results. Further work on sheep breeds is still necessary and use for pigs scarcely seems justified. Its principal use is clearly as a mineral and vitamin supplement.

2.4 SEAWEED AS MANURE

Although algae were known and prized at a very early date in the Orient, several centuries elapsed before there was any mention of the use of seaweeds in western lands.

It is mainly the large brown algae, wracks and oar weeds that are used for manure. Other species have been employed if they are washed up in sufficient quantity, e.g. the sea-lettuce *Ulva*, which is rich in nitrogen. Driftweed that collects on the shore is never or but rarely composed solely of the brown weeds and it usually contains an admixture of red and green algae.

2.4.1 France

The greatest use of seaweed for the land is probably to be found in the north-west of France, where the coastal region is known as the 'Ceinture d'ore'. Here, 400 miles (643·7 km) of coastline are involved to a depth of about 500 m from the sea, and throughout this strip the peasants apply annually 30–40 m^3 of weed per hectare. The recent (1978) oil spill of the tanker Amoco Cadiz caused great consternation because of the damage, among other effects, done to the potential seaweed harvest. An early worker (Mangon, 1859) calculated that the addition of 30 000 kg of weed is equivalent to adding 49 kg of nitrogen to the same area. The seaweed, however, in addition adds significant amounts of trace elements and growth substances.

Driftweed, known as 'goémon épave', 'goémon de dérive' or 'goémon d'échouage', is primarily employed, though the value of this material depends to some extent upon the length of time it has been floating in the sea. Its condition, i.e. degree of fragmentation, will also depend upon whether it has been exposed to much wave action. The French peasants also cut and collect the brown rockweeds which grow below mid-tide mark; this material is known as 'goémon de dérive' or cut seaweed and is said to be better than the oarweeds (Bergeron, 1949).

At Roscoff very considerable quantities of the button weed (*Himanthalia*) are collected in the autumn and used as manure on the artichoke fields. In certain parts, e.g. the Ile de Ré, goémon épave is put on the barley fields just before the seed is sown, and then later a layer of goémon de dérive is put on the young seedlings. The effect is said to be perfect, the fields of barley are magnificient and the crop much in demand by brewers. The steady use of the seaweed manure completely obviates any necessity for a rotation of crops, which would otherwise have to be practised. This fact, though, is somewhat surprising in view of the deficiency of phosphates in the algae (see p. 47).

In France the collection of the different kinds of goémon for manure is very strictly controlled. Anyone is allowed to collect goémon épave, but special permission has to be obtained for other types, and it is only given to those persons whose land adjoins the beach or to maritime municipalities.

At present the cutting and collecting of seaweed in France is controlled by a decree published in 1890 together with some of the earlier decrees.

When the seaweed has been collected it is either dug into the ground fresh or else quickly dried in the sun and built into a stack.

Apart from the brown algae, certain of the red seaweeds which produce lime (Calcium carbonate) are used in some areas for a special purpose. These particular red seaweeds are principally *Phymatolithon calcareum* and *Lithothamnion coralloides* and they always grow submerged in 3–25 ft (1·0–7·6 m) of water. They are collected on the French coast, mainly from the mouths of rivers and protected bays, where they are known as 'maerl'. Neither species will grow in sea-water with a temperature below 13°C. The seaweeds are of special value because of the high calcium carbonate content (up to 80%), and they are therefore employed instead of ordinary lime in order to 'sweeten' humus-rich acid or peaty soils. In France it is again transport costs that prevent such material from being used more widely. The action of the maerl is slow if it is used in the form of lumps and in order to get the best results it should be finely ground.

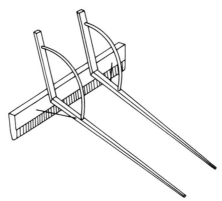

Fig. 2.1 Rake used in collecting 'maerl'.

An early account of the maerl industry, which is largely confined to Brittany, has been given by Pierre (1853). The coralline sand or mud is collected either by dredging or digging, or by the use of a special rake called a 'havel' (Fig. 2.1). The method of employing the tangue depends on the nature and quality of the mud, the nature of the soil, the proximity of supplies and the type of crop involved. The local peasants used to make compost heaps consisting of alternate layers of tangue and farmyard manure. The tangue or maerl acts on the soil mechanically and chemically but it is not necessary to apply it every year (Bergeron, 1949). Fields growing lucerne receive a dose every second or third year, whilst it is only applied every third or fourth year to pasture meadows.

2.4.2 Eire

Seaweed manure is still used fairly extensively on the west and south-west coasts of

Ireland, and in one or two localities the species concerned have been cultivated. Potato sets are placed over what is virtually a layer of fresh seaweed, whilst in the autumn the manure is put on the stubble after the oats have been cut and it is also used on pasture land. The peasants think so highly of it that it may be conveyed seven or eight miles inland. *Fucus vesiculosus* is the species most commonly cultivated in those places where the farmers cannot find enough. Cotton has given an account (1912) of how, in sheltered sandy bays, the farmers bring down stones, in size about one foot cube, and lay them in rows about three ft (0·91 m) apart across the sand, leaving special paths for the carts. When the stones sink into the sand they have to be raised, but they must not be tilted because growth takes place best on the old exposed surface. It is only when there are too many unusable 'weed' algae present that the stone is turned over. The weed is cut after two years' growth, and is made into stacks six feet (1·83 m) high. These are tied together with ropes and towed ashore when the tide is coming in.

2.4.3 Scotland

In south-west Scotland the driftweed is more highly esteemed as a manure, but in north Scotland the crofters prefer cut rockweed. In the Hebrides driftweed is mainly used as manure, and it is regarded as specially valuable for barley crops. The ground is covered to a depth of 3–4 in (7·6–10·2 cm), but it is not ploughed in, which is a somewhat wasteful method.

2.4.4 England

Agricultural practice in southern England was rather different from that of Scotland, because in Cornwall the seaweed was first mixed with sand and allowed to rot before being used at the rate of about 10–12 tons per acre (10·2–13·9 tonnes per 0·4 ha). In the Isle of Man seaweed is still used for potatoes, broccoli and other green crops. The seaweed is usually put in a pile alternating with layers of stall manure and then left until it has composted.

In both the Scilly and Channel Isles seaweed manure also forms an integral part of their agriculture and has done so for many centuries. In the various islands each species has its own particular value to the farmers, e.g. in Jersey *Fucus serratus* is the species most highly regarded. In both groups of islands the usual practice is to put on about 45–50 tons (45·7–50·8 tonnes per 0·4 ha), per acre in the autumn where it is intended to grow early potatoes.

It has been found that seaweed manure dug into the soil to a foot below the surface increases the fruiting period of tomatoes and it has also been claimed that it renders them free from blight. A claim such as this requires substantiation from carefully designed experiments (see p. 52). There is also a report that potatoes grown on land manured with seaweed are less susceptible to scab disease and also to the virus disease known as leaf curl, but this requires confirmation.

2.4.5 America

On the Pacific coast of North America the big kelps are collected and after being chopped up are used wet. There is therefore no transportation to a drying centre. This technique was used by a firm known as the Pacific Kelp Mulch Company. Other firms attempted to dry weeds in the sun, but it was found to be difficult because of the viscous exudate. The weed is now transported to a central factory and dried in large horizontal driers, about 50 ft (15·24 m) long and 5 ft (1·52 m) in diameter, which are capable of handling four tons (1·02 tonnes) of wet weed per hour. The dried kelp still contains about 15% of water. Farther north, in Alaska, *Alaria fistulosa* is regularly employed as a manure for potatoes and the results are extremely satisfactory.

2.4.6 New Zealand

In the southern hemisphere, coast dwellers in New Zealand have made use of algae as manure. *Macrocystis pyrifera, Lessonia variegata* and *Ecklonia radiata* are the bottom weed species that have been used, whilst *Carpophyllum, Cystophora (Blossevillea)* and *Sargassum* are the principal low-water forms. The cost of driftweed removal in New Zealand is, however, expensive. The principal objection to a more extensive use of seaweed as a manure in New Zealand, and indeed anywhere, is that it is heavy bulky material (it contains 90% water), and, unless dried, it must therefore be utilized near its source. It is, therefore, only profitable to establish an industry in places where large quantities are likely to be continually available. Rapson *et al.* (1942) reported on attempts to use dried and ground *Macrocystis* as a manure but nothing eventuated. Grimmet and Elliott (1940) prepared a dry manure from New Zealand *Macrocystis* which contained 17% of potash and 2% of nitrogenous material. The bull kelp, *Durvillea,* which is also abundant in these waters, would not be a satisfactory alga for manurial purposes since it only contains about 1·75 % potash and 0·7% nitrogen (Aston, 1916). Field experiments carried out in New Zealand have shown that plots fertilized with dried *Macrocystis* yield crops only slightly inferior to plots treated with an equivalent amount of artificial manure containing 30% of potash salts.

Little (1948) studied the rate of decomposition of certain large New Zealand brown seaweeds when dug into the soil. *Macrocystis, Ecklonia* and *Durvillea* decompose completely within four months, but *Carpophyllum* is still recognizable after one year. The bulk of the sodium, potassium and chlorine is released in the first fourteen days, so that these elements are made available very rapidly. It is likely therefore that the minor elements will be made available equally rapidly.

2.4.7 Other countries

In Japan algae which are not used for other purposes (pp. 72–86) are used for manure, either alone or with other materials (Davidson, 1906). Species of *Sargassum* are used in China (Liang-Ching-Li, 1934) and they are usually collected from driftweed. It is used fresh, or dried, or burnt and the ash then applied to the soil. Peanuts and sweet potatoes are the crops for which it is primarily used (Tseng, 1933). In Hainan, and also in India and Ceylon, it is employed, together with *Gracilaria* and fish offal, for coconuts and coffee brushes (Kirby, 1953; Chidambaram and Unny, 1953). Seaweed as coconut manure is also recorded from Brazil where the red alga *Hypnea* is most used (Schmid and Hoppe, 1962) and the West Indies (Honduras, Grenada). Coastal farmers in Brazil not only use *Hypnea* but *Ulva* and *Enteromorpha* also. In Chile green, brown and red seaweeds are all used as manure whilst in South Africa a mixture of *Ecklonia maxima* and *Laminaria pallida* is sold as a soil conditioner.

2.5 LIQUID MANURES

In recent years dried seaweed meal and liquid extracts have been increasingly employed by horticulturists, market gardeners, farmers and orchardists. With the gradual exhaustion of presently known mineral fertilizer supplies over the next century, it is likely that even more use will be made of this annually renewable source.

Dried meal, comparable to that used for animal fodder, obviously takes longer to be effective than a liquid extract because it must undergo bacterial breakdown. It is therefore better used where consistent, long-term effects are desired. After application there will be a temporary decrease in available soil nitrogen which is utilized by the bacteria breaking down the meal. Later there will be an increase in the available soil nitrogen. The dried meal, with its slow release of alginic acid and other polyuronides, represents a better soil conditioner than does a liquid extract. According to Stephenson (1974) its use is well justified over the years for crops such as potatoes, asparagus, flowers, fruit and hops but not for cereals.

The liquid extracts operate more rapidly because the compounds are already dissolved and in a state that the plants can use. There is absorption also through the leaves as well as through the roots.

In recent years various liquid extracts of brown algae have appeared on the market. United Kingdom brands are 'Maxicrop' and 'Alginure' whilst 'Seagro' is manufactured in New Zealand. 'Maxicrop' is primarily used for gardens and glass-house crops whereas the others are mainly promoted for pastures, orchards and field crops. 'Maxicrop' is very wide-spread and is exported to Australia, Bahrein, Barbados, the European Continent, Bermuda, British Guiana, Canada, Ceylon, Curaçao, Falkland Island, Finland, Ghana, Iceland, Jamaica, Kenya, Malawi, Mauritius, New Zealand, Nicaragua, Nigeria, Rhodesia, Singapore, Thailand, Trinidad, USA and Zambia. 'Maxicrop' is used on citrus in British Guiana, on citrus and grapes in Greece, on orchids in Belgium, on garden crops in Thailand and on glass house crops in Iceland (Stephenson, 1974). 'Seagro' in New Zealand is primarily used on pasture but it has an additional use on orchard crops.

Table 2.5 Analysis of typical 'Maxicrop'

Dried material	Moisture	5·2%	CaO	0·44%
	Solids	94·8%	Fe_2O_3	0·34%
Solids	Organic matter	51·2%	Al_2O_3	0·23%
	Ash	43·6%		
Ash	Nitrogen (N)	0·72%	Copper	40 ppm
	P_2O_5	2·0%	Cobalt	4 ppp
	SO_3	6·1%	Nickel	24 ppm
	Cl	6·7%	Zinc	100 ppm
	S_1O_2	0·2%	Molybdenum	10 ppm
	Iodine	0·9%	Manganese	40 ppm
	Bi	0·8%	Boron	1 ppm
	Na_2O	18·9%		
	K_2O	3·0%		
	Mgo	0·58%		

'Maxicrop' is manufactured by a process of alkaline hydrolysis (full methods are given in British patents Numbers 664989 and 909563) mainly from *Ascophyllum*, With some formalin added as a preservative. 'Seagro' is also manufactured from Norwegian *Ascophyllum* by a cold alkaline process whereas 'Alginure' is manufactured from oarweeds. A typical analysis of 'Maxicrop' is given in Table 2.5. Because seaweeds are low in concentrations of phosphorus some manufacturers add more of this elements to the seaweed extract. On the other hand there are experiments which indicate that seaweed extract has the effect of releasing phosphorus that is normally bound within the soil.

The exact composition of a liquid seaweed extract or of dried ground seaweed can vary with locality, season and latitude, but in the case of extracts such as 'Maxicrop' and 'Seagro' there is likely to be only a small seasonal difference (see also below).

A somewhat unusual algal manure is that known as ficoguanoide or algal guano. This is composed of the blue-green alga, *Microcoleus chthonoplastes,* which occurs on the Atlantic coast of Spain in sufficient quantity to be collected. The total nitrogen content of this manure varies between 0·029 and 0·04% (Guerrero, 1954).

2.6 MANURIAL VALUE

It is evident from the above account that there is a fairly widespread, though locally restricted, use of algae as manure or as a plant nutrient, though the latter use is spreading. All the algae are relatively high in nitrogen and potash but they are low in phosphorus content. Indeed the continued use of seaweed manure, without some balancing with artificial phosphate, is increasingly less satisfactory as all the phosphate becomes extracted from the soil. However, phosphate, if added, reportedly need not be applied necessarily every year. Table 2.6 compares the composition of one ton of average wet seaweed with one ton of average farm manure from western Scotland (Anon, 1940). More recent analyses, if they were available, would not show any great difference.

Table 2.6

	Nitrogen	Phosphoric acid	Potash	Common Salt	Organic matter
	(lb)*	(lb)	(lb)	(lb)	(lb)
1 Ton† wet weed	11	2	27	35	400
1 Ton farm manure	11	6	15	—	380

* 1 lb = 0·454 kg.   † 1 ton = 1·016 tonnes.

Unfortunately, although there is plenty of nitrogen present it is not immediately readily available (see p. 48). The available potash and organic matter in seaweed manure are rapidly used up by the plants, and for this reason the Hebridean crofters reckon that one cartload of farm manure is equivalent to two and a half loads of seaweed, because the effect of the farm manure lasts longer. Seaweed, however, has one advantage over farmyard manure on account of its freedom from weeds and fungi. Because of its high potash content seaweed manure is particularly suitable for root crops and cabbage, and in the case of mangolds the high salt content makes it even more valuable.

A rather better comparison with normal farm manures is provided in Table 2.7. It will be seen that the algae and wet kelp are not so good as the other manures in nitrogen and phosphate content, but that they excel nearly all in potash content.

Analyses from America (Table 2.8) show that weeds from northern California are richer in nitrogen, potash and phosphates than the same species growing father south, but that the reverse is true for the organic matter and iodine in the fronds.

The variations in the chemical composition of seaweeds in relation to change of latitude in the Pacific are extremely provocative. It is probable that the phenomenon is more widespread, but it will require additional analyses of samples on a far larger scale than heretofore before any interpretation becomes possible. Table 2.8 shows that the composition of the different organs varies, but this is typical of nearly all plants. The time when the dried weed is applied to the land is obviously of some importance, but at present our information can hardly be regarded as adequate. The evidence available suggests that the mineral and nitrogen content is at its maximum between January and May (Fig. 2.2) (Lunde, 1937; Black, 1948 et seq.; Macpherson and Young, 1949). These happen to be the months when there are good casts. In British Columbia mineral and nitrogen content in *Macrocystis* and *Nereocystis* is also maximal in December, January and March (Wort, 1955).

In Tables 2.9 and 2.10, figures are provided for fresh weight and dry weight analyses of the more important European algae.

Table 2.7 (After Hoagland, 1915)

	Moisture	Nitrogen	Potash	% Phosphates (dry wt)
Horse manure (fresh)	—	0·45	0·35	0·2
Stable manure	73·3	0·5	0·6	0·3
Green alfalfa	75·3	0·7	0·45	0·15
Cow peas	78·8	0·25	0·3	1·0
Garbage tank	—	0·7	0·3	0·6
Street sweepings (Washington D.C.)	—	0·85	0·55	0·55
Wet kelp (species unspecified)	85	0·3	0·5	0·2
Laminaria digitata*	75–82·9	0·22–0·3	1·2–1·83	—
Ascophyllum	69·6	0·3	0·8	—

* Hendrick (1916).

Table 2.8

	Organic matter	% Nitrogen (fresh wt)	% Phosphates (fresh wt)	% Potash (dry wt)	% Iodine (dry wt)
Macrocystis. Leaves					
N. (Pacific Grove)	7·9	2·2–2·67	1·03	12·55	0·14
S. (San Diego)	9·5	1·25	0·73	10·71	0·22
Macrocystis. Stem					
N. (Pacific Grove)	5·5	0·75–1·11	0·57	22·01	0·13
S. (San Diego)	6·5	0·55–0·71	0·55	19·49	0·12
Nereocystis					
Leaves		2·85	0·85	18·65	0·10
Stipe		1·23	0·52	26·37	0·07
Pelagophycus					
Leaves		1·55	0·83	18·65	0·32
Stipe		1·00	0·56	29·52	0·13

A comparison of the fresh weight analyses with comparable analyses of Pacific algae (Table 2.8) shows that the European species are not so rich in nitrogen and phosphates, but they are apparently richer than *Macrocytis* in organic content. This last feature may not be of great significance because it depends upon the proportion of organic material that can be converted in the soil into a suitable humus.

It is important to note that only fresh material should be used for analysis

Table 2.9

	L. digitata		L. saccharina			% of fresh weight F. serratus	F. vesiculosus	Ascophyllum	Range
	Stipe	Leaf	Stipe	Leaf					
Water	82.9	75.0	83.2	78.5		76.3	68.1	69.6	68.0–83.2
Organic matter	11.0	19.7	11.0	16.7		20.75	25.5	24.15	11.0–25.5
Nitrogen	0.22	0.3	0.3	0.2		0.3	0.3	0.3	0.2–0.3
Potash	1.83	1.2	1.9	0.95		1.0	0.95	0.80	0.8–1.9
Phosphates	—	—	—	—		—	—	—	0.2–0.17

Table 2.10 Percent composition of dry matter of seaweeds (after Hendrick, 1916)

	Laminaria hyperborea*		Laminaria digitata*		Fucus vesiculosus	Fucus serratus
	Stipes	Fronds	Stipes	Fronds		
Organic matter	64.25	77.99	65.27	77.53	79.7	78.63
Nitrogen	1.9	1.19	0.98	4.0	0.99	1.19
Water sol. ash.	28.08	16.92	28.64	17.91	15.92	15.87
K_2O in Ash	10.66	5.19	11.85	4.54	2.76	3.77

*Augier and Santimone, 1978

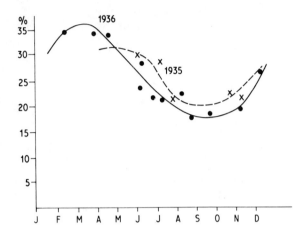

Fig. 2.2 Variations in the ash content of frond of *Laminaria digitata*.

Table 2.11 Minor elements (ppm) (after Johnston, 1966)

	Fe	Mn	Cu	Zn	Mo	Co	Sr
Laminaria digitata frond	138	9	3–4	64	0	0·29	4000
Laminaria digitata stipe	293	10	5	62		0·92	4000
Laminaria hyperborea frond	159	10	14	76		0·25	
Ascophyllum	283	27	1·1–4	60	0·29	0·73	2600
Fucus spiralis	638	104	6	62	0·29	1·39	
Fucus serratus	375	155	5–17	70	0·65	0·84	2800
Fucus vesiculosus	221	116	3·4–7·4	60	0·34	0·65	

because washing, either by rain or artificial means, removes some of the salts and results in false values. Beckmann (1917) found that thorough washing reduced the ash content of *Ascophyllum* by 14% and of *Laminaria digitata* by 13%. Vincent (1924) reported that this lost fraction consists to a very large extent of carbohydrates, nitrogenous material and salts. On the other hand, the effect of washing serves as an indication of the rapidity with which the mineral salts become available.

The amount of potentially available nitrogen in the giant kelps is 1–2%, but it is only liberated slowly into the soil. The availability of the nitrogen appears to depend upon the type or condition of the weed, e.g. that of *Nereocystis* is readily available whilst that of *Pelagophycus* is not. In the case of *Macrocystis* the availability of the nitrogen decreases on drying, so that it is desirable to use this alga when wet. It has been estimated that the organic matter of these three seaweeds is sooner or later converted to valuable humus and that it increases the humic content of the soil to an extent comparable with a crop of alfalfa or stable manure and straw.

Table 2.12 Trace elements in seaweed meal

Element	Grass (g ton^{-1})*	Ascophyllum meal (g ton^{-1})	Max. amount lost from soil (g acre^{-1})†	Amount added 7 lb art. supply (g acre^{-1})
Vanadium	0·06	0·6	0·12	340
Cobalt	0·14	1·4 (1–10)	0·28	190
Iodine	0·50	500 (–1200)	1	2430
Molybdenum	0·82	(0·3)–1·25	1·64	1740
Copper	4·6	61 (1–10)	9·2	810
Boron	10	167 (40–100)	20	362
Iron	56	1132 (150–1000)	112	640
Zinc	56	110 (50–200)	112	724
Manganese	95	45 (10–50)	190	785

Figures in brackets from Jensen *et al.*, 1968.
* 1 ton = 1·016 tonnes.   † 1 acre = 0·4 ha.

Many plants require minute quantities of certain elements if they are to grow properly. If sufficient of these trace elements is not present the plants suffer, in the same way as animals (see p. 37), from what are known as deficiency diseases. The symptoms of these diseases are varied, but they frequently involve changes or 'burning' of the leaves. Seaweed manure may be valuable because it contains a number of these trace elements, e.g. manganese, boron and barium (Table 2.11). These vary seasonally, as do the major elements. In the oarweeds they are more abundant in the stipe than in the lamina and are less than in the rockweeds (Black and Mitchell, 1952). This aspect of seaweed manuring has, however, been more or less neglected.

The great capacity of these brown algae to accumulate strontium is highly significant in relation to radio-active strontium fall-out from atomic explosions. Fish and other marine animals eating these algae will also accumulate the strontium. Seaweeds appear to be highly sensitive to the uptake of radioactive ions and it has been suggested they could be used to measure the radioactivity of the surrounding water (Czapke, 1966).

A comparison (Table 2.12) of the trace element content of *Ascophyllum* meal with grass, the amount lost per acre from the soil and the amount supplied by artificial fertilizer, shows that seaweed in the liquid form (p. 43) is probably more efficient (Booth, 1966).

Whilst much has been made of the trace elements present in dried seaweed or in the liquid extract as a basis for the results observed when either are used for plant growth, recent studies tend to suggest that it may really be the growth hormone content that is responsible for the results. Booth, back in 1969, stated that the manurial value of either meal or spray was not related to the nitrogen, phosphorus or potash content but was more likely to be related to hormonal or trace element content in view of the unusual properties – e.g. promotion of seed germination, increased frost hardiness, increased resistance to fungal and insect pests. It has been known for some time that auxins and gibberellins are present in the large brown seaweed *Laminaria, Ascophyllum, Fucus* spp. (Mowat, 1964; Augier, 1976) and both these compounds are known to be important in the growth of terrestrial plants as well as that of the algae. Furthermore Augier (1976), Fries (1979) and Taylor and Wilkinson (1977) have reported the presence of phenylacetic acid (PAA) as a growth agent in large brown macroalgae, as well as a closely related compound p-OH-PAA. These may well be the two extra auxins referred to by Stephenson (1974) as occurring in *Laminaria* and *Ascophyllum*. These two compounds are regarded as encouraging growth by division of plant cells rather than by elongation, as mainly occurs with the auxin indole-acetic acid (IAA). Plant physiologists consider that there is considerable evidence to indicate that auxins (IAA, PAA, indolyl-3-acetic acid, didolyl-3 carboxylic acid) participate in the growth control of the various plant organs.

The later discoveries of the gibberellins (and at least two of the known 23 members of the group are regarded as present and available in dried seaweed or liquid seaweed extract) provides the presence of compounds that promote growth and cell division in higher plants as well as causing resumption of growth in dormant seeds and buds, both of which are features following liquid seaweed application.

Indole-Acetic Acid (IAA)   Gibberellic Acid (GA$_3$)

Still later (1954–56) came the discovery of the cytokinins and in particular the compound kinetin.

These growth hormones are apparently released fairly steadily from the algae because Fries (1973) has reported their presence in coastal waters. In the higher plant it is the interaction of all these growth compounds, together with some growth inhibitors (Hussain and Boua, 1973) that controls growth (Black and Edelman, 1970). It is evident, therefore, that addition of any of the growth promoters is likely to affect growth.

Blunden (1977) has experimentally demonstrated that it is indeed cytokinins that increase tuber yield in potatoes by 112%, increase protein content in grass by 0·6–0·7% and increase the keeping time of limes (time to yellowing) (Table 2.13).

A study by Kingman and Senn (1977) on the effect of liquid seaweed extract on *Helianthus annuus* (sunflower) indicated that the responses were mainly due to

Table 2.13 Mean time taken after immersion in diluted seaweed extract or in kinetin solutions for limes to become yellow (after Blunden, 1977).

	Mean time (days) for limes to become yellow			
	Cytokinin conc. (ppm) calculated as kinetin			
	30	0·15	7·5	0
SM3	75·1	67·9	61·2	—
Marinure	71·7	50·8	57·7	—
Kinetin	83·6	58·4	47·9	—
Water	—	—	—	55·6

Table 2.14 (After Stephenson, 1974)

Crop	% gain or loss 250 lb* per acre	% gain or loss 500 lb per acre
Sweet pepper	31 gain	13 gain
Tomato	37 gain	11 gain
Sweet corn	104 gain	75 gain
Okra	3 loss	7 gain
Peas	6 loss	4 gain
Sesame	17 loss	4 gain
Cotton	8 loss	78 loss
Soya bean	23 gain	20 gain
Lima bean	12 gain	zero change

* 1 lb per acre = 0·454 kg per 0·4 ha.

the hormone groups, but that the gibberellins and cytokinins were the most important, the latter affecting nitrate uptake. Unfortunately these compounds have only a limited life in the liquid extract and therefore it should not be kept but used soon after manufacture.

As may be expected the use of liquid seaweed extract varies with the type of plant for which it is used. In the case of soya bean there is no increase in yield but there is an increase in the protein content (Senn and Kingman, 1977) and the same increase in protein has been reported for pasture grass. This increase is reflected in the meat quality of the animals that graze the grass.

Over the last 30—40 years numerous trials have been conducted with dried seaweed meal or liquid extract on a range of hops. McIntosh apple trees produced larger and more abundant fruit with a hay or seaweed mulch than with the use of sawdust or grass.

A further study at Clemson University in South Carolina, using a range of crops

Table 2.15 Potato yields from control plants and plants treated with seaweed extract* (after Blunden, 1972)

Row number	Potato yield (kg) per row	
	Control plants	Seaweed-extract-treated plants
1	1238·3	1524·1
2	1047·8	1587·6
3	1079·6	1492·3
4	1111·3	1428·8
5	1079·6	1587·6
	5556·6	7620·4

* Yield calculated from number of 31·75 kg bags filled.

and two alternative seaweed meal (Norwegian) treatments of 250 lb per acre (114 kg per 0·4 ha) and 500 lb per acre (228 kg per 0·4 ha) with a basal dressing of 1000 lb per acre (456 kg per 0·4 ha) artificial fertilizer produced both gains and losses (Table 2.14).

A study of residual fertility showed that application at 250 lb per acre was more effective. It was obvious that a crop such as cotton does not respond to the application of seaweed.

Blunden (1972), using a product SM3 of Chase Organics Ltd, reported that bananas gave an increase in bunch weight from an average of 14·57–17·69 kg, that gladioli corms increased in weight, and that there were increased yields with potatoes (Table 2.15), maize (Table 2.16), peppers, tomatoes, pineapples and oranges (Table 2.17)

Experiments with tomatoes have generally resulted in an increase in yield but in this case a 1 in 25 solution was found to be better than a 1 in 50 (Stephenson, 1974). Further study of the crop is perhaps necessary because Senn and Kingman (1977) reported no significant difference on tomato crops. Heavier crops of grapes, cucumbers and gherkins have been reported from France and Praha (Algea Produkter). Povolny (1971) stated that the marketing quality of stored cucumbers was prolonged by 14–21 days after spraying with algifert. Turnips grown on a sandy soil in Norway over a period of three successive years showed that the use of seaweed at 125 or 250 kg per decare produced a much greater yield than basic artificial fertilizing over the first two years but that in the third year the difference was not nearly so great.

Other crops that have responded to liquid seaweed treatment include Brussels sprouts, potatoes, carrots and beetroot. With the Brussels sprouts there was an estimated increased yield of 12 cwt (0·42 tonnes) on a 5 ton per acre (5·1 tonnes per 0·4 ha) crop and with potatoes an average increase of 6·8% with standard 'Maxicrop' and 18·9% with extract plus chelated iron (Stephenson, 1974).

Turning to fruits, application of liquid seaweed extract has given increased yields

Table 2.16 Ripe ears of corn collected from control plants and plants treated with seaweed extract (after Blunden, 1972)

		No. of ripe ears collected		
		Control plants		Seaweed-extract-treated plants	
Date of collection		Row 1	Row 3	Row 2	Row 4
July	22, 1963	1	2	1	6
	23, 1963	1	4	8	11
	26, 1963	2	7	7	7
	28, 1963	11	7	14	16
	30, 1963	4	7	17	9
	31, 1963	4	9	6	9
August	1, 1963	2	8	5	6
	3, 1963	4	2	3	5
	5, 1963	11	0	2	1
		39	46	63	70

Table 2.17 Pineapple and orange yields from control trees and trees treated with seaweed extract (after Blunden, 1972)

		Average orange yield per tree (boxes)		
Variety	Year	Seaweed-extract-treated trees and plants	Control trees and plants	% increase
Hamlin and Parson Brown	1966	11·85	11·30	4·9
	1967	11·75	11·14	5·5
	1968	10·60	9·77	8·5
Pineapple	1968	5·60	5·29	5·9
	1969	8·48	7·51	12·9
	1970	5·53	6·72	12·1
	1971	9·99	8·89	12·4

in strawberry crops (19–133%), peaches and blackcurrants, the last named giving increases of 12 and 27% in two experiments with the use of a very dilute solution (1 in 400) three times at fortnightly intervals (Stephenson, 1974).

Seaweed extract can also be used on flowers though here it is difficult to secure quantitative data. Trials at Clemson University with poinsettias have shown that there is an increase in the number, weight and quality of the flowers with low concentrations of extract or meal (Table 2.18).

Other features associated with the use of seaweed include improved germination of seeds of grasses, (Senn and Skelton, 1969) cereals, vegetables and flowers.

Table 2.18

Control	Average number flowers 3·8	Diameter flowers (ins)* 14	Average weight flowers (mg)
1 in 50 seaweed solution	4·7	14	106
1 in 25 seaweed solution	4·5	13	92
1 in 5 seaweed solution	4·4	12	81
Meal at 300 lb per acre†	4·5	14	123
Meal at 600 lb per acre	4·9	14	126
Meal at 900 lb per acre	4·5	14	119

* 1 in = 2·54 cm.   † 1 lb per acre = 0·454 kg per 0·4 ha.

Table 2.19

	Number of rotten fruit								
	Days after picking								Total
	4	5	6	7	8	9	10	11	12	
3 Untreated plots	11	12	3	0	0	2	0	9	0	37
3 Treated plots	3	0	1	1	0	1	0	1	1	8

Improved seed germination is not necessarily wholly related to the seaweed extract but also may depend upon the soil type. Thus Goh (1971) reported that use of seaweed extract promoted germination of white clover on mineral pasture soils of low fertility but had no effect on garden soils of high fertility. From Europe and the USA there are consistent reports of the reduction in red spider mites on both fruit trees (apples), cucumbers (Povolny, 1971) and flowers (chrysanthemum), possibly caused by interference in the breeding cycle. Experience in New Zealand has shown that use on pasture reduces the incidence of grass grub, possibly because the uptake of iodine may make the roots impalatable.

The use of seaweed extract also discourages aphids, especially the black bean aphis, as well as those occuring on sugar beet and tomatoes. In the case of cucumbers, red mite is controlled (Povolny, 1971). It is also very useful in controlling certain fungal infestations — in particular *Botrytis* on strawberries, the treated plants not only showing a great reduction in infestation but also giving a higher yield. Sprayed plants are much less susceptible to attack by mildew, e.g. turnips, or Verticillium wilt (carnations).

Two final features associated with the use of liquid seaweed extract are an increase in frost hardiness and in shelf life of fruit. Tomatoes, citrus and celery all overcome frost periods without damage whilst untreated plants undergo great loss. It is thought that the vitamins present in meal and extract may be responsible for this effect.

Peaches sprayed with extract may have a shelf life up to four times longer than those not sprayed (Table 2.19) and the same applies to apples, apricots, oranges and gherkins (Povolny, 1969, 1972).

Skelton and Senn (1969) reported that in the case of peaches early spraying by liquid seaweed extract during full bloom definitely increased the quality and shelf life but had no effect on firmness or soluble solids. Spraying later in the season was not so effective. However, not all varieties of fruit respond equally well. Thus Povolny (1969) reported that spraying improved the storage quality of Cox's Orange apples, but was less effective with Nonnetit and had no effect of Goldparmane.

In the same way that seaweed meal appears to exert a more beneficial effect with poor animals (see p. 38) so there is evidence that the application of seaweed meal as a manure is more effective on special or poor soils than on soils that are already very good.

Both meal and extract can be used in the process of making compost. Peat can be soaked in the extract before it is mixed in compost or in a potting mixture. Poor soils, especially sandy soils, can be greatly improved by the addition of seaweed meal or even of raw seaweeds, and, because of its composition, seaweed meal improves the physical state of soils derived from chalk. Blunden and Woods (1969) at the Sixth International Seaweed Symposium pointed out that the carbohydrates of the large brown algae (e.g. algin, mannitol, laminaran) were all very important when seaweed was used as a fertilizer. These carbohydrates all exert an effect upon the physical nature of the soil and are really insignificant so far as plant food value is concerned.

Although it is evident that quite a considerable body of information exists about the composition of seaweeds as a manure, and their effect on the chemical composition of the soil, we still know very little about their effect on the physical properties. The presence of algin and the existence of algin-decomposing bacteria (*B. terrestralginicum*) that produce a form of lignin, must have an effect on the physical condition of the soil, consolidating and binding more sandy soils, and loosening the more clayey soils. However, we need to know much more about this aspect and it is here that future research should be directed.

2.7 KELP FOR ENERGY

The pressure to use biomass for energy production, human food and stock feed and to provide plant nutrients has in the past few years resulted in what is probably the most exciting concept concerned with organic chemicals from the sea. This is an energy-food-fertilizer farm using kelps and established in the open ocean, each farm occupying up to 100 000 acres (40 000 ha) (American Gas Association, 1977). The first quarter acre (0·1 ha) unit was established late in 1978.

The conversion of large brown algae to methanol presents no great problem and this can be converted economically to gasoline (Wise and Silvestri, 1976) (Fig. 2.3). It is likely that a similar process will enable the conversion of ethanol to gasoline

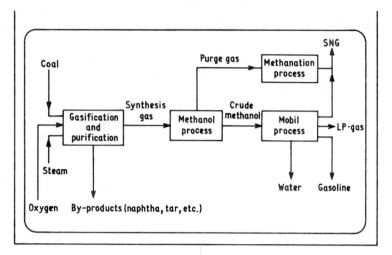

Fig. 2.3 Coproduction of gasoline and S.N.G. (from Oil and Gas Journal, 22 November, 1976).

to be equally economic or cars will be developed to run on ethanol. Fig. 2.3 shows the co-production of high quality gasoline and S.N.G. by the Mobil process with a thermal efficiency of 65·5%.

The concept of the kelp farm (Wilcox and Leese, 1976) has been developed to ensure adequate supplies of algae that can be harvested economically. At the present time attention is centred around the large Pacific kelp, *Macrocystis pyrifera*, (Fig. 1.8), which grows in extensive beds that are regularly harvested off the Pacific USA. The algal genus also occurs off the coasts of South Africa, South America, Australia and New Zealand. Experimental work in California (North, 1974) has shown that plants can be cultured and nets innoculated with sporelings, or larger plants can be readily transferred to nets. However, *Macrocystis* is not the only large brown alga that can or could be harvested mechanically. Other algae that should be considered and experimented with now are *Pelagophycus porra* (Pacific USA), *Ecklonia buccinalis* (South Africa) and *Durvillea antarctica* (Australia, New Zealand and South America).

In the case of *Macrocystis*, which has been unsuccessfully exploited in Tasmania for algin production, experiments have shown that it can be grown successfully outside its present distribution area in the West Indies (La Croix) and in the English Channel. In the case of France the plants were removed before maturity and reproduction because of opposition to its introduction. The use of any of the large algae for farming immediately raised the problem of a potential 'weed' species that could change an existing ecosystem. For this reason it will be wisest to restrict kelp farms to ocean areas where the alga grows naturally. There is clearly a big potential for both *Macrocystis* and *Durvillea* in ocean areas off southern Australia and New Zealand.

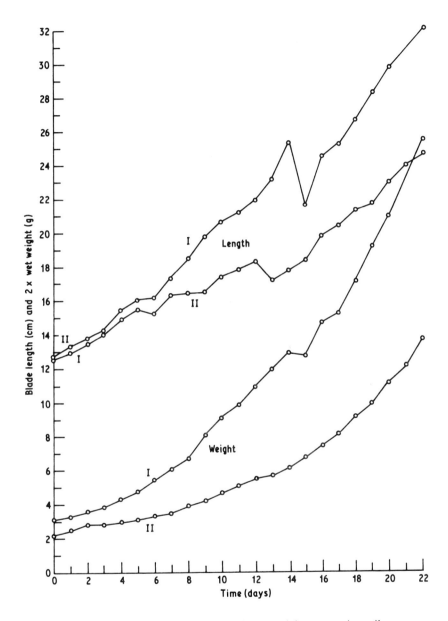

Fig. 2.4 Daily changes in blade length and wet weight among juvenile *Macrocystis* sporophytes grown in up-welled Atlantic water from 870 metres depth at St. Croix (I) and in water from Newport Bay, California (II).

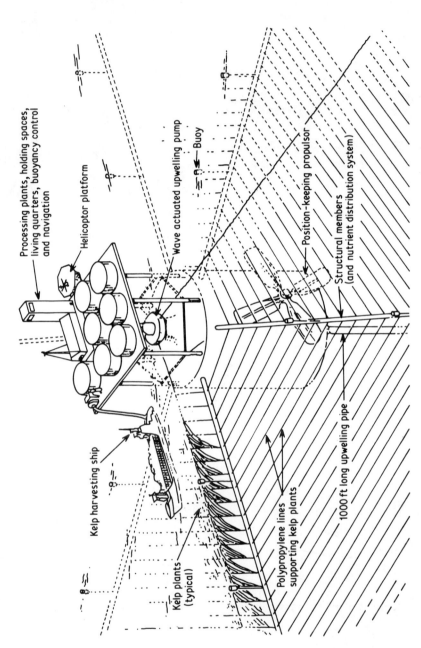

Fig. 2.5 Conceptual design 1000-acre ocean food and energy farm unit.

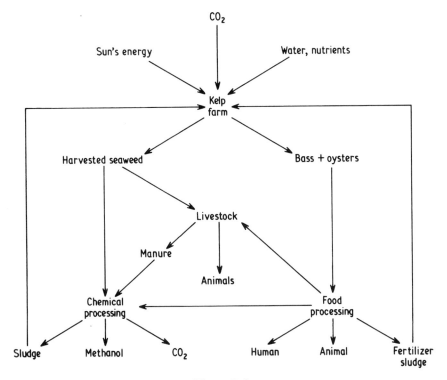

Figure 2.6.

Initial experiments using *Macrocystis* on nets attached near the shore resulted in problems with ocean currents, wave action and shipping. However, they did demonstrate the value of locating farms in areas where there was natural upwelling of nutrients or in areas where upwelling could be generated artificially. Jackson (1977) showed that low nutrients in surface waters provide the primary limiting factor. At St. Croix the test plants grown in upwelling water grew better than comparable plants in California (Fig. 2.4).

It is now proposed to use the solar energy falling on the open ocean and employ a buoyancy controlled structure (Fig. 2.5). 'Potentially a square mile of ocean farm could produce enough energy and other products to support more than 300 persons at today's US per capita consumption level, or more than 1000 to 2000 persons at today's world average per capita consumption level' (Wilcox and Leese, 1976).

The advantage of *Macrocystis* and other large brown algae is that they are immune to frost or drought and in the case of *Macrocystis* individual frond life is 6 months so that 2–3 crops per year are feasible. It is estimated that a 100 000 acre (40 469 ha) farm would produce $15-45 \times 10^9$ ft^3 ($42.48-126.62 \times 10^7$ m^3) methane per year at a cost of \$ 3–9 per 1000. Fig. 2.7 illustrates the general flow pattern. Initial cost and productivity estimates make the project roughly comparable with future sources of energy.

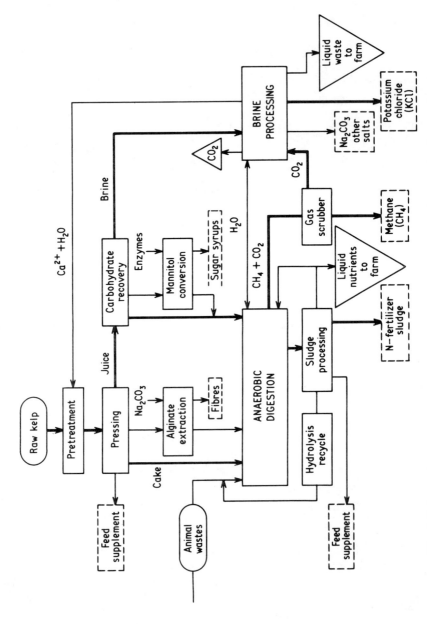

Fig. 2.7 Ocean farm project: process chart for production of methane and other products.

The present base line design envisages a farm of 100 000 acres (40 469 ha) located 100 miles (1·61 km) off the California coast and divided into 10 acre (4 ha) units set a 100 ft (30·38 m) down with provision for upwelling from 300 ft (91·47 m) down. Production is estimated at 340 tons (345·4 tonnes) wet weed per acre per year. This would yield 42·5 tons solid material per acre per year (43·2 tonnes per 0·4 ha per year) of which 23·8 tons (24·2 tonnes) would be volatile, with a conversion rate of 49% volatiles to S.N.G. One could also farm kelp bass and oysters concurrently. Mineral salts would be recovered and sold as fertilizer whilst animal food and fertilizer would be produced from the sludge. Wastes would be sprayed over the kelp as fertilizer for the seaweed. Much of the work would be carried out on the floating platform, but S.N.G. production would be land based involving an area of about 580 acres (235 ha) and a work force of about 850 (Fig. 2.7).

It is estimated that the annual outflow from such a farm would be as follows:

Synthetic Natural Gas (S.N.G.)	$22·1 \times 10^9$ ft^3 ($62·6 \times 10^7$ m^3 at $3·65 per 1000 ft^3 (per 28·3 m^3)
Fertilizer Salts	$4·9 \times 10^5$ tons ($49·78 \times 10^5$ tonnes) valued at $30 per ton (per 1·016 tonnes)
Kelp Bass	$1·84 \times 10^5$ tons ($1·87 \times 10^5$ tonnes)
Oysters	$0·36 \times 10^5$ tons ($0·37 \times 10^5$ tonnes)
Feed/fertilizer ex sludge	$9·3 \times 10^5$ tons ($9·45 \times 10^5$ tonnes) valued at $30 per ton (per 1·016 tonnes)

New Zealand is currently interested in this possibility and I believe Australia should be also. Russia has set up its first experimental plantation, using *Laminaria japonica* off Priomorye, using ropes upon which to attach the algae. The average harvest is between 50–60 tons (50·8–60·96 tonnes) per hectare (Buyankina, 1977). There seems therefore to be a big future in such ocean farms.

3
Sea Vegetables
(Algae as Food for Man)

Seaweed as a staple item of diet has been used in Japan and China for a very long time. To a lesser extent various species have been employed in Europe and North America. On the Continent and larger islands where normal agriculture can be supported, there is no great demand for seaweed as food. There are, however, many islands where conventional agriculture cannot meet local demands, and it is here that people have, of necessity, relied on the sea as a major source of food.

A summary of the seaweeds eaten in Europe and North America is given in Table 3.1, whilst Table 3.2 sets out the main country of usage, the common local name and the way in which the seaweed is used.

Since large quantities of seaweed are available around the coasts of many countries all that is required is the labour to gather them. The fact that few are eaten in Western countries is probably mainly due to palatability. However, the supply of food for the increasing world population will become an increasingly serious problem, and use of seaweeds may have to become more widespread (Chapman, 1978).

In 1977, because of a growing interest in mariculture (see p. 241), more people gathering fresh seaweed for eating and more sea vegetables becoming popular, a committee was set up in the USA to standardized the names of the sea vegetables and their products. These have been incorporated in the subsequent text and also in Table 3.2 (Shurtleff, Madlener, Green and Schooner, personal communication).

3.1 EUROPE AND AMERICA

3.1.1 Chlorophyceae

Among the green algae the sea lettuce or green laver (*Ulva lactuca*) used to be eaten as a salad and it has also been used in soups (Sauvageau, 1920). It is, however, a food with a taste that needs to be acquired. Species of *Ulva* are eaten today throughout the West Indies (Jamaica, Trinidad, Grenada, Antigua, St. Kitts) and in Barbados they are employed to make a 'bush' tea (Michanek, 1975).

Table 3.1 Sea vegetables consumed in the world

	Western Europe	England and Wales	Scotland	Ireland	Mediterranean	Iceland	West Indies	Eastern Canada	Eastern USA	Alaska	Western USA	Chile
Ulva lactuca	×		×									
Alaria esculenta	×	×	×	×		×				×		×
Durvillea antactica												×
Durvillea utilis												×
Fucus vesiculosus	×		×	×								
Laminaria digitata	×	×	×									
Laminaria saccharina	×		×									
Laminaria bongardiana										×		
Nereocystis luetkeana										?	×	
Porphyra columbina												×
Porphyra laciniata	×	×										
Porphyra perforata	×									×	×	
Porphyra umbilicalis		×										
Chondrus crispus	×			×				×	×			
Gigartina stellata				×								
Gracilaria compressa		×					×					
Iridaea edulis			×			×						
Laurencia pinnatifida	×		×		×							
Palmaria palmata	×		×	×	×	×		×	×			

Table 3.2 Algal species, country, common name and principal usage

Species	Country	Common name	Usage	Special features
Annfeldtia gigartinoides	Hawaii, Japan	Sea Nibbles	Cooking	
Alaria esculenta	USA, UK, Greenland, Iceland	Alaria	Cooking or eaten raw	High Vit. B_6, K
Alaria crassifolia	Japan	Japanese alaria	Cooking	
Analipus japonicus	Japan	Sea fir	Cooking	
Ascophyllum mackaii	USA	Wrack	Steamed	
Asparagopsis taxiformis	Hawaii, Indonesia	Supreme limu	Cooking	
Bangia fusco-purpurea	China, Taiwan	Bangia	Salad	
Campylaeophora hypnaeoides	Japan	Ego nori	Cooked	
Caulerpa racemosa	Japan, Philippines, Indonesia, China, Taiwan	Sea grapes	Raw or as salad	
Chaetomorpha crassa	China, Taiwan, Philippines	Sea sphagetti	Salad, cooking	
Chondrus crispus	USA, Iceland, France	Irish Moss	Cooked	High Vit. A
Chondrus ocellatus	Japan	Japanese moss	Cooked	High Vit. A
Chorda filum	Japan	Mermaid's line	Salad	
Codium fragile	Japan, Korea	Codium	Sweet, soup, salad	High Fe
Ecklonia cava	Japan	Kajime	Soups	
Eisenia bicyclis	Japan	Arame	Soups	
Enteromorpha clathrata	USA, Taiwan, China, Tobago	Light green nori	Raw, salad	
Enteromorpha compressa	Japan, China	Yellow green nori	Raw, salad	
Enteromorpha intestinalis	Japan, China, Malaya, Philippines	Green nori	Raw, toasted, steamed	

Table 3.2 Algal species, country, common name and principal usage (continued)

Species	Country	Common name	Usage	Special features
Enteromorpha prolifera	Japan, China	Dark green nori	Raw, salad	
Eucheuma	Philippines, USA	*Eucheuma*	Jellies, raw	
Fucus vesiculosus	USA, Alaska	Rockweed	Tea	High Vit. A, Mg, protein
Gelidium amansii	Japan, China, Indonesia	*Gelidium*	Jellies	
Gigartina papillata	Iceland	Grapestone	Puddings	
Gloiopeltis furcata	Japan, China,	Funori	In soups,	High Vit. C
Gloiopeltis tenax	Taiwan	Purple funori	Fried, fresh	
Gracilaria verrucosa	Vietnam, S.E. Asia, Japan, Philippines	Ogo	Raw, salad, cooked	Very high Mn
Halosaccion glandiforme	USSR	Sea Sac.	Raw	
Hijikia fusiforme	Japan, China, Korea	Hijiki	Raw, cooked	High protein
Kjellmaniella gyrata	Japan	Sea banner	Soup	Very high Na
Laminaria angustata	Japan	Tender kombu	Vegetable, soup	Very high Fe
Laminaria japonica	Japan, Korea	Royal kombu	Soup, pickled, cooked	High protein
Laminaria longicruris	USA	Atlantic kombu	Cooked, soup	High protein
Laminaria angustata var. longissima	Japan	Barner kombu	Vegetable, soup	High protein
Laminaria japonica var. ochatensis	Japan	Soup-stock kombu	Soup	High protein, Vits.A,C
Laminaria saccharina	UK, France, Eire	Sugar kombu	Stip-fresh	High protein
Laurencia pinnatifida	USA	Pepper dulse	Seasoning	
Monostroma latissimum	Taiwan, Japan	Jade nori	Cooked, soup, seasoning	

(continued on the next page)

Table 3.2 Algal species, country, common name and principal usage (continued)

Species	Country	Common name	Usage	Special features
Nemacystus decipiens	Japan	Mozuku	Eaten raw	
Nemalion helminthoides	Japan, Italy	Sea Noodles	Raw, soup, salad	
Nereocystis luetkeana	USA	Bull whip kelp	Pickle, candy	
Nostoc spp.	China, Japan, S. America	Nostoc	Cooked, fresh, soup	High protein
Palmaria palmata	W. Europe, Iceland, Canada, USSR	Dulse	Raw, relish, drink	
Petalonia fascia	Japan	Sea petals	Soup	
Pleurophycus gardneri	USA	Tender kelp	Steamed, salad	
Polyneura latissima	USA	Polyneura	Soup, salad	
Porphyra miniata	USA, (Japanese)	Red nori	Soup, roasted	High protein
Porphyra nereocystis	USA (Chinese)	Rose nori	Soup, stews	High protein
Porphyra perforata	USA, Canada, (Indians)	Fluted nori	Cakes, dry, toast	High protein, Vit. C
Porphyra suborbiculata	USA, China, Korea	Korean nori	Soups	High protein
Porphyra tenera	China, Korea, Japan	Asakusa nori	Cooked, soups	High protein, Vit. A
Porphyra umbilicalis	UK, Eire	Purple nori	Fried, stew, salad	High protein
Porphyra yezoensis	Japan	Open sea nori	Cooked, soups	
Postelsia palmaeformis	USA (stipes only)	Sea palm	Candy, boiled	
Sargassum enerve	Japan	Japan sargassum	Soups	
Sargassum fulvellum	Japan, Korea	Sargassum	Cooking	
Scytosiphon lomentaria	Japan	Sugara	Soups	
Ulva lactuca	China, S.E. Asia, W. Indies, Philippines, Chile	Sea lettuce	Soups, salad, raw cooked, 'bush' tea	Very high Fe
Ulva linza	W. Indies	Slender sea lettuce	Soups, salad, raw, cooked, 'bush' tea	Very high Fe
Ulva pertusa	W. Indies	Lacy sea lettuce	Soups, salad, raw, cooked, 'bush' tea	Very high Fe
Undaria pinnatifida	Japan, Korea	Wakame	Soups, baked, raw	High protein

Table 3.3 Dulse produced (lb* dry wt) in Nova Scotia and New Brunswick (after Macfarlane, 1964)

	Nova Scotia	New Brunswick	(Value $ 000)
1940	3 000	68 000	—
1941	8 000	47 000	—
1942	8 000	38 000	—
1943	3 000	54 000	—
1944	2 000	50 000	—
1945	2 000	54 000	—
1946	6 000	70 000	—
1947	9 000	36 000	—
1948	11 000	36 000	—
1949	13 000	—	—
1950	8 000	64 000	14·3
1951	15 000	94 000	20·8
1952	—	86 000	25·9
1953	—	78 000	24·9
1954	1 000	96 000	24·3
1955	2 000	95 000	25·0
1956	1 000	76 000	17·8
1957	6 000	85 000	22·6
1958	2 000	101 000	32·6
1959	1 000	74 000	24·4
1960	3 000	99 000	38·9
1961†	562 000		35·4
1962†	76 000		27·7
1963†	78 000		29·0
1964†	100 000		39·0
1965†	63 000		22·0
1966†	724 000		50·0

* 1 lb = 0·454 kg.
† Canadian Bureau of Statistics.

An anonymous writer (1944) stated that the Germans collected and used algae in Norway, where they erected two bakeries in order to make 'bread' of dried, ground and desalted algae. It is also recorded that the Indians in Alaska made soups from the giant kelps of the Pacific coast of America. In the Commander Islands of the Behring Sea, *Fucus* stems are eaten raw whilst *Laminaria bongardiana* is a favourite course of its own.

3.1.2 Rhodophyceae

Among the red seaweeds there are a few which have been used for food. One of the more important of these in Eire is dulse, or *Palmaria (Rhodymenia) palmata* (Fig. 1.7). Dulse has also been employed in Iceland, where it is known as 'sol', since the eighth century. In the old days it was eaten along with dried fish, butter and potatoes, and in times of famine was even baked into bread. In Canada it is used as a 'nibble' with beer (Madlener, 1977), whilst in the USSR it is fermented into an alcoholic beverage.

In the United States, dulse grows in abundance in the Bay of Fundy (Wilson, 1943) and it is also collected on the shores of Nova Scotia (see p. 256). Today the industry in the Bay of Fundy exports the packaged dulse to Vancouver as well as selling it locally. Amounts harvested in some recent years are given in Table 3.3.

The harvest season lasts from mid-May to mid-October. The weed is dried by sun and wind and this must be done as soon as the plants are picked, because otherwise they perish. The process takes about six hours and when first dried the plants become black, but under proper conditions they return later almost to the natural colour (Kirby, 1953).

Ulva, Porphyra and *Palmaria* were all at one time commonly used on both the Atlantic and Pacific coasts of North America as well as Japan. It is said that Chinese residents at California still collect purple laver, *Porphyra perforata*, and use it for food. Three hundred thousand pounds of dried *Porphyra* were collected in 1929, some of which was exported to China and some used in Chinese restaurants in 'seaweed soup'.

In the West Indies jellies are prepared in Trinidad and Jamaica from species of *Gracilaria. Eucheuma isiforme* is said to be eaten in Antigua and Barbados, and another species in the Philippines.

There are two other red seaweeds that used to be eaten, principally in Scotland. These are 'pepper dulse', or *Laurencia pinnatifida*, which is a small tufted bushy plant that grows in rocks about mid-tide, and *Iridaea edulis,* which is also sometimes known as dulse because it is very like *Rhodymenia*: it differs from true dulse by being thicker and brighter red in colour. Sauvageau considered that it tasted better than dulse.

Mrs. Griffith, an eminent British algologist of the last century, pickled plants of *Gracilaria compressa* and she considered that they were excellent. Sauvageau, who appears to have sampled most of these edible algae, also recommended the brown seaweed *Dictyopteris polypodioides* as making good eating.

A very delectable product, which is sold in north-west America where it is commercially known as 'Seatron' is prepared by removing the salt, adding flavouring extracts, and candying portions of the stipes and bladders of the giant *Nereocystis.*

Summing up the position in the northern hemisphere one may say that apart from carragheen (see p. 130) and laver (see p. 98) the general use of seaweed as a food dropped out of use in Europe, except perhaps for Iceland, about fifty to a hundred years ago.

3.1.3 Cyanophyceae

Two blue-green terrestrial algae, *Nostoc commune* or Ke-Sien-Mi and *Nematonostoc flagelliforme* (Elenkin, 1934) or Fa-Tsai, are still used by the Chinese of the interior for food. The second species is more or less confined to north-western districts, e.g. Kansu and Sinkiang. In Mexico, two algae from Lake Texcoco in the Valley of Mexico are eaten for food. The two species involved are *Phormidium tenue* and *Chroococcus turgidus* (Ortega, 1972). Recently the use of blue-green algae has also been reported from the Republic of Chad in Central Africa (Anon., 1968; I.F.P. 1967; Léonard et Compère, 1967). The alga concerned (*Spirulina platensis* and *S. maxima*) grows on the surface of shallow fresh water ponds with water containing high amounts of HCO_3^-. In practice the alga is collected and dried out in a sheet in the sun. The Institut Français du Pétrolle has shown that the *Spirulina* can be grown in troughs in the open and the algae separated by simple filtration. In the experimental area in the South of France the yield was 98 lb (44·5 kg) of dry matter per acre per day and this could increase to 124 lb (55·2 kg) at $86°F$ ($30°C$) and long summer day length. Since the alga is very rich in protein (62–68% of dry weight) it forms a potentially very rich food base (Clement, 1978). It has been estimated that one acre (0·4 ha) will yield 10 tons (10·16 tonnes) of protein as compared to 0·16 tons (0·162 tonnes) for wheat and 0·016 tons (0·016 tonnes) for beef.

3.2 SOUTH AFRICA

The red seaweed, *Suhria vittata*, was employed traditionally for jelly-making by the early Cape colonists. It is usually collected from the stems of *Ecklonia maxima* where it grows epiphytically. It is equally good whether fresh, bleached, wet or dry. The recipe is as follows: 'Take a handful of the seaweed, wash well and boil to a pulp in three to four pints of water. When at the boil leave the lid off for about fifteen minutes to allow the fumes to escape. Then strain through cloth and add sugar, lemon or orange juice and brandy or sherry to taste. Additional flavouring can be added by using cloves, cinnamon, and lemon peel, but these must be put into a muslin bag and dipped into the boiling seaweed for about fifteen minutes'.

3.3 HAWAII

In the east Asiatic–Australasian area seaweed foods occupy a rather more important place in the life of the people. Thus the inhabitants of the Hawaiian Islands eat large quantities of seaweed, which are collectively known as 'limu'. The name 'limu' is also employed in Samoa, where a number of species are eaten, but in Tahiti and Mangaia the algae are known as 'rimu' whilst in Guam they are called 'lumut'. It is interesting to note that the Japanese in Hawaii apparently do not make much use of the local algae, but import their own products (see p. 72) from Japan (Moore, 1944). About seventy-five different kinds of seaweed were originally used as 'limu', forty of them being in general use, whilst the remainder were less frequently employed. Doty's list of 149 is excessive as 86 species have no known use (Abbott and Williamson, 1974). The following list contains the names of some of the commoner forms used in Hawaii.

Limu papahapapa, palahalaha	=	*Ulva fasciata*	Chlorophyceae
Limu ele'ele (black limu)	=	*Enteromorpha flexuosa*	Chlorophyceae
		Enteromorpha intestinalis	Chlorophyceae
Limu wawae'iole or aalaula	=	*Codium muelleri*	Chlorophyceae
Limu lipoa	=	*Dictyopteris plagiogramma*	Phaeophyceae
Limu kala	=	*Sargassum echinocarpum*	Phaeophyceae
Limu lo-loa (long limu)	=	*Gelidium* spp.	Rhodophyceae
Limu ko'ele'ele (dry limu)	=	*Gymnogongrus vermicularis*	Rhodophyceae
		Ahnfeldtia concinna	Rhodophyceae
Limu huna	=	*Hypnea nidifica*	Rhodophyceae
Limu kohu, koko, nipaakai (red limu)	=	*Asparagopsis sandfordiana*	Rhodophyceae
Limu lipe 'epe'e	=	*Laurencia* spp.	Rhodophyceae
Limu manauea	=	*Gracilaria coronopifolia*	Rhodophyceae
Limu hula hula waena	=	*Grateloupia filicina*	Rhodophyceae

The same plants were often called by different names in the various islands. None of the algae are normally eaten alone, but they are usually finely chopped up in the raw state and used as a relish in combination with other foods, e.g. the kernels of kukui nuts mixed with salt and limu. Another favourite application is to use them as a vegetable with fish and soya sauce. The different species tend to have different culinary uses and they may also require different treatment. Limu kohu, for example, has to be soaked for 24 hours before using in order to remove the bitter iodine taste; limu manauea is chiefly employed for thickening chicken broth; limu huna is boiled with squid or octopus; limu fuafua (*Caulerpa clavifera*) and limu lipoa are used as relishes because of their penetrating spicy flavour, whilst limu ele'ele is eaten uncooked. After collection, the algae are cleaned by hand and then rapidly in fresh water, after which they are bagged. When required for use they are chopped, pounded and soaked in water overnight (Abbott and Williamson, 1974). The majority of the seaweeds can be dried and they are then kept tied up in leaves. In some parts of Hawaii the algae were so highly esteemed that up to 1940 they were cultivated by the nobility in algal gardens. With the advent of abundant western food, seaweed eating is now mainly reserved for special occasions. An unusual use is that of *Sargassum* as a cleansing for priests after conducting a burial (Abbott and Williamson, 1974).

3.4 SOUTH AMERICA

In the south Pacific in Chile, and especially on the island of Chiloe, the giant bull kelp, *Durvillea antarctica,* and *Ulva lactuca* (sea lettuce) were collected and eaten extensively by the natives, though not by the white population, in soups and as a vegetable, when it was known as cochayugo. In the case of *Durvillea* it is the stipe that is primarily used in cooking or in salads. Hoffmann (1939) records that this practice was still being followed in 1937. The algae are dried and sold in bundles as salted 'cochiyugo': they can be bought everywhere, even in the markets of cities such as Santiago. The stems of certain seaweeds, species of *Phyllogigas* and *Durvillea,* are also chewed in districts where goitre is prevalent, and so they are known as 'goitre sticks'.

3.5 AUSTRALIA

There are several references in the literature to the use of *Sarcophycus potatorum* (= *Durvillea potatorum*) as a food by the aborigines in Australia. *Durvillea antarctica* was also used as a food by the same people: it was prepared by drying and roasting, after which it was soaked in fresh water for about twelve hours before being eaten.

3.6 NEW ZEALAND

In this country the Maoris used to employ certain of the green seaweeds said to be very palatable in salads and soups. They still make use of the red seaweed *Porphyra* (see p. 98). It is recorded that seaweed meal made from *Macrocystis,* together with milk from a seaweed-eating cow, produced a very striking speed-up in the development of a four year old child who, at the beginning of the treatment, was not able to sit up and talk.

3.7 CHINA

Seaweeds have always been regarded by the Chinese with great esteem. In 600 B.C. Sze Teu wrote that 'Some algae are a delicacy fit for the most honourable guest, even for the king himself'. It might be surprising, as Johnston (1966) has pointed out, that harvesting has never developed to any major extent but this is because there is a paucity of indigenous marine algae in China.

A final feature is the high water temperature caused by the Pacific North equatorial current.

Table 3.4 lists those algae or their products eaten in China and their reported localities of occurrence. Michanek (1975) reported that on latest information the

Table 3.4
(After Johnston, 1966)

Alga	Locality
Enteromorpha	Amoy, Pei-tai-ho, Wei-hai-wei
Monostroma	Amoy
Ulva lactuca	Amoy, Pei-tai-ho, Swatow
Ulva pertusa	Amoy
Ecklonia cava	
Laminaria spp.	Used as kombu
Sargassum fusiforme	Amoy
Sargassum serratifolium	
Sargassum spp.	Hong Kong, Macao, Wei-hai-wei
Undaria pinnatifida	Chusan Island
Porphyra dentata	
Porphyra suborbiculata	Amoy
Chondrus elatus	
Eucheuma papulosa	
Eucheuma spinosum	
Gelidium amansii	Pei-tai-ho
Gelidium divaricatum	Amoy, Macao
Gloiopeltis coliformis	North China
Gloiopeltis furcata	
Gloiopeltis tenax	North China
Gracilaria verrucosa	Amoy, Pei-tai-ho, Wei-hai-wei
Grateloupia filicina	Swatow, Wei-hai-wei

Chinese consumed up to 2400 tons (2438 tonnes) of kombu (see below) in 1959 and that the *Laminaria* species were then under cultivation.

3.8 JAPAN

The principal users of the marine algae as food are the Japanese, who gather the harvest of the wild species from the shore by hand, whilst those from the sublittoral are collected by divers.

Some 21 varieties of sea vegetables are used in everyday cookery, six of them being in use since the eighth century. According to Shurtleff and Aoyagi (1979) these account for 10% of the Japanese food requirements. Apart from the Nori (see p. 98) the most important items of seaweed food in the Japanese diet are Kombu (kubu, konbu) which refers to species of *Laminaria* and wakame (*Undaria*) (see p. 81). According to Smith (1904) the production of kombu dates back to about 1730, and the methods differ very little from those used in the eighteenth century. In Japan, Osaka is the principal centre, with less important ones at Tokyo and Hakodate (Fig. 5.1).

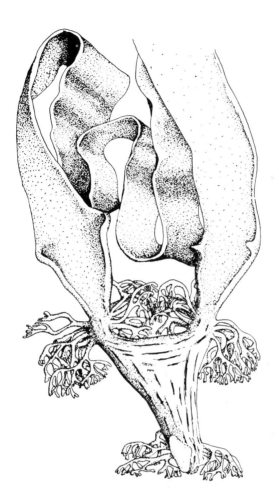

Fig. 3.1 *Arthrothamnus bifidus.*

The seaweeds used for the manufacture of kombu all belong to the *Laminariales* and are collected almost entirely from Hokkaido, the most northern of the main islands. The best plants are found on the north-eastern coast which is bathed by the cold Arctic current. According to Davidson (1906), nine (Miyabe, 1904, records twelve) different species of *Laminaria* are collected, but it is more likely that only some half dozen species and varieties are really involved (see Table 3.2). Other species involved include *Kellmaniella gyrata,* sea banner or chigaiso, and two species of *Arthrothamnus.* One species is *A. bifidus* or Nekoashi-kombu (Fig. 3.1), and the second *A. kurilensis* or 'Chishima-Nekoashi-kombu'. The number of species of

Fig. 3.2 Gathering kelp with poles and grapnels.

Laminaria depends on the taxonomic treatment. At present the principal species appears to be *L. japonica* (royal or makombu) and its varieties *ochatensis, religiosa* and *diabolica,* treated as species by some workers and all yielding products of high quality. The other important species is *L. angustata* (tender or kizami kombu) and its variety *longissima* which exists in two forms, ovate and lanceolate (Hasegawa, 1962).

The gathering of the kelps, which begins in July and terminates in October, is carried out from open boats. Hooks of various types on long handles, or weighted and attached to ropes, are used to tear or twist the weed from its home on the rocky bottom (Figs. 3.2 and 3.3). This is a very primitive method of collection, and when

Sea Vegetables (Algae as Food for Man)

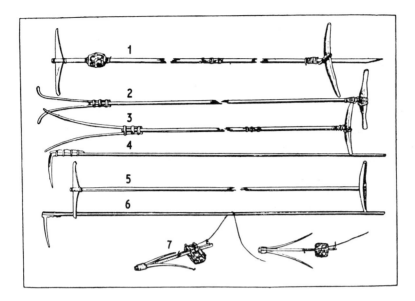

Fig. 3.3 Forms of hooks used in gathering Hokkaido.

it is remembered that great quantities of the same algae also used to be collected for the iodine industry (p. 19) it may well be wondered how the fishermen managed to harvest it all. The method also commonly results in the wrenching off of the holdfast so that no regeneration is possible. When cut near the base of the blade it results in a new shoot that grows more rapidly than a new plant.

There is evidence that the harvest from wild sources is not adequate because in some places the kelp is even cultivated. Sandy ground below five fathoms, where there is no wave action that would disturb the sporelings, is made suitable by the planting of stones. Ten fathoms is usually the maximum depth for cultivation because these algae do not often grow below that depth. In order to secure the greatest possible surface the size of the stones is such that they would just not be washed away when the plants on them are fully grown. Mature plants offer a considerable resistance to the currents and if the stones are not heavy enough plant and stone are both swept away. The stones are usually laid down at the time of year (September) when the plants are reproducing, so that the spores will settle on the stones at once. The type of rock composing the stones is also important: andersitic or basaltic rocks are best because they possess minute cavities in which the young spores can settle. Sandstone is not a good substrate for *Laminaria* because when the plants become large the rock is not sufficiently resistant and the plants are torn off if wave action is vigorous. The best growth of *Laminaria* is actually found in places where rocks are scattered in the sand, because then the competition for light between the plants is not too severe.

Sometimes the *Laminaria* beds become invaded by the marine phanerogam *Phyllospadix*, and when this happens the beds have to be weeded, a special tool being used for the purpose.

When the kelp has been gathered it is spread out carefully on the beaches in parallel rows to dry. When almost dry the stipe and rooting portion are cut off. It is then left to dry further until a gloss appears on it, after which the fronds are tied in bundles, each bundle containing about 50–70 lb (22·7–31·8 kg), and sent to a factory. At the factory the kombu is treated in one of several ways, one of the commonest being the preparation of shredded ('kizami') or green ('Ao') kombu. This is carried out as follows: the dried algae are immersed in boiling water in large iron or copper vats containing a strong solution of the aniline dye, malachite green. Formerly slats of copper used to be employed, but because of their poisonous properties their usage has been forbidden for a long time. The function of the dye is to impart a uniform deep-green colour to the product. The algae are boiled for 15–30 min with occasional stirring and then are drained and hung out in the air to dry. When the surface of the kelp is dry they are rolled up and sent to women who unroll them and lay them out flat in long wooden frames. When the frames are filled the weed is tightly compressed by means of four cords. These piles may weigh up to 125 kg and they are cut into four equal parts, each portion being held together by one of the cords. These smaller bundles are put into presses and when the press is full the bundles are sprinkled with water, in order to make them adhere better, and then they are compressed by means of levers, wedges and ropes. Finally one of the sides of the press is removed and the mass is shredded by means of a plane. The shavings are laid out on mats and dried in the open air until the surface is dry, when they will still contain enough moisture to keep them pliable. The kombu is now ready for packing, and in this state it will keep for a year or two without deterioration. For local use it is made up in paper packages, but for export it is put in boxes or tins.

From the thicker species, especially *Laminaria longipedalis* (= *L. diabolica*) ('oni kombu'), other and finer preparations are made. The various grades represent successive steps in the treatment of a single sample. The frond is first soaked in weak wine-vinegar until it is quite pliable. This imparts a flavour and no doubt has a preserving effect. It is then drained and dried and the blue-green surface layer is scraped off from both sides, leaving the thick white core of the front. The pieces scraped off are called 'kuru-tororo' or black pulpy kombu, which is often inferior in quality because it also contains sand and dirt. If the shredding is continued with a saw-edged knife a white stringy mass, known as 'shiro-tororo' kombu shreads or white pulpy kombu, is produced. Instead, however, a sharp-edged knife may be used to obtain thin and delicate filmy sheets, which are known as 'oboro' kombu veils or filmy kombu. The remaining middle thin pieces of the alga, which can no longer be cut up are pressed together with other similar pieces, divided into lengths and planed down as in shredding green kombu. The shavings are like coarse hair, and so it is known as 'shirago-kombu' or white hair kombu. When prepared less carefully the

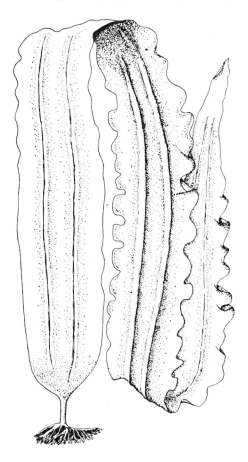

Fig. 3.4 *Laminaria japonica.*

product is called hanori- or mobozoroi-kombu, which is not regarded as being of good quality.

Fronds of the thinner species of *Laminaria* (e.g. *L. japonica* (Fig. 3.4), from which black kombu has first been prepared, are later cut into small pieces of various shapes, e.g. strips, squares, circles, etc. and dried over a fire until they become quite crisp. Such pieces are sold as 'kiri' or dried-on-the-fire kombu. *L. japonica* is high in alcohol-sugar content and so is used for sweet-meat products. The algae can also be coated with white or pink icing when they are called 'hoiro' or sweet-cake kombu. Another method of treatment is to grind the dried pieces into a fine, greenish or greyish poweder. According to Davidson (1906) *L. religiosa* is especially used for this purpose. The product is known as 'Saimatsu' (finely powdered) kombu, and it can be compressed into small cakes which are often coated with sugar. There is also a form of kombu which is known as tea or 'kombucha'. This is prepared in the same way as green kombu, but as the shavings are cut a second time they are similar to rolled tea leaves. Other forms of kombu are 'dashi kombu' soup (stock kelp), 'hayani kombu' (precooked kelp), abita kombu (green flat kombu) and 'sosei kombu' or crude kelp as used for fertilizer manufacture. Each type of kombu is described by its name, e.g. a superior variety of kombu used to make 'dashi' stock is *I.* makombu (*L. japonica*) grown off the coast of Hakodate and harvested in its second year. Because this is a highly valued species, the price varies with locality and quality of product, e.g. 1000–2290 yen per kg. Already experiments have been carried out to try and transplant sporelings from high priced to low priced areas (Sansbonsuga and Torii, 1973). A great number of products are obtained from the various kombu algae, and Shurtleff (personal communication) lists some 32 names of different products.

The various kinds of kombu find different uses in cooking. Green kombu is boiled with meat, fish and soups, and is also used as a vegetable or with boiled rice. Powdered kombu is employed in sauces and soups or is added to rice in the same way as curry. These two forms, together with tea kombu, are also used for making a tea-like liquid. There is also a series of condiment products. Kombu further plays an essential part in marriage ceremonies and New Year celebrations, as well as being a popular gift.

The amount of kombu used in Japan is as much now as it was at the beginning of the century, though about half the total yield is exported to China. Hoffman (1949) quotes Japanese statistics which show that the weight of *Laminaria* collected for kombu between 1934–36 was as follows:

    1934   475 316 tonnes yielding 57 087 tonnes of trade product.
    1935   333 423 tonnes yielding 52 057 tonnes of trade product.
    1936   293 284 tonnes yielding 48 044 tonnes of trade product.

Between 1955 and 1960 the average annual harvest was estimated as 142 000 tons (144 272 tonnes) (Hoppe, 1966). Later figures (Hasegawa, 1976) show some variation, and also the development of cultivated *Laminaria*, which is likely to increase (Table 3.5) The yield of dry product is around 30 000 tons (30 480 tonnes) per year and is valued at some $ 30 000 000 US.

Gloess (1919) estimates that the huge quantity of one million tons of kombu was gathered in 1901, but in view of the later figures given above this seems very unlikely because the iodine industry had barely started (see p. 19). These data do provide some idea of the enormous quantities of oarweed that are handled each year by the Japanese, but as it is most unlikely that they can collect all the available seaweed, the figures also serve to give some idea of the tremendous quantities available around Japan (see p. 269). At the Ninth Seaweed Symposium (1977) it was suggested that there was a potential harvest of about 700 000 tons (711 100 tonnes) wet weight.

A species of *Laminaria* is also used in Japan to yield a material known as 'kan-hoa'. It may be *L. japonica* or *L. longipes,* but in another capacity it has been

Table 3.5 Supplies (tons* wet weight) of *Laminaria* collected
(after Shurtleff and Iwamoto, personnal communication)

Year	Wild source	Cultivated source	Total
1967	175 884	–	175 884
1968	169 874	–	169 874
1969	147 580	30	147 610
1970	110 780	282	111 062
1971	151 725	666	152 391
1972	155 415	3 338	158 753
1973	130 537	7 648	138 185
1974	119 405	10 177	129 582†
1975	157 760	15 696	173 456
1976	159 162	22 087	181 249

* 1 ton = 1·016 tonnes.
† 138 000 tons (New Age Food Study Centre leaflet).

referred to as *L. bracteata* (see p. 235). The same edible product is used in China, where it goes under various names, e.g. hai-tai, kouanpon, hai-houan, yan-tsai, or chai-tai. It differs very little from kombu and is prepared by first washing the alga, after which it is dried and cut up into fragments.

The Japanese have cultivated *Laminaria* species since 1718 and the Chinese have done so since 1952. In Japan, cultivation is carried out in areas where stones are laid or reefs blasted to open them up (Hasegawa, 1971). Attempts are also now being made to propagate *Laminaria* plants on plastic pipes laid in sandy sea beds so that rocky shores may no longer be necessary (Torii and Kawashima, 1977). Further experiments have been carried out to grow *L. japonica* under a form of forced cultivation in order to reduce the period of harvesting from 2 years to one year. Hasegawa (1971a) has demonstrated this successfully and shown that the forced plants are equal in weight to those grown naturally over two years. The success of this technique has resulted in the great increase of cultivated crop to wild crop in recent years. Forced cultivation does, however, present some problems. Deformities such as a spiral or spoon-shaped blade can appear, epizoa become abundant on the laminae, there is an interaction between natural and cultivated plants in the same area with respect to nutrient demand, and the quality may not be quite so good, e.g. chemical differences and losses on storage (Hasegawa, 1979). With time it is likely that these difficulties will be overcome. Apart from these problems it is known that ordinary environmental factors can also affect the blade features of wild and cultivated plants (Sanbonsuga and Torii, 1974).

The cultivation of Hai-Tai takes place in China at Tsingtao, Chefoo and Darien where fresh water inflow reduces the salinity. In such a place the alga grows very quickly and between November and June the plants can reach 3–4 m long and 25–30 cm wide; the plants are a chestnut brown. In more saline areas growth is less

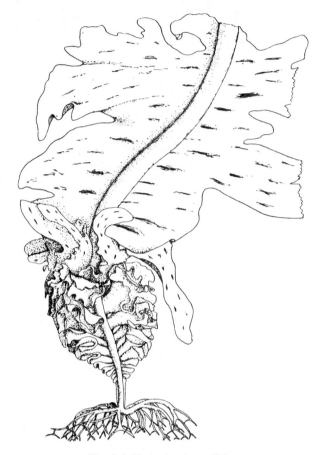

Fig. 3.5 *Undaria pinnatifida*.

rapid, plants may not be more than 1 m long, they are golden brown in colour, and the product is not satisfactory.

In Japan in the early days, sporing fronds were placed in the open sea in suitable localities, at the end of October or in November as soon as the sea temperature fell below 20°C. Rocks and stones on which the spores settled also formed a suitable substrate for *Ectocarpus* and other algae and these 'weeds' frequently hindered the growth of the Laminarias or even eliminated them. In order to surmount this problem fertile fronds are now allowed to liberate spores into artificial tanks where the water temperature is kept below 20°C, and then later when new sporelings have developed from the gametophytes in culture they are put out into the sea and they are then of such a size that the other 'weed' algae cannot inhibit their subsequent growth.

Table 3.6 gives a list of various Japanese species of *Laminaria* and their products.

Table 3.6

Species	Product	Notes
L. angustata Kjellm	Kizami-kombu Shredded kombu Green kombu	Mostly exported to China
L. diabolica Miyabe	Kuro-tororo kombu White-pulpy kombu Oboro kombu	Regarded as of inferior quality
L. fragilis Miyabe		High salt and I_2 content. Mostly exported to China
L. japonica Aresch	Ma-kombu Kuru-tororo kombu Hoiro-kombu	Made into sweetmeats
L. longissima Miyabe (= *L. angustata* var. *longissima*)		Mostly exported to China
L. ochotensis Miyabe	Rishiri-kombu	Gives a high quality product
L. religiosa Miyabe	Hosame-kombu Saimatsu-kombu	Gives a high quality product
L. yezoensis *L. cichorioides*	Chiimi-kombu	

Since 1959 Russians living on the Pacific coast have also used species of *Laminaria* in cooking and as sweetmeats (Schmid and Hoppe, 1962).

3.8.1 Wakame

The seaweed from which this food product is made is primarily the large brown *Undaria pinnatifida* (= *U. distants,* Miyabe, 1904) about 40–50 cm long (Fig. 3.5). Two other species, *U. undarioides* and *U. peterseniana* may also be collected or cultivated on rocks in places where there is a strong current of water, and at depths of 20–40 ft (6·09–12·18 m). There is a close correlation between growth rate and water temperature (Saito, 1975). It is collected between February and June by means of long handled rakes (Fig. 3.3) or by diving, the weeds being torn from their attachment by a twisting motion.

Because of the demand for this sea vegetable and the need to increase output there has been increasing use of cultivated areas of the alga. After selection of a suitable area stones are laid on the sea bed and can be 'seeded' naturally or by laying of sporophylls or by using a zoospore suspension obtained from the sporophylls. Another method is to blast rock reefs in order to increase the ground area and then proceed as above. Yet another techniques is the use of nets with

Table 3.7 *Undaria* quantities (tons, wet weight) from wild and cultivated sources (after Shurtleff and Iwamoto, personal communication)

Year	Wild source	Cultivated source	Total
1967	63 533	53 080	116 613
1968	48 263	76 698	124 961
1969	38 016	59 821	97 837
1970	45 574	76 358	121 932
1971	38 480	95 155	133 635
1972	21 364	109 678	131 042
1973	26 340	113 211	139 551
1974	20 098	153 762	173 860
1975	19 200	102 058	121 258
1976	19 337	126 901	146 238

sporophylls deposited in the rope twists or a zoospore suspension spread over the nets. This inoculation of the nets can also be done in tanks prior to the nets being placed in the sea. Attempts have also been made to produce more productive strains and ones that respond even better to water temperature by hybridization experiments. Hybrids have now been obtained from crosses between *U. pinnatifida* and the other two species (see above) and both hybrids exhibit some better features than either of their parents (Saito, 1975).

The treatment of harvested plants is very simple as it only involves drying and baling. Sometimes, however, a rather more complex method of preparation is employed. The alga is first washed in fresh water and dried and then resoaked in fresh water and the mid-rib removed. When the leafy parts are half dry they are kneaded together and dried out rapidly (Kusakabe, 1967). The dried seaweed, known as 'Hoshi Wakame', can also be cut up into short lengths and sold as 'Kizami-wakame' or wakame chips. In the province of Shizma the chips are tinned, after being coated with sugar, and are then known as 'Ito-Wakame'. In northern Japan the ripe fronds, that is those bearing the sporangia, are cut off and compressed into a slimy liquid which is mixed with boiled rice and eaten. A toasted form of the fronds is known as 'Yaki Wakame'. The thick rooting portion has been given a different name and is known as 'Mehibi'; this is also dried and then cut or shaved into slices and eaten with sauce. Extracts of this alga appear to increase the intestinal absorption of calcium and thus stimulate bone formation (Hong and Cruess, 1979).

Large quantities of this seaweed are collected annually: in 1936, 44 600 tons (45 300 tonnes) were collected, and from them 9000 tons (9141 tonnes) of finished product were produced. From 1950–60 the annual harvest was 50 000 tons (51 100 tonnes) (Hoppe, 1966), and from 1960–69 the annual crop of wild plants amounted to 55 100 tons (55 980 tonnes) (Saito, 1975).

In more recent years there has been a change in pattern but the overall amount collected has increased substantially. There has been a decline in the quantity collected from wild sources but this has been more than offset by the amount gathered from cultivated areas (Shurtleff, personal communication). Table 3.7 sets out these changes in more detail. Cultivation commenced in 1962 (2100 tons) (2133 tonnes), but did not reach significant values until 1967 (66 900 tons) (67 950 tonnes), (Saito, 1975).

It is believed that there is a potential harvest here of 60 000 tons (60 960 tonnes) dry weight annually which would involve a harvest of nearly 350 000 tons (355 600 tonnes) wet weight.

3.8.2 Other Japanese food products

There are a number of other species which are used in home consumption, and are therefore prepared by the householders themselves. They are employed for making jellies or as condiments, or else are eaten as vegetables and salads. 'Arame' (*Eisenia bicyclis*, Fig. 3.6), is eaten as an ingredient of soups or mixed with soya sauce. The dried alga is known as 'Hoshi arame' and when cut into slivers it is sold as 'Kizami arame' and in the powdered form as 'Kona arame'. 'Brick-arame' is a product made in the Isle of Sado from *Ecklonia stolonifera* (Noda and Kitami, 1964). 'Hijiki', or *Hijikia fusiforme,* is a brown seaweed that grows on the rocks near low-water mark. It is collected from January to May, the best period being the first two months of the year, when the plants are small and tender. It is dried in the sun and later used after boiling in fresh water, sauté with vegetables or added to soups or cooked with soya beans.

'Ao-nori' or Aonoriko is the collective name given to flakes of various large species of the green seaweed genus *Enteromorpha* (green nori). These grow at the mouths of rivers, where they are subject to the alternating influence of fresh and salt water. In some places they are even cultivated by placing bundles of twigs in suitable localities. The crop is harvested during the winter and spring and is preserved by drying in the sun. For eating it is gently heated over a charcoal fire and powdered: it is used primarily as a condiment. It is the richest source of iron in Japanese foods. 'Aosa' or 'Awosa', the sea-lettuce (*Ulva* spp.) as well as species of *Enteromorpha* (Ao-nori) form of garnishing for meats and fish and are used in salads. Species of *Monostroma* (Jade nori) are also used as food. It is mainly used in the production of three condiments, Wasabi nori, Shutake nori and Nori-no-Tsukudani (Shurtleff *et al.* personal communication). These algae can be cultivated in the sea by using twigs or nets as for Asakusa nori (p. 100) (Arasaki, 1949), cultivation being especially practised in Ise, Mikawa and Tokyo Bays. The annual harvest was around 9000 tons (9141 tonnes) wet weed between 1955–60, yielding about 1400 tons dry product (Hoppe, 1962). Further details are given in Levring, Hoppe and Schmid, 1969. Whilst separate figures are not available in recent years, the total harvest of wild sea vegetables (algae) other than *Laminaria* and *Undaria* has been as follows:

1967   56 056 wet tons (56 940 tonnes);   1968   46 202 tons (46 920 tonnes);

1969	46 477 tons (47 210 tonnes);	1970	41 846 tons (42 500 tonnes);
1971	44 702 tons (45 400 tonnes);	1972	43 819 tons (44 510 tonnes);
1973	44 104 tons (44 790 tonnes);	1974	44 165 tons (44 850 tonnes);
1975	41 444 tons (42 090 tonnes);	1976	36 326 tons (36 890 tonnes)

(Shurtleff, personal communication). This indicates a steady decrease in consumption.

A red seaweed, also utilised as a garnishing as well as for agar (p. 153) is known as Ogo-nori (*Gracilaria verrucosa*) but this is usually first treated with lime-water or dipped into hot water in order to change the colour from purple to green.

Fig. 3.6 *Eisenia bicyclis*.

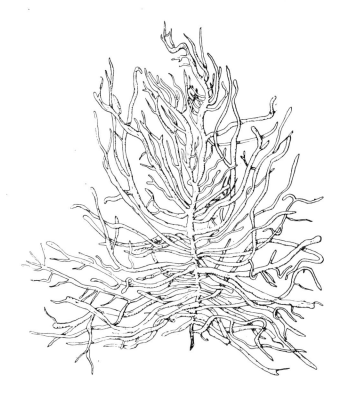

Fig. 3.7 *Mesogloia crassa*.

'Miru' is the collective name given to species of the green genus *Codium*, e.g. *C. fragile* (Miru) and *C. divaricatum* (Kirumiru). The plants grow near low-water mark on the beach and are collected in April and May, and when they have been dried they are preserved in ash or salt. For culinary purposes they are boiled in water and put in soups, or alternatively, after washing in water, they are mixed with soya-bean sauce and vinegar. A thin brown leafy seaweed looking very like *Porphyra* and known as *Petalonia (Ilea) fascia* (this also grows in Europe) is treated and used like Amanori. Another brown seaweed is the slippery, gelatinous *Mesogloia decipiens* or 'Mozuku' (Fig. 3.7). This is gathered in April or May whilst young and preserved by salting. It is also used in soya-bean sauce or, after the salt has been washed out, in vinegar. It is possible that the related species, *M. crassa* or Futo-mozuku, is additionally used. Young plants of *Sargassum enerve* ('Hondawara') are eaten in soup or soya-bean sauce.

A small brown alga called *Heterochordaria abietina*, which looks like a spray of fir, is very abundant in northern Japan where it is known as 'Matsumo'. It is preserved by packing in salt and is cooked with soya-bean sauce. A somewhat curious use is also found for it in the preservation of mushrooms. The mushrooms are first washed in salt water and are then packed in barrels in layers alternating with layers of the salted matsumo.

Among the red seaweeds there is 'Umi-somen' or *Nemalion vermiculare*, a small slimy species 5–12 in (12·7–30·5 cm) long, that grows on rocks, being specially abundant in San-in, Hoku-riko and the north-eastern districts. The related species, *N. multifidum*, is known as 'Tsukomo-nori'. They are both preserved by drying or by mixing in salt or ash. Like many of the others they are eaten in soup or with vinegar and soya-bean sauce. The crest-like seaweed *Eucheuma papulosa* (= *Kallymenia dentata* (Smith, 1905)), or 'Tosaka-nori', grows on submerged reefs at depths of eight feet (2·4 m) or more off the island of Kozu and along the provinces of Ize, Shima and Higo. After storms in August and September it is thrown up on the beaches, where it is collected and dried for use as a condiment or in soya-bean sauce. It is also collected by divers from depths down to ten fathoms (18·3 m).

Irish Moss occurs on the shores of Japan, where it is known as 'Tsunomata'. Another name recorded in the older literature is 'Hosokenomimi'. The species involved is *Chondrus ocellatus*, but there are two other species, *C. armatus* ('Toge-tsunomata') and *C. elatus* ('Naga-tsunomata'). Davidson (1906) refers to *C. elatus* as 'Kotojitsunomata' but Takamatsu (1938) uses this Japanese name for *Gymnogongrus pinnulatus*. On the other hand Takahashi and Shiragama (1934) refer to *Chondrus elatus* as 'Tsunomata'. It is possible that the usage of these names varies in different parts of Japan. The plants after collection are dried in the sun: they are then boiled in order to make a jelly, which can be eaten as a food or form a starch for use in laundries (Tokida, 1954).

There are a number of other Japanese red seaweeds which are dried and eaten in various ways. These include 'Toriashi', 'Yuikiri' or *Acanthopeltis japonica*; 'Katanori' (*Grateloupia divaricata*) and 'Shikin-nori' (*Gigartina teedii*, which also occurs in Europe); 'Kome-nori' (*Carpopeltis flabellata*), which is first immersed in hot water when it turns bright green and is then used for garnishing fish dishes. Hoffman (1939) and other earlier writers give 'Kome-nori' as the Japanese name for *Grateloupia affinis*. The Japanese plant is now known as *Carpopeltis affinis* accoding to Japanese algologists, and they quote the corresponding indigenous name as 'Matsunori', Kome-nori referring to the species *C. flabellata*. It is probable that both Matsunori and Kome-nori are used. 'Mukade-nori' (*Grateloupia filicina*, also European), 'Yego-nori' (*Campylaeophora hypneoides*), 'Okitsu-nori' (*Gymnogongrus flabelliformis*), 'Tosaka-nori (*Meristotheca japonica*) and 'Atsuba-nori' (*Sarcodia ceylanica*) are further species. In spite of all these numerous seaweeds and the great quantities that are collected, it is somewhat surprising to find that they only form abou 0·5% of the total vegetable food of Japan (see also Levring, Hoppe and Schmid, 1969).

For those interested in the use of sea vegetables, full detailed information and recipes are printed in the *Book of Sea Vegetables* (Shurtleff and Aoyagi, 1979) and *The Sea Vegetable Book* (Madlener, 1977).

On the adjacent continent, in Cochin China, there is apparently only one reference to an algal food. This mentions the use of the red alga *Griffithsia* sp., which is eaten with sugar after it has been bleached, compressed and cut up.

Table 3.8 Indonesia food algae

Latin name	Type of seaweed	Native name	Place	Region (*See above*)
Turbinaria sp.	Brown		Bangka	6
Sargassum sp.	Brown	Bebojot	Lombok	5
Gelidiopsis rigida	Red	Sangan?	Lingga	6
Sarcodia montagneana	Red	Bebiroe	Lombok	6
Gymnogongrus javanicus	Red	?	Bangka	6
Gracilaria lichenoides	Red	Doejoeng	Lingga	6
Gracilaria taenoides	Red	Doejoeng Djanggoet	Bangka	6
Corallopsis minor *Corallopsis salicornia*	Red	Boeloeng	Bali	5
Hypnea cervicornis	Red	Boeloeng Djadja	Bali	5
Laurencia obtusa	Red	Sangan?	Lingga	6
Acanthophora spicifera	Red	Boeloeng Bideng	Lombok	5
Caulerpa laetevirens	Green	Boeloeng	Bali	5
Caulerpa peltata	Green	Lata	Bangka	6
Caulerpa racemosa	Green	Lelato	Lombok	5
Caulerpa racemosa var. *clavifera*	Green	Lai lai	Ternate	Batjan Island
Codium tomentosum	Green	Soesoe lopek	Lombok	5

3.9 INDONESIA

Here algae are usually eaten raw, especially the green algae (*Ulva, Chaetomorpha, Caulerpa*), or else after being plunged into boiling water for a minute. They may be eaten with a sauce of allspice, or used with sugar or eaten as a relish, for which purpose they are cooked in sugar obtained from palm trees or soya beans. The brown alga, *Padina australis,* is an example of a seaweed that is made into a gelatine-like sweetmeat. Other brown algae, *Dictyota, Sargassum* and *Turbinaria,* are commonly cooked with coconut milk (Michanek, 1975).

In the above list (Table 3.8), taken from Tondo (1926), it will be observed that many of these seaweeds belong to genera that are also used as food in Europe, Hawaii or Japan. However, few of these are now used.

3.10 OTHER COUNTRIES

In Korea, to the north of Japan, *Undaria* is harvested for food, and in 1970 38 400 tons (38 470 tonnes) were obtained. Some 2200 tons (2235 tonnes) of *Sargassum* were also collected (Michanek, 1975). On the mainland of Asia, in Burma, *Catenella nipae*, collected probably from mangroves on the Tenasserim coast, was regularly on sale in the markets of Rangoon (Post, 1939), whence it was used in salads after boiling water had been poured on it. *Bostrychia* was also used in the same area. *Cladophora*, obtained from the Mekong River, is dried and sold in markets in Laos, whilst in Thailand *Caulerpa racemosa* and *Gracilaria crassa* are common in Phuket market where 10–20 kg per day are sold for use in spicy sauces.

In the Philippines, especially in the north (Luzon), seaweeds are also boiled and mixed with vegetables. A number of algae are also eaten raw; they include the green algae, *Caulerpa racemosa, Chaetomorpha crassa, Enteromorpha intestinalis*, the brown alga *Hydroclathrus cancellatus* and the Rhodophytes *Gracilaria verrucosa* and *Eucheuma* spp. *Agardhiella* is boiled with sugar and spices and made into a sweet, whilst *Sargassum siliquosum* is used as a vegetable (Johnston, 1966). *Chnoospora pacifica* is eaten raw in Vietnam, whilst on the island of Guam (see p. 70) some of the gelatinous algae are used for making blancmanges.

3.11 NUTRITIONAL VALUE

The principal components of the edible algae are carbohydrates (sugars or vegetable gums), small quantities of protein and fat, ash, which is largely composed of sodium and potassium (Ishibashi *et al.*, 1960), and 80–90% of water. Table 3.9 (abridged from Tressler, 1923) gives the composition of a certain number of the edible seaweeds.

Apart from these major constituents sea vegetables are rich in essential trace elements (Table 3.10). The distribution of these in kombu blades has been studied by Yoshimura *et al.* 1976. Many can be successfully frozen, though drying is better. When required for use, the dry algae are soaked for a few seconds up to 30 min, depending on species (Madlener, 1977). In recent years some special attention has been given to the general components (Hosada, 1970) and flavouring compounds (Hosada, 1975) of *Laminaria longicruris*.

3.11.1 Proteins

From Table 3.9 it can be seen that apart from *Nostoc* (a gelatinous blue-green alga that is eaten in China), Amanori and *Ulva linza*, which all have high nitrogen contents, the food value must lie principally in the carbohydrates. Even in the case of the three species which have a high nitrogen content, it is not known how far the nitrogen is soluble and, therefore, available for digestion. Konig and Bettels (1905) give some figures (Table 3.10) for the solubility of the nitrogen, but they do not say how far the soluble products can be utilized in digestion.

Table 3.9

Seaweed	% Water	% Raw protein	% Fat	% Starch sugar	% Fibre	% Ash
Nostoc commune f. *flagelliforme*	10·6	20·9	1·2	55·7	4·1	7·5
Enteromorpha compressa	13·6	12·4		53·0	10·6	10·4
Ulva linza	13·5	19·3	1·73	46·2		19·2
Ulva lactuca and *U. fasciata*	18·7	14·9	0·04	50·6	0·2	15·6
Laminaria spp. (av.)	23·5	5·85	1·15	41·95	6·7	21·1
Arthrothamnus bifidus	24·4	5·8	0·7	45·6	6·4	17·0
Undaria pinnatifida	18·9	9·74–15	0·3	37·8–57		31·3
Hijika fusiforme (av.)	16·1	6–9‡	0·5	35‡–56·9		16·9
Porphyra tenera* (Amanori) (av.)	17·1	25–30	0·8	40·1		10·3
Gracilaria coronopifolia	12·85	7·9	0·05	58·4	3·0	17·8
Ogo-nori (*Gracilaria* sp.)†	—	4·3	—	24·3	4·3	3·55
Ego-nori (*Campylaeophora*)	—	13·65	—	32·2	12·25	3·0
Heterochordaria‡	21·7	22·4	—	46·5	—	—

* Tressler incorrectly uses the names *P. vulgaris* and *P. laciniata*.
† Probably mainly *G. confervoides*.
‡ Levring, Hoppe and Schmid, 1969.

Table 3.10

Species	% Water soluble nitrogenous substances	Price in sen of 100 g of dry asakusa-nori	Protein content
Wakame	5·31	75	5·05
Arame	7·50	53·8	4·25
Enteromorpha	5·5	53·9	3·7
Hijikia fusiforme (av.)	3·89	77·5	5·65
Laminaria japonica	5·44	68	5·55
Porphyra tenera (av.)	21·85	55·5	4·8

The problem of the nitrogen utilization of most of these algal foods had not been experimentally tested up to 1939, but most of them are now regarded as highly digestible (Madlener, 1977). It is evident that *Porphyra* is really protein rich, and 75% of this material is said to be digestible (see p. 108). The nature of the protein also requires to be studied because some proteins are much more digestible than others; also people can become conditioned to them with persistent use. Matsui (1916) also claimed that the market price of hoshi-nori reflected the value of the contained food materials, but this is certainly not borne out by all his figures.

Pentegow (1929) gave some detailed figures (Table 3.12) for the composition of Japanese algae used in the manufacture of kombu. The free amino acid content of Makombu (*Laminaria japonica*) grown under forced culture conditions has been studied by Oishi and Kunisaki (1970) at different stages of growth. Further work of this nature would be of considerable importance, since culture methods vary for a number of sea vegetables.

The variations in protein content of one and the same species from different localities is very striking, whilst the values for fat content show even greater variation. The locality of the material is therefore of great importance. *L. angustata* var. *longissima* differs from all the other species in having a much lower content of the nitrogen-free substance (carbohydrates mainly), but this is perhaps compensated to some extent by the high protein and fat.

3.11.2 Carbohydrates

In considering the digestibility of the sugars and starches, it is important to distinguish between the different kinds of carbohydrate, because like the proteins some are more digestible than others. If a seaweed contains 50% of carbohydrates it does not follow that all of it is available food. On the other hand, those carbohydrates classed as pentosans are readily soluble and hence can be regarded as digestible. All the seaweeds in the list above contain from 4 to 11% of pentosans.

Oshima (1905) considered that about two-thirds of the total sugars could be digested. His experiments, however, lasted for too short a time and were not

Table 3.11 Analysis of edible seaweeds, expressed on a dry-weight basis (after Johnston, 1966)

		Monostroma latissimum	Heterochordaria abietina	Alaria crassifolia	Undaria pinnatifida
Nitrogen	g%	1·40	3·1–3·48	3·35	1·56–2·05
Crude protein	g%	8·76	21·75	19·05	9·74
Sulphur	g%	6·27	2·33	0·80	1·0
Potassium	g%	0·65	3·9	1·6–5·7	1·6–4·2
Magnesium	g%	3·05	1·98	1·82	3·05
Calcium	mg%	576	869–890	906	1162–300
Phosphorus	mg%	53	457–550	488	195–260
Iron	mg%	278	10–11	8	13–43
Manganese	mg%	14	N.T.*	N.T.	T.†
Molybdenum	p.p.m.	3·8	1·7	N.T.	N.T.
Selenium	p.p.m.	0·076	0·085	0·128	0·059

* N.T. = no trace. † T = trace.

Table 3.11 Analysis of edible seaweeds, expressed on a dry-weight basis (after Johnston, 1966) (continued)

		Laminaria angustata	Laminaria japonica	Laminaria religiosa	Kjellmaniella crassifolia
Nitrogen	g%	1·07–1·55	1·17–1·49	0·96–1·94	1·68
Crude protein	g%	9·71	9·32	12·12	10·5
Sulphur	g%	0·54	0·68	1·04	1·18
Potassium	g%	3·3–4·9	50–10·6	2·9–6·7	5·4
Magnesium	g%	1·96	1·67	1·65	1·70
Calcium	mg%	850–1035	800–954	1096–1200	969
Phosphorus	mg%	283–550	150–358	171–280	292
Iron	mg%	4–10	7·9	10–12	5
Manganese	mg%	N.T.	N.T.	N.T.	N.T.
Molybdenum	p.p.m.	1·9	2·2	N.T.	2·1
Selenium	p.p.m.	0·031	0·035	0·031	0·041

* N.T. = no trace. † T = trace.

Table 3.12 Composition of Kombu species

Species	Locality	% Composition (water excluded)				
		Protein	Fat	N-free subs.	Fibre	Ash
L. angustata	Ezan	6·48	1·16	52·72	7·21	32·41
	Hakuro	7·04	1·82	63·63	4·88	22·62
	Mitsuishi	8·10	1·99	60·15	6·55	23·19
	Urakawa	7·55	1·61	63·54	5·24	22·06
	Horotzumi	6·44	3·24	68·68	4·98	16·65
	Kunashiro	7·05	1·95	56·84	6·49	27·66
L. angustata var. longissima	Kujiro	8·32	3·06	47·94	7·61	33·51
	Ochiishi	8·26	1·3	42·98	8·59	38·87
	Hanasaki	10·71	2·68	38·63	9·81	38·16
L. japonica	Ofudgun	6·44	2·06	61·15	7·57	22·27
L. cichorioides	Tomomale	9·2	1·56	60·09	6·70	22·44
	Rijivi	9·85	0·91	53·65	8·63	26·95
	Tsunashiro	8·3	1·42	53·89	7·49	2·82
	Menashi	7·51	0·60	55·40	8·95	29·05
L. religiosa	Fukugama	6·11	1·06	55·46	13·2	24·16
L. fragilis	Muroran	5·26	0·85	52·52	9·3	32·07
Arthro. bifidus	Kujiro	7·71	0·97	60·31	8·52	22·48

sufficient in number. Saiki (1906) came to the conclusion that carbohydrates were not attacked to any extent by digestive juices, and that under a quarter was absorbed. In his experiments he employed raw algae and common Japanese algal products, e.g. kombu, asakusa-nori. In addition to the failure of the digestive juices to attack the foods, Saiki found that bacteria had little effect upon them.

A similar conclusion was arrived at some years later by Swartz (1914). In the meantime Lohrisch (1908) had carried out experiments in which a man was fed on soluble agar: under these conditions he was found to use about 50% of the carbohydrates, which were probably mainly reducing sugars. The utilization of pentosans and galactans in algal foods was confirmed to some extent by Swartz (1914), as shown in Table 3.13, and the complete utilization by the man in the case of dulse is outstanding. Apart from this exception the amount of carbohydrate absorbed is very low. In Swartz's experiments the algae were administered in the form of attractive jellies or blancmanges.

In many of the experiments quoted above people unaccustomed to a seaweed diet were used and, in view of the success obtained earlier with horses after a period of conditioning it is possible that better results would have been reported if the persons undergoing the trials had been on a seaweed diet for some time

Table 3.13

Material	Seaweed	% Utilization	
		Dog	Man
Pentosan	Dulse (*Rhodymenia*)*	73(2)†	100(2)
Pentosan	*Enteromorpha*	35(2)	9(2)
Pentosan	*Ulva*	–	34(1)
Pentosan	*Haliseris pardalis* (brown alga)	16(2)	–
Galactan	Irish Moss (*Chondrus*)	33(2)	6(2)
Galactan	*Gracilaria coronopifolia*	33(2)	30(3)
Galactan	*Hypnea nidifica* (red alga)	56(2)	10(1)
Galactan	*Ahnfeltia concinna* (red alga)	–	60(1)

* According to Colin and Gueguen (1930) the sugar of *Rhodymenia* is a complex of glucoside and galactose which they call Floridoside. This exhibits a regular seasonal variation with a maximum in the summer and a minimum in winter. More recent work reports the polysaccharides as xylans (Bjorndal *et al.* 1965; Percival, 1969).
† The figures in brackets refer to the number of experiments.

previously. Oshima's results may perhaps be accounted for partially in that he used Japanese subjects for his experiments, and they were probably accustomed to eating seaweed. Another valid criticism that has been put forward is that the experiments did not last for a long enough period. An interesting point is raised by the last example in Swartz's group because the subject had chronic constipation and in his case 60% of the material was utilized. A similar result was also obtained by Lohrisch (1908), and there was also the case of the effect of seaweed on a poorly developed child (see p. 71). There would therefore seem to be a definite indication that seaweed foods are of more value to sick animals and persons than to those in normal health.

Nothing is known at present about the fate in the human body of the carbohydrate laminaran (see p. 226), which is abundant in the oarweeds at the time of year when they are collected for the manufacture of kombu, and it would seem desirable that some research should be carried out on this subject. Gloess (1919 and 1932) thinks of kombu only in terms of the algin (see p. 197) which it contains, but there again there is little or no information concerning its utilization.

If there is no food value in the nitrogen and carbohydrates of algal foods because of their indigestibility, it has been suggested (Oshima, 1905) that some food value may exist in the mineral salts, and in this respect they could play some part in the control of deficiency diseases (see p. 49). Swartz (1914) and Perrot and Gatin (1912), on the other hand, adopt the view that the roughage of the algae compensates for the one-sided rice and fish diet. The value of agar in human diet is also largely ascribed to roughage. This interpretation, however, does seem to be too naroow: it would appear, from the evidence available, that dulse, amanori and kombu may have some food value, mainly on account of the carbohydrates they contain. There is little doubt that further experimental work is necessary, and future digestion experiments should be accompanied by a simultaneous estimation of the energy metabolism. It should be noted that at present only one seaweed is regarded as dangerous if consumed as a food. This is the blue-green alga *Lyngbya* spp. (Madlener, 1977).

3.11.3 Iodine and goitre

There are, however, other aspects which require consideration. For example, the use of algae and algal products in the Orient is regarded as the principal reason for the lack of goitre. This may have been achieved unconsciously, but on the other hand it does represent a real contribution to national health. The incidence of goitre in Japan has been placed at one in a million; this is probably due to the iodine that the algae contain, and analyses have been carried out in order to determine not only the iodine content but also the state in which it occurs.

With regard to the actual content of iodine, McClendon (1933) found that *Laminaria* species used in manufacture of kombu have high values, e.g. *L. japonica* contained 0·07–0·45% of its dry weight and *L. religiosa* 0·08–0·76%. Among the edible green algae *Codium intricatum* (Motsure-miru) contained a considerable quantity of iodine, 0·13–0·16% of the dry weight, whilst red algae such as *Gelidium* and *Grateloupia* contained a medium amount. The Chinese have also carried out some analyses of the iodine contents of the principal algae used as foods, but the results of independent investigators show a wide divergence (cf. *Digenea* in Table 3.14), and it is evident that reliable data have yet to be obtained.

Table 3.14 Iodine content of algae
(after Tang and Whang, 1935)

Alga	I_2 in parts per million of the dry material	
Gloiopeltis furcata	53	
Hijikia fusiforme	320	
Digenea simplex	260	(2000 according to another source)
Ulva lactuca	31	
Gelidium amansii	1600	
Laminaria religiosa	240	(11 580 according to another source)
Porphyra tenera	18	

3.11.4 Vitamins

At the present time the supply of food is becoming a serious problem and in particular there is the daily need of man for certain important minor items. An adult requires some 66 g of protein daily and a major portion of this could be met by eating 100 g of either dulse or amanori, providing 42 g and 30 g respectively. 100 g of algae provide more than the necessary daily intake of Vitamins A, B_2 and B_{12} and 67% of Vitamin C. They are in fact an excellent source, being particularly rich in Vitamins A and E. With the exception of calcium and phosphorus 100 g of algae provide all that a human being needs in respect of sodium, potassium and magnesium (Loose, 1965). One could indeed develop a very strong case for the use of compressed seaweed food packets, each of 100 g, for daily consumption in any space journey of the future.

Vitamin A is found in *Ulva lactuca, Laminaria digitata, Undaria, Hijikia,* in *Codium tomentosum* and in kelp meal from Pacific coast kelps, the quantity in the first-named being comparable to that found in cabbages. Vitamin A is also extremely abundant in plankton diatoms. Early work has shown that the diatom *Nitschia* is extremely rich in this material, and as it forms the food of fish it suggests that this alga is probably the source of Vitamin A in fish-liver oils. From this point of view the use of plankton as food could perhaps be strongly recommended. Vitamin B_1 is found in sea-lettuce (*Ulva*), *Enteromorpha, Monostroma, Alaria valida, Laminaria* spp., *Hijikia, Undaria, Porphyra nereocystis, Wildemannia perforata* (= *Porphyra perforata*), *Rhodymenia pertusa, Chondrus crispus* and *Rhodomela subfusca*. The amount of Vitamin B_{12} is low, 0·004—2·8 µg, but it resists decay during storage. In *L. hyperborea* (dry) there are 40 mµg, in *L. digitata* 50 mµg and in *Alaria esculenta* 60 mµg. Seasonal variation does not follow any pattern (Karlstrom, 1963). In the case of Vitamin C there is a wide range: 3—135 mg per 100 g dry wt: the principal species containing it are sea-lettuce (*Ulva*), *Enteromorpha, Alaria valida, Undaria,* and the species of laver mentioned above, together with *P. naiadum, Wildemannia perforata* and *Gigartina papillata*. Apart from *Ulva* all these algae are Oriental or Pacific species, but for comparison there is the following information from Lunde and Lie (1938) about European species (Table 3.15).

From the figures it can be stated that, weight for weight, dulse contains half as much Vitamin C as oranges, whilst some of the Fucoids and *Porphyra* are even richer. In Greenland the algae, which are eaten raw or dipped in boiling water or eaten with blubber oil, may be of considerable value because Hoygaard and Rasmussen (1939) consider that the Angmagssalik Eskimo obtains 50% of his Vitamin C from algal foods.

Riboflavin (Vitamin B_2) in amounts ranging from 0·84—23·08 $\mu g\ g^{-1}$ dry weight occurs in a number of algae including *Chondrus ocellatus* and *Porphyra tenera*. Niacin also is present in marine algae in quantities ranging from 1 to 68 $\mu g\ g^{-1}$ dry wt. Other vitamins which have been detected in marine algae include Pantothenic acid (0·18—12 $\mu g\ g^{-1}$ dry wt), folic acid (0·4—8·5 $\mu g\ g^{-1}$ dry wt), Vitamin D and δ-Tocopherol (1—34 mg per 100 g dry wt) (Levring, Hoppe and Schmid, 1969).

Apart from the vitamins, dulse has been studied for its desmosterol content (Idler and Wiseman, 1970) though the sterol content may be more important for other possible uses (see p. 236).

There is no doubt that whatever may be the direct food value of algae they are useful because of their iodine content, which serves as a protection against goitre, whilst the bulk and water prevent constipation.

Table 3.15 Vitamin content of european algae

	Ascorbic acid in fresh seaweed	
	(mg per 100 g fresh wt)	
Laminaria digitata	3 (Winter)	− 15 (Spring)
Laminaria hyperborea (tangle)	10 (February)	− 47 (March)
Laminaria saccharina (sugar wrack)	4 (February)	− 24 (May)
Himanthalia	28 (Winter)	− 59 (May)
Alaria esculenta (Murlins)	11 (December)	− 29 (May)
Alaria sp. (Greenland)	45	
Fucus serratus (Black wrack)	11 (December)	− 48 (September)
Fucus vesiculosus (bladder wrack)	13 (December)	− 77 (September)
Fucus sp. (Greenland)	13	
Ascophyllum nodosum (knobbed wrack)	30 (Winter)	− 62 (September)
Ascophyllum nodosum (knobbed wrack) (Greenland)	11	
Palmaria palmata (dulse)	24 (March)	− 49 (October)
Palmaria palmata (Greenland)	17	
Gigartina mamillosa	26 (February)	− 63 (May)
Porphyra umbilicalis	44 (February)	− 83 (May)
Ulva lactuca	27−28	

4

Laver or 'Nori' Industry and Carragheen or Irish Moss

4.1 THE LAVER OR 'NORI' INDUSTRY
(Based on species of porphyra)

The thin delicate red seaweed *Porphyra laciniata* still forms a culinary dish in certain parts of Great Britain. It is eaten either as a salad or, more usually, is cooked and made into a breakfast dish. It requires to be fried in a great quantity of fat. Yarham (1944) reported that the miners of south Wales are the biggest laver eaters in Great Britain. The latest figures (Schmid and Hoppe, 1962) suggest that about 200 tons (203·2 tonnes) are still manufactured annually.

There are numerous local methods of preparing laver, but the usual procedure is given by Hill, 1941.

In Eire the laver is collected in spring and stewed or boiled to a jelly and then kept until required. It is principally used fried with butter. Only the small form of *Porphyra* is used, because it is regarded as more tasty.

In New Zealand another species, *P. columbina*, known as 'Karengo' is much relished by the Maoris. Quantities were sent to Maoris serving in the Middle East during the World War II, where it was reported to be more thirst quenching than chewing gum on desert marches.

Recently *Porphyra nereocystis* has been examined in California as an alternative to the importation of nori from Japan, but at present its exportation is regarded as scarcely economical (Woessner, Sorenson and Coon, 1977).

Japan is the most important user of *Porphyra*. Amanori or Hoshi-nori is made mainly (85%) from *Porphyra yezoensis* (Susab-nori) though in the past *P. tenera* (Asakusa-nori) was more important. Other species are also collected and used and they include *P. onoi* (Ono-nori), *P. okamurai* (Kuro-nori), *P. seriata* (Ichimatsu-nori), *P. pseudo-linaris* (Uppurui-nori), *P. kunieda* (Maruba-nori), *P. dentata* (Oni-amanori), *P. akasakai* (Murone-amanori) and *P. angustata* (Kosuji-nori). Recently two very useful strains have been selected from *P. tenera* and *P. yezoensis* respectively (Miura, 1975).

In areas of wild cultivation a mixture of species generally occurs. Suto (1950a) reported that seven different kinds of nori were grown in Tokyo Bay. Yoshida (1966)

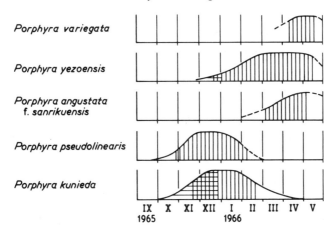

Fig. 4.1 Diagrammatic representation of growth of *Porphyra* species observed in the season from Sept. 1965 to May 1966, on the coast of Shukunohe, Iwate Pref. Area with vertical lines represents the sexual maturation, and that with horizontal lines liberation of monospores (after Yoshida, 1966).

mentions that six species were found in the cultivated grounds (Fig. 4.1). The main species grow in bays and near river mouths, but today the supplies are almost all obtained from cultivated grounds or from artificial cultures (see later).

The exact date of the beginning of the seaweed culture is not known, but the practice is quite old and probably began in Tokyo Bay between 1624 and 1651 (Miura, 1975) near the estuary of the Sumida River. Before that time *Porphyra* grew naturally at the mouth of the River Sumidagawa, but as this river continually brought down gravel and silt, land developed at its mouth and the fresh-water current was diverted farther out into the sea. The resulting increasing extension of the fresh-water influence gradually reduced the crop of *Porphyra*, and so in order to maintain supplies cultivation commenced. Some years ago about 1000 acres (4·04 km^2) were under cultivation, but in 1901 there were over 2000 acres (8·08 km^2) yielding nearly 5 000 000 lb (2268·5 tonnes) of dried seaweed. Apart from Tokyo Bay the other large area of cultivation is Hiroshima on the inland sea.

In Tokyo Bay, where the principal culture area used to be at Omori, the grounds were prepared in October or November by fixing bundles of bamboo or oak brushwood, known as 'hibi', into the mud in places where the water is 10—15 ft (3·05—4·57 m) deep. These bundles have now been replaced by nets or blinds (see below).

The best grounds for growing Amanori were in great demand, and so the local governments lease the planting privileges. In Tokyo Bay there used to be five classes of licence, the cost of each depending upon the yield of the ground.

In recent years the life history of Japanese *Porphyra* species has been fully worked out (Kurogi, 1953; Suto, 1954; Yamasaki, 1954). The carpospores produced in the

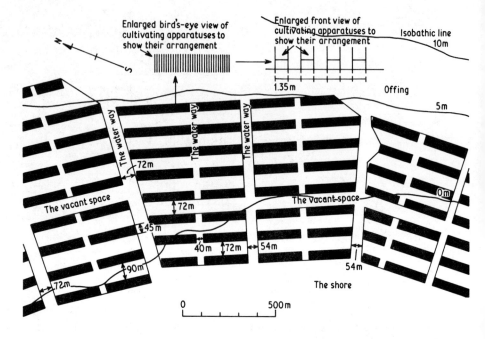

Fig. 4.2 Bird's-eye view of an example of the Nori cultivation ground to show the arrangement in orderly rows of the Nori cultivating apparatus of pole system (Fixed type) (after Miura, 1975).

adult thallus between November and May develop into a shell-boring, filamentous or 'conchocelis' phase. When mature, this phase produces conchospores, mostly between September and November. These germinate and develop into the large commercial leafy thallus which, in some species, is also capable of reproducing itself by means of neutro-spores. Nursery nets can be placed in seeding grounds (Tane-ba) to catch natural seeding of conchospores. This does not, however, result in uniform coverage, so that now artificial seeding is mainly practised.

The planting out of the twig bundles was a slow operation and their use has been compounded by dredging and pollution of the inshore waters. At the present time, therefore, cultivation is largely based on extensive floating nets ('hibi') of coconut fibre or other synthetic material with meshes of about 15 cm. Alternatively 'blinds' constructed of bamboo are used. Both can be located in more open water than previously used. Nets are usually 1–2 m wide and 18–45 m long, whilst the blinds measure 1–2 m x 18 m (Hoppe, 1966). The nets or blinds are spread out horizontally, either attached to poles (Fig. 4.2) or anchored (Fig. 4.3) in order to catch conchospores naturally or after they have been seeded (see below). They are so arranged that at low tide nets are exposed for about 4–4½ hours. Poles can be used where water depths do not exceed 10 m and where there is a tidal range of 1–2 m (Miura, 1975).

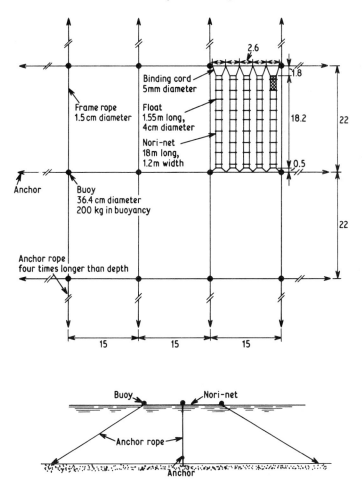

Fig. 4.3 Diagrams showing a plane figure (above) and a profile (below) of the Nori cultivating apparatus of floating system used in Seto Inland Sea. Numerals are in metres (after Miura, 1975).

The nets are exposed daily in order to kill off algal infections, especially species of Chlorophyceae. During mild winters, when the water temperature remains high, fungal infections can seriously reduce the yield. Such diseases can also be controlled by 3–4 hours of exposure to the air every day. With floating nets a special ladder-shaped arrangement allows for exposure (Fig. 4.4). Bacteria can bring about a green spot rotting-like deterioration of the thallus (Nakoa *et al.* 1972), and there is also a disease known as Cancer Disease (Watanabe and Kato, 1972). It appears that two compounds present in the local muds, 2-chloroanthoquinone and reduced dibenzanthrone, are the responsible agents (Ishio *et al.* 1972). In very recent years

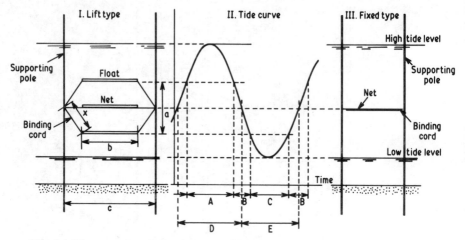

Fig. 4.4 Diagrams showing two types of Nori cultivating apparatus of pole system (I and III) and the position of Nori-nets in relation to the change of level (II) (after Miura, 1975).

a pathogenic *Pythium* responsible for a laver red-rot has appeared in the Ariake Sea farm (Fujita 1978; Fujita and Zenitani, 1977). With increasing oil transport by sea, and hence the increasing possibilities of oil spills, a study has been made by Tokuda (1977) of the effect of oil spill emulsifiers and surfactants upon *Porphyra* growth. This ought to be extended.

Attachment of the carpospores to the shells or other artificial substrate is either by sprinkling a carpospore suspension over the shells (flat upper part) or by adding fragments of carpospore-bearing adult thalli. The carpospore suspension is prepared by inducing spore liberation from reproductive thalli that have been dried overnight in shade and immersed next day in sea water for 4–5 hours. The amount of thallus required is between 15–150 g per 3·3 m^2 varying with species and maturity of thallus. Shells are either laid upon the tank bottom or linked with a string into a row of shells and then hung from a pole. In the latter case, because of uneven growth, the shells are reversed once or twice a month. The water in the tank is kept at a temperature below 28°C and a specific gravity lower than 1·028 (Miura, 1975). The carpospores then grow into the conchocelis phase of the life history.

The conchospores commence to be released in the autumn when the water temperature falls to 17–21°C, and continues into December. From filaments covering 1 cm^2 of shell, 200–1000 monospores will be liberated daily every two to three days (Shitanaka and Suto, 1954). It has also been shown that conchospore shedding takes place usually just after sunrise, and spore settlement is optimal just after shedding (Suto *et al.* 1954; Yamasaki, 1954; Miura, 1975).

In order to 'seed' the nets in the sea, the shells are attached to the nets and placed in vinyl bags or bamboo troughs so that they do not become exposed at low tide. More recently bag-shaped nets have been used for the shells with the nori net spread over them. A number of nori nets can be placed one upon another and then put over the conchocelis nets. Five to six shells are sufficient to seed one net 18 m long by 1·2 m wide, and it usually takes three days for satisfactory seeding. When nets have been seeded the shells are then removed.

Seeding of nori nets can also take place without removing the shells from the tanks. The nets can either be moved about in the tanks in order to stir up the spore-containing water, or the water itself can be stirred, leaving the nets motionless. In the former case a rotating frame with the net attached is rotated in the tank, whilst in the latter case compressed air is bubbled through the water from pipes laid on the bottom. With 1000 shells in a tank, up to 1000 nets (1·8 x 1·2 m) can be seeded in 24 hours. After seeding the nets are kept in seawater for up to 24 hours and then spread in the sea on the poles or as floating rafts.

Some 15—20 days after nets have been seeded young plants become visible as black spots on the netting. After a further 30 days the plants are 2—3 cm long and soon will reach a length of 15—20 cm. The number of plants range up to 1000 per 10 cm length of net strand. The relationship between density and final nori yield has been studied by Yoshida (1972). One can understand the size of the final crops (Table 4.3), when it is realised that there are now about 8—7 million nets covering 191 million m^2 and 171 000 blinds covering 18 million m^2 (Michanek, 1975). Once on the nets young *Porphyra* thalli can be stored at low temperatures providing they have had some preliminary drying. Generally storage lasts for 1—4 months and frozen nets are thawed out by immersion in seawater and not in the air.

This kind of cultivation is quite as complicated as that of any of our land crops and is just as dependent upon certain external factors. For example, changes in the direction of the fresh-water currents may bring about a disastrous failure of the laver crop: on one occasion along the east coast of Izu the loss incurred worked out at £1200 per mile. Such an effect is known as 'reef-burning' and the cause was not discovered until 1905. On very rare occasions reef-burning can be brought about by changes in water temperature, and one or two such cases have been recorded. In Mikawa Bay the harvest is apparently correlated with December and February temperatures and March rainfall (Hamail and Hotta, 1949). Some bays, e.g. Matsushima, are known to produce laver of low quality. This may be related specifically to the water conditions during the winter (Matsudaira *et al.* 1953).

A considerable volume of work has been carried out on the conditions necessary for the cultivation of this crop (Iwasaki and Matsudaira, 1958; Iwasaki 1961; Kurogi, 1956, 1959, 1961, 1963; Saito, 1956; Imada, Saito and Maeki, 1970; Miura, 1973). Plants grow better with sunlight than with artificial light and best results are obtained with sunlight of less than 25 000 lux that has passed through glass. In the case of floating nets, thallus colour can fade if there is little or no wind to maintain water circulation. Where there is the pole system of cultivation, if the nets are too dense, water circulation is reduced. The ground then has to be arranged in a regular manner (Fig. 4.2). Day length is also significant because maximum growth takes place with a long day photoperiod but spore-shedding takes place under short day conditions (Kurogi, 1959; Iwasaki, 1961). Other important factors are pH and dissolved oxygen in the sea water (Imada et al. 1970).

It is evident that as far as natural conditions are concerned, the pattern has changed from the old 'hit and miss' procedure with twigs to nets and blinds with the conchocelis phase being cultivated artificially in tanks in order to impregnate the hibi more efficiently. Even with this improvement, conditions in many of the bays have become increasingly unsatisfactory because of pollution. Floating nets further off-shore are one palliative but the other solution is algal cultivation in bulk under completely artificial conditions in factories away from the sea.

In these factories the conchocelis phase is grown in filtered or artificial sea water with the addition of potassium phosphate and potassium nitrate, but it does even better if Fe–EDTA is added. Alternatively, sodium nitrate and sodium hydrogen phosphate can be used (Kurogi and Sato, 1967). Excessive sunlight and desiccation kill the plants though they will survive 90 min in the air of a very high relative humidity (Takenchi et al., 1956). Growth is most successful at temperatures between 10–22°C (Saito et al., 1954; Kurogi, 1956) and spore liberation is also related to water temperatures as well as day length etc. With *Porphyra angustata* the maximum increase was obtained at 20°C and a light intensity of 4000 lux and a 12 hour day period (Liaw and Chiang, 1979).

The procedure is to allow carpospores from mature thalli to be deposited in deep tanks on plates or clean shells. In commercial cultures of *P. yezoensis* the medium is maintained at 15°C under a light intensity of 1500 lux and a 12 hour day period. Later the conchocelis plants are transferred to tanks where the water is at a rather higher temperature. When there is good growth they are then put in tanks for about ten days under light with air bubbling through continuously. At this stage sporangira are produced with conchospores and when these latter have germinated to young plantlings they are scraped off into large tanks containing 100 l of artificial seawater medium continuously aerated with CO_2-enriched air and slightly illuminated. The water in these tanks is changed weekly and adult plants are ready for harvesting in about 80 days.

More recently the Kyowa Hakko Kogyo Co. has developed a free-living conchocelis culture method. The culture liquid, which is autoclaved at 120°C for 15 min, is set out in Table 4.1.

Pieces of mature frond with carpospores are cleaned by dipping in agarised media several times and then placed in the medium at 15°C and 1500 lux. After 10 days the parent fronds are removed. After 3–4 months there are filamentous conchocelis colonies which are then raised at 25°C at 1000 lux using an 8 hour day. Under this regime maturity is reached after a further two months. The mature plants are transferred to 40 litres of new medium (Table 4.2) which is stirred and air blown through it ($2 \cdot 8$ l min^{-1}). After a few days conchospores are liberated and caught on a nylon thread and they grow up to young fronds. These are removed into 50–100 litre vessel with further stirring and air blown through at 3–5 litres per mi

Table 4.1

Composition of culture liquid (1 litre water)			
NaCl	28 g	*S II metals (in 1 litre)*	
$MgSO_4.7H_2O$	7 g	NaBr	1·2 g
$MgCl_2.6H_2O$	4 g	$AlCl_3.6H_2O$	1·2 g
$CaCl_2\ 2H_2O$	1·5 g	$SrCl_2.6H_2O$	600 mg
KCl	0·7 g	$Na_2MoO_4.2H_2O$	120 mg
Nitrilo-Triacetic acid	100 mg	LiCl	120 mg
$NaNO_3$	100 mg	RbCl	30 mg
K_3PO_4	10 mg	KI	1·5 mg
B-Glycerophosphate Na	10 mg	*P II metals (in 1 litre)*	
Tris (oxymethyl) aminomethane	1 mg	EDTA Acid	1 g
S II metals	10 ml	H_3BO_3	1 g
P II metals	10 ml	$MnCl_2.4H_2O$	140 mg
Vitamin B	100 μg	$FeCl_3.6H_2O$	50 mg
Biotin	1 μg	$ZnCl_2$	10 mg
B_{12}	0·2 μg	$CoCl_2.6H_2O$	4 mg
		$CuSO_4.5H_2O$	5 mg

Table 4.2 Artificial media for buds and fronds

	For buds		For fronds
H_2O	1 l		1 l
NaCl	24 g		24 g
$MgSO_4\ 7H_2O$	8 g		8 g
$CaCl_2.2H_2O$	1 g		1 g
KCl	750 mg		750 g
$NaHCO_3$	250 mg		250 g
H_3BO_3	50 mg		50 g
CTM*	5 ml		4−2·5 ml
NPS†	1 ml		3−7·0 ml
β-Alanine	2·5 mg		2·5 mg
Ornithine−HCl	2·5 mg		2·5 mg
Guanine−HCl	0·2 mg		—
$Na_2EDTA.2H_2O$		2·0 mg	
$FeSO_4.7H_2O$		0·8 mg	
$MnSO_4.nH_2O$		0·32 mg	
$Na_2MoO_4.2H_2O$		0·16 mg	
$ZnSO_4.7H_2O$		0·08 mg	
$CoSO_4.7H_2O$		0·032 mg	
$CuSO_4.5H_2O$		0·008 mg	
$NaNO_3$		67 mg	
$Na_2HPO_4.12H_2O$		16 mg	

* CTM 1 ml
† NPS 1 ml

During this 8 hour day period 0·1–15% CO_2 is added to the airflow. At this stage of growth the temperature is maintained at 11–13°C and the light intensity at 8500–10 500 lux. The plants reach full growth between 50–60 days depending on the species.

Another method of increasing the yield is to introduce species from one area into another and prolong the harvesting period. Experiments are also being conducted with species of *Porphyra* from other parts of the world in order to find those species that grow best under artificial conditions.

Porphyra umbilicalis is the species most commonly cultivated or collected in Hokkaido. When cultivated farther south in north-east Honshu spores could be liberated in middle January to early March and harvesting of fronds then takes place in late May and June long after the normal local harvest of *P. tenera* has occurred (Kurogi and Yoshida, 1966). By employing two species of differing temperature requirements it is therefore possible to obtain two harvests a year instead of one. Fixation of the monospores is rapid and takes place within 1–2 min (Suto, 1950b; Yamasaki, 1954) so that unless there is a strong current most spores should become attached to some surface.

In *P. tenera* spores are formed under short-day conditions at 20°C (Kurogi and Hirano, 1956; Kurogi, 1959; Kurogi, Akiyama and Sato, 1962; Kurogi and Sato, 1962; Kurogi and Akiyama, 1966). This contrasts with the behaviour of the 'conchocelis' phase of *P. umbilicalis* where spores are formed under long day conditions at temperatures below 20°C. In *P. tenera* monospores are liberated under light conditions ranging from 200–4000 lux (Kurogi, 1965). There is abundant liberation at 10°C (Kurogi and Sato, 1967) from autumn plant 'conchocelis'. 'Conchocelis' obtained from dioecious spring plants liberated spores abundantly only at temperatures below 5°C (Kurogi, Sato and Yoshida, 1967). It is quite evident that the species vary in spore production and liberation, both adults and 'conchocelis' phase, and it would seem that by making use of suitable species cultivation and harvesting could be a year long process.

Cultivation needs to be watched because unfavourable factors may affect the fronds. Thus on the concrete laver nursery at Moheji, yellow discolouration and fading of the fronds appeared to be associated with low nitrogen content in the sea water (Sawasaki, Torii and Nakamura, 1965).

When maximum growth has been attained (varying with the species), harvesting is commenced by thinning the plants. With repeated harvesting successive collections are comprised of smaller and smaller thalli. Harvesting from a single net can be repeated up to five times when the net is then replaced by a new nursery net from low temperature storage. From a single net of 18 x 1 m the farmer will obtain 35–105 kg of fresh alga that will process into 1000–3000 sheets (Hoppe, 1966).

Whilst in the old days the manufacture of Hoshi-nori was carried out in a simple primitive way (Fig. 4.5), today all the manufacturing processes from cutting to drying are done by machinery. The plants, after harvesting, are first washed with sea water in order to remove contaminants (especially diatoms) and discoloured plants or parts

Fig. 4.5 The preparation of *Nori*.

of plants. Diatoms can best be removed after dehydrating plants and storing them at low temperatures for three days. After washing, the plants are chopped into small pieces by a machine and then put in fresh water to give a suspension. The suspension is poured into a small rectangular frame and a sheet or nori is formed. The frame is then put into a hot air chamber and this then gives a sheet of dry nori. The standard size sheet is 20·2–21·2 cm long by 17·5–19 cm wide. If harvested plants exceed the manufacturing capacity they can be kept for about 80 days at low temperature ($-20°C$) after centrifugal dehydration.

After preparation, nori is sold as paper-thin purplish black sheets and packed in bundles of ten. Toasted nori (Yaki-nori) is ordinary nori pre-toasted and sold in an airtight package. Toasted sheets brushed with a mixture of soya, sugar, sake and seasonings form sea-zoned nori (Ajitsuke-nori). Flake-like trimmings comprise momi-nori or crumbled nori whilst strips form Pine-needle nori or matsuba-nori.

When the Asakusa-nori or hoshi-nori is required for use it is first of all baked or toasted over a fire until the colour changes to green. It can then be broken up and added to sauces, soups or broths to which it imparts a flavour. Sometimes it is just soaked in sauces and eaten. In 1903 it was even being put up in tins for boiling with soya bean sauce. In Japanese railway station buffets, hotels and restaurants it takes the place of the inevitable sandwich, being offered to the public under the name of

Table 4.3 Changes in net numbers and total harvest of nori

Year	Net nos	Harvest (wet wt. tons)*	Sheets (10^8)	Yield (1000 yen)
1957	–	–	15	–
1959	20	–	18	–
1960	58	–	35	221
1061	112	–	34	802
1962	250	–	41	2424
1963	400	–	32	4686
1964	700	–	45	4498
1965	900	–	30	3700
1966	–	–	37	4000
1967	–	157 550	35	5570
1968	–	144 969	29	3333
1969	–	134 320	55	–
1970	–	231 460	60	–
1976	–	291 000	–	–
1977	–	–	70	–

* 1 ton = 1·016 tonnes.

'sushi'. This is prepared by placing boiled rice and strips of meat or fish on a sheet of hoshi-nori, which is then rolled up and cut into slices. It is also cut up into small fragments and used in biscuit manufacture.

The following figures, taken from Hoffman (1939), will give some idea of the extent of this product. In 1913 the industry was valued at £800 000. In 1927 the culture area was 34 km^2 with a yield of 23 827 tons (24 210 tonnes) of wet weed, whilst in 1936 the area involved comprised 50 km^2 with a corresponding yield of 31 540 tons (32 040 tonnes) that gave 2400 tons (2438 tonnes) of dried nori. To this, however, must be added the naturally growing seaweed which is also collected; this wild material is known as 'Iwa-nori' and in 1936 it amounted to 2244 tons (2279 tonnes). In that year the total quantity of weed amounted to 33 784 tons (34 310 tonnes) and it yielded about 2750 tons (2794 tonnes) of hoshi-nori.

In 1960–61, some 133 000 tons (135 128 tonnes) were harvested and yielded about 3500 million sheets of hoshi-nori. In 1969 the yield was 5·5 billion sheets, whilst the latest figure (1977) is 7 billion sheets (Shurtleff). It should be noted that almost all the sheets produced in a single year are consumed in that year. Some 70 000 workers were engaged in the industry and it will be more now. Some indication of the growth of the industry is given in Table 4.3.

The food value of the nori lies in the high protein content (25–35% of the dry weight), vitamins and mineral salts, especially iodine (Table 4.4) (Noda, 1971). It appears that the protein content of laver in some localities is related to the velocity of the inflowing water (Iwasaki and Matsudaira, 1954). The Vitamin C content is about 1·5 times that of oranges and nori is also rich in Vitamin B and riboflavin

Table 4.4 Composition of Hoshi-nori per 100 g (from Miura, 1975)

Quality	Water (g)	Crude protein	Fat (g)	Sugars (g)	Ash (g)	Vitamin A activity
Superior	11·4	35·6	0·7	39·6	8·0	11 000
Medium	11·7	34·2	0·7	40·5	8·7	10 000
Inferior	13·4	29·0	0·6	39·1	10·9	5 600

Quality	Carotene	Vitamin B_1 (mg)	Vitamin B_2 (mg)	Niacin (mg)	Vitamin C (mg)
Superior	33 000	0·25	1·24	10·0	2·0
Medium	30 000	0·21	1·00	3·0	2·0
Inferior	17 000	0·12	0·89	3·0	2·0

(Schimizu, 1971). Humans digest 75% of the protein and carbohydrate (Schachat and Glicksman, 1959) and in this respect it is much better than other seaweeds (see p. 93). The specific taste of nori is provided by the glutamic acid, glycine, alanine and volatiles that it contains (Tsuchiya and Suzuku, 1955; Kasahara and Nishibori, 1975; Noda *et al.* 1975). The carbohydrate content of nori is a mixture of sulphated galactans, made up of 3, 6 anhydro-L-galactose, with D- and L-galactose units, and β, 1,4 mannan. In this respect the carbohydrate extract is intermediate in characters between agarose (p. 186) and κ-carrageenan (p. 122). A study of galactolipids in *Porphyra tenera* has been carried out by Sato (1971), and Kaneda and Ando (1972) who demonstrated then antioxygenic activity. No figures relating to the chemical composition of lavers can be too specific because changes in the components of the artificial seawater culture medium can result in changes in the chemical composition of the lavers (Fujikawa *et al.* 1971).

Whilst the laver industry is a feature of Japanese culture there is also a laver industry in South Korea where the species of *Porphyra* are cultivated in much the same way as they are in Japan. The food preparations made from the Korean lavers are said to be of a very high quality. The most important ones are Doug-, Phoo- and Poug-laver. *Porphyra* or gamet is also cultivated at Luzon in the Philippines and is sold after drying as a food (Michanek, 1975).

4.2 CARRAGEENAN

Certain red seaweeds give rise to a group of natural gums or mucilages with a great variety of uses, most of which cannot be replaced by synthetic products. The gums or hydrocolloids (Smith and Montgomery, 1959) of red algae are related to each other chemically and in most cases they consist of at least two major compounds, one of which is responsible for the gelling properties and the other for the viscous properties. Some workers divide these compounds, which are polysaccharides, into three major

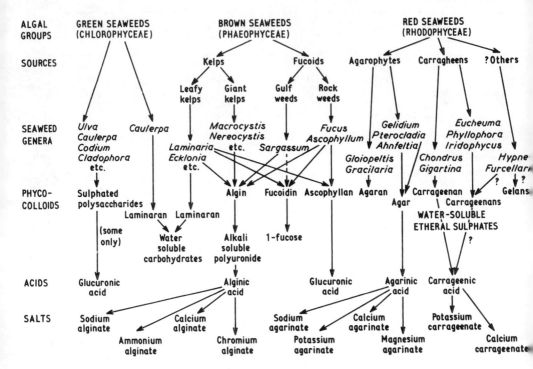

The above diagram gives an indication of the sources and interrelations of the three groups of phycocolloids (algal polysaccharides) (Tseng, 1945a).

groups, the agars, the carrageenans and the gelans. All of them have some properties which are similar to those of polysaccharides from larger brown algae, including the alginates (p. 212), and collectively these algal polysaccharides are known as phycocolloids. Most gums and mucilages are polyuronides, but alginic acid is the only polyuronide from the algae. The suffix colloid has been added to all these compounds because they form a colloidal solution when dispersed in water.

There are three groups of these phycocolloids:

(a) The water-soluble ester sulphates such as carrageenan, agar (p. 48), and fucoidin (p. 232), with some properties similar to mucilages.
(b) Water-soluble reserve carbohydrates such as laminaran (p. 226).
(c) Alkali-soluble polyuronides as represented by algin (p. 194).

A tentative arrangement of these phycocolloids showing their sources and interrelations is shown in the scheme above (Tseng, 1945a).

Seventy-eight red algae, representing 2% of the total known species, yield phycocolloids. They are distributed among 30 genera, which form 6% of the total number of Rhodophyte genera. Within the three orders Gelidiales, Cryptonemiales and Gigartinales, 4% of the species and 12% of the genera are involved.

These algae and their products can be classed in three groups as follows:

(1) Agars derived from the agarophyte genera *Gelidium, Ahnfeltia, Pterocladia, Gelidiella,* and *Acanthopeltis.* The genus *Gracilaria* is generally included in this group, but some workers call its product agaran and not agar. Funoran from *Gloiopeltis* probably belongs here.

(2) The carrageenans derived from the algae *Chondrus crispus* and *Gigartina stellata* and other species of *Gigartina.* These two species represent the true carragheen or Irish moss. Stoloff (1959, 1962) adds *Eucheuma, Phyllophora,* and *Gymnogongrus,* and also the products from *Endocladia, Yatabella, Aeodes orbitosa, Iridaea,* and *Rhodoglossum.* Other species of *Gigartina* belong here, as does *Iridophycus*. In New Zealand *Pachymenia* is a potential source yielding a product similar to carrageenan though with less sulphate and no reaction with milk proteins (Luxton, 1977). Usov (1977) adds to this group six Russian species as well as *Tichocarpus crinitus.*

(3) The gelans derived from species of *Furcellaria* (furcelleran), *Hypnea* (hypnean), *Dumontia incrassata, Dilsea* and *Aeodes* sp. Whether funoran and aeodan are true gelans is debatable, especially if Stoloff's definition of a gelan is used (similar to κ-carrageenan, but with less ester sulphate). Furcelleran and hypnean may eventually be shown to belong to a particular type of carrageenan. In this event the term gelan would become superfluous. For the moment it can be retained.

4.2.1 Raw materials

The raw materials that are used for the production of carrageenans are set out in Table 4.5. In the past the products from *Hypnea, Eucheuma, Furcellaria* and *Phyllophora* have been given specific names and these can still be used when it is convenient to indicate the source material, but technically the names should be abandoned (Stancioff, 1968). Table 4.6 from Moss (1977) gives figures for exports of the various carrageenan algae and the importing countries.

4.3 IRISH MOSS

One of the more important algal foods, especially on both sides of the North Atlantic, is Irish moss, carragheen or goémon blanc. The industry associated with this plant originated in Ireland and was introduced into America in 1835. The principal seaweed involved is *Chondrus crispus* (Fig. 1.6), though in some places *Gigartina stellata* (= *G. mamillosa*) is also employed or the two are mixed together. The name carragheen refers specifically to *Chondrus* though in New Zealand it is applied incorrectly to species of *Gigartina* (see p. 120). *Chondrus crispus* is extremely variable and a large number of forms have been described. It is further a species that exhibits a significant

Table 4.5 Raw materials for carrageenan (after Nielsen et al., 1977)

Extract	Algae division	Principal species	Main growth areas
Carrageenan	Rhodophyte (red)	*Chondrus crispus*	Canada, USA, France, Spain, Portugal
		Chondrus ocellatus	Korea, Japan
		Gigartina stellata	Canada, Southern Europe
		Gigartina acicularis	Southern Europe, Morocco
		Gigartina pistillata	Southern Europe, Morocco
		Gigartina scottsbergii	Argentina
		Gigartina canaliculata	Mexico
		Gigartina chamissoi	Chile
		Gigartina radula	South Africa
Furcelleran		*Furcellaria fastigiata*	Denmark, Russia, Canada
Eucheuman		*Eucheuma cottonii*	Far East, East Africa
		Eucheuma edule	Far East, East Africa
		Eucheuma serra	Far East, East Africa
		Eucheuma spinosum	Far East, East Africa
Hypnean		*Hypnea cervicornis*	Caribbean, Far East
		Hypnea musciformis	Brazil, Senegal, Philippines
Phyllophoran		*Phyllophora*	Russia

Table 4.6 Estimated international shipments dry seaweeds for 1976*

Carrageenan	Exporting country	Total dry tons† exported	Price per dry ton for exporting country ($)	Total value to exporting country ($)	Estimated dry tons received					
					England	France	Denmark	Japan	USA
Chondrus crispus	Ireland	500	500–550	250 000	250	250			
	Canada (Maritimes)	9000‡	500–550	4 500 000		500	4000		4500
Other *Gigartina*	Mexico	600	500–550	300 000					600
	Peru	400	500–550	200 000					400
	Morocco	300	500–550	150 000		300			
	Portugal	500	500–550	250 000		500			
	Korea	900	500–550	450 000				100	
Iridaea	Chile	3600	450–500	1 800 000		?			3000
Eucheuma cottonii	Philippines and adjacent areas	4500	270–300	1 260 000					
E. spinosum	Philippines and adjacent areas	2500§	320–350	850 000		1000	1000	1000	1500
E. spp.	Tanzania	700	—	—					
Hypnea	Brazil	?	400–425	?		700	700	400	700
				$ 10 010 000					

* James R. Moss, President, Agro Mar Inc., 'Some Essential Considerations in Establishing a Seaweed Extraction Plant', paper prepared for the Ninth International Seaweed Symposium, Santa Barbara, California, August, 1977.
† 1 ton = 1·016 tonnes.   ‡ Recorded harvest 51 000 000 lb (23 133 tonnes) which would not meet this amount.
§ Probably too high.

genetic variability (Cheney and Mathieson, 1979).

In all areas the weed after collection is washed in sea water to remove sand and is then spread out to dry and bleach, a process which may take 10–14 days. Bleaching can be done artificially by sulphur dioxide but this has proved unsatisfactory, particularly when used for edible purposes. The bulk of commercial material today is of unbleached (even by sun) black weed. The principal regions where *Chondrus* is collected are Ireland, France, Canada and the USA. The total output of carrageenan weeds in 1975 was around 11 000 metric tons and Moss (1979) estimates that by 1985, using artificial cultures, the output could be 20 000 tonnes.

4.3.1 Ireland

Here a narrow form is generally used for edible purposes, whereas a wide form is used for preparing a size for dressing manilla ropes and linen. *Gigartina stellata* is collected as well. Most of the weed, after bleaching, is packed in bales and exported. Before exporting the weed is classified into one of four grades:

1st quality	= pure, well-bleached *Chondrus*.
2nd quality	= well-bleached *Chondrus* and pure, well-bleached *Gigartina*
3nd quality	= as above but not so well treated.
4th quality	= weedy and badly prepared material.

Table 4.7*

Year	Quantity exported (tons)†	Value (£)
1936	216	4 183
1937	439	10 181
1938	288	5 989
1939	322	6 469
1940	606	15 744
1941	559	24 209
1942	625	45 747
1943	531	39 203
1944	416	26 133
1945	515	29 726
1946	410	28 802
1947	403	26 372
1948	325	17 560
1949	352	19 302
1950	289	16 795
1951	410	27 815

* Kindly supplied by Department of Lands, Dublin.
† 1 ton = 1·016 tonnes.

According to Kirby (1953) the amount of 1st quality material is around 80 tons (81·3 tonnes) annually or 25% of the total crop. Table 4.7 gives the export figures from 1936. Eire ranks third among the world's producers of this crop.

After 1951 other inedible seaweeds, probably for production of alginates, were included and the figures cannot be separated. It is said that production since 1950 has varied between 300–350 tons (304·8–355·6 tonnes) annually. During the 1939–45 war the price ranged from £24.50 a ton in 1939 to around £70 a ton in 1944–5. The price actually varies according to quality and purpose, e.g. whether for food or sizing of textiles.

Great care is necessary in harvesting the weed if regeneration is to occur. Damage to the hold-fast must be avoided and only the top half of each plant should be collected (Marshall, Newton and Orr, 1949).

4.3.2 France

Large quantities are collected on the coasts of France from Fregastel, where it was first harvested, to Finisterre and southwards to Noirmontier, but Brittany is the chief collecting region. The industry is controlled by laws like the other goémon (p. 39), and harvesting occurs from 1 May to 25 October. After a rough sorting it is transported to the drying grounds, washed in fresh water and laid out to bleach. The French product is said to be of a better grade than the Irish product despite the low price paid the collectors and it ranks second in the world's production.

The annual harvest for 1919 was recorded by Gloess (1932) as 2000 wet tons (2032 tonnes), of which 1000 tons (1016 tonnes) were collected around Finistere. 100 lb (45·4 kg) of wet weed generally yield about 20 lb (9·08 kg) of dry product. After 1920 the quantity of Irish moss collected decreased and by 1932 was not more than 400 tons (406·4 tonnes). A certain amount of the bleached product was exported to England, the price depending on the quality. In 1939 industrial moss was worth £26 a ton (1·016 tonnes) whilst the top quality grade was £56 a ton (1.016 tonnes). Subsequently there was a further decline in the industry (Lami, 1941) and in the production of *Chondrus,* and it was suggested that the beds might need to have some legislative protection. In recent years, because of the increased demand, the amount collected and exported has greatly increased and the latest figures give an annual output of about 2000 tonnes in 1965 which rose to 3190 tonnes in 1970. *Chondrus* also grows abundantly in the Channel Islands and during the last war it was gathered and sold in shops. It was used to thicken vegetable stews and was cheap at a time when bread was 15s a loaf.

Elsewhere in Europe *Chondrus,* together with *Gigartina,* is harvested in Spain where the amount is about 1100 tonnes per year.

4.3.3 Canada and the USA

The principal manufacturing centre of the Irish moss industry is America, and used to be at Scituate (Massachusetts), but the main factory, Marine Colloids, is now

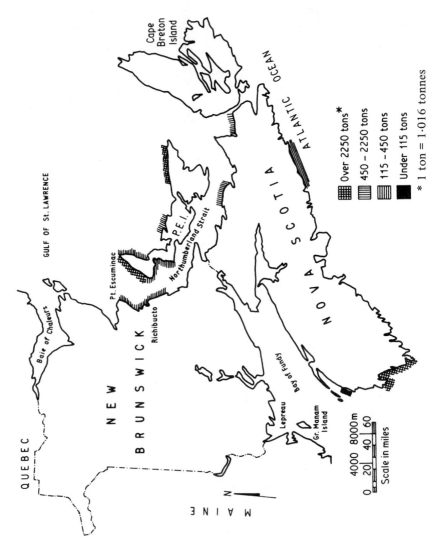

Fig. 4.6 Estimated size of prevailing Irish moss beds (after Ffrench, 1970).

at Rocklands, Maine. Stauffer Chemicals have also recently entered the same field.

In the early years of the industry considerable quantities were collected and processed. The weed reached maximum biomass between 0 and 3 m of mean sea level, which is the level on the shore where the main collecting is carried out. In 1880, 204 000 kg were handled and in 1898, 349 000 kg, but by 1912 this had dropped to 90 600 kg, and in 1924 there was a further drop to 50 200 kg. Consumption had increased to 227 000 kg by 1944 (Tseng, 1945a), and to around 1220 tonnes per annum by 1953 when weed had to be imported from Portugal, France, Ireland and Canada. The last country has now become a very important supplier of the raw material, coming from Nova Scotia, Prince Edward Island and New Brunswick (Fig. 4.6). In 1942 production from P E I was just over 673 000 kg and nearly 227 000 kg came from Nova Scotia as well. Up to 1943 harvesting also took place in Newfoundland. In 1945 the total production was over 1 020 000 kg (Macfarlane, 1964). By 1960, 2 540 000 kg of carragheen were produced in Canada and in 1961, 9000 tonnes were exported and this had risen to 11 080 tonnes dry weight in 1966. In 1968 harvesting commenced again on the west coast of Newfoundland leading to yet a further increase in production.

Since 1966 professional moss harvesting, using raked moss and mechanical dryers, stimulated the industry, and the mean annual harvest 1967–1976 has been 34 000 000 wet kg. In south-west Nova Scotia some 700 harvesters crop 65 distinct beds (Pringle, 1979). In recent years, hand operated rakes have become replaced by basket rakes attached to boats. Such rakes are regarded as damaging to beds (see later) and they have been banned since 1977. Pringle (1979) has now designed a better basket rake that does less environmental and crop damage.

Until 1971 Canadian production formed 80–85% of the world's harvest. Since then it has dropped to 30–35%, half of which comes from Prince Edward Island where some 2000 families are directly or indirectly involved (Anderson *et al.* 1977). 1974 was a 'boom' year in *Chondrus* production but this gave an over-supply and subsequent years have seen a great slump in production, and many 'mossers' lost their occupation (Table 4.8). In 1975 in order to assist, some 250 000 kg were exported to France at about 60c per kg.

Part of the slump was brought about by cheap weed produced in Chile and bought by Marine Colloids. In this case the weed supplied was a species of *Iridaea*. There has also been increasing competition from *Eucheuma* farms in the Pacific (see p. 138). In order to meet this competition, cultivation of *Iridaea* and of *Chondrus* may be started in Canada or the USA. If so, this will obviously affect the wild industry. Experiments by Hanic and Pringle (1977) have shown that *Chondrus* can be seeded on to pottery units. There appear to be two distinct genetic varieties of *Chondrus*, an estuarine broad form and a deep water narrow form which could have separate uses. A cloned 14 strain has also been grown and propagated vegetatively. Growth in cultivation is substantially improved by the addition of ammonium nitrate and phosphorus but not by urea (Simpson *et al.* 1979). Yields are maximal if the pH is controlled at 7·0 by means of CO_2 and if there is a low plant density (Simpson *et al.* 1978).

Table 4.8 Canadian *Chondrus* harvests (lb* dry wt)

Year	Nova Scotia	Prince Edward Island	New Brunswick	Value ($000)
1940	10 000	0		000
1941	53 000	208 000		
1942	490 000	1 490	26 000	
1943	155 000	722		
1944	489	733	26 000	
1945	573	666		
1946	566	2 430		
1947	552	1 488		
1948	560	754		
1949	1 174	1 126		
1950	–	980		122·9
1951	94 215	1 253 617		172·4
1952	2 750	2 000		227·8
1953	4 831	2 117 800		278·7
1954	3 130 600	2 354		319·2
1955	3 477 260	2 316 720		353·2
1956	3 023 280	1 700		292·5
1957	3 054 711	2 131 800		334·4
1958	3 506 722	2 990 200		399·0
1959	2 930 751†	2 364 300	1 773	322·3
1960	3 086 352	2 498 650‡	1 856	372·7
		Total wet wt landings‡		
1961		39 461 000		772·2
1962		43 496 000		783·2
1963		36 855 000		632·0
1964		28 207 000		475·0
1965		39 291 000		604·0
1966		51 009 000		972·0
1973		33 721 000		1995
1974		59 375 000		5829
1975		79 600 000		3378
1976		51 600 000		1600

* 1 lb = 0·454 kg.
† After season of severe storms.
‡ From Canadian Bureau of Statistics.

In New Hampshire and Massachusetts, and probably elsewhere, there is a vigorous spring growth in March and April leading to maximal size and biomass in late summer or early autumn. Where harvesting takes place, regrowth and reproduction is dependent on the time of previous harvesting and the intensity of harvesting. Thus moderate harvesting in summer leads to regeneration of the original biomass within

5–6 months and plants will reproduce after 9 months. If harvesting takes place in winter, recovery is substantially longer (Mathieson and Burns, 1975). It is perhaps worth noting that growth rates in the sea are lower than those of plants in culture (Harvey and McLachlan, 1973). The growth of spores is a function of light intensity, light duration and temperature (Prince, 1971).

In recent years attempts have been made to see whether mechanical harvesters could replace hand-gathering (Pringle and Semple, 1976; Pringle, 1979). Plants and animals are removed by hydraulically powered, rotating brushes and the material pumped to the boat. One of the problems is that 'mossers' are generally involved also in lobster fishing, so that there is concern about the effect of this type of *Chondrus* harvesting on the lobster crop. Also, any harvesting techniques should have a minimum impact upon holdfasts in order to promote regeneration.

Use of the experimental harvester yielded a biomass of *Chondrus* (57%), *Phyllophora* (26%), *Fucus* (9%) and other algae (7%). In the *Chondrus* biomass, 59% comprised immature plants (58% by raking), though only 21% of the total weight. Up to 26% of the plants had holdfasts attached compared with 33% using rakes. Lobsters were also captured at the rate of 1·6 per hour with 80% of them suffering injury. The traditional method of drag-raking also brings in lobsters at a rate of 1·6–2·5 per hour. It would seem that the experimental harvester is no worse in its ecological impact than traditional methods. Further work may produce a better harvester. An underwater open submarine vessel with harvesters using SCUBA may well be the answer. SCUBA has been used previously as means of monitoring and comparing undisturbed and exploited beds (Taylor, 1972).

Concern about the future of the industry in Prince Edward Island has resulted in a lengthy report by Anderson *et al.* (1977). The report raises many questions and does not pretend to have all the answers. The main issues are the influence of the outside multinational corporations that buy the harvested weed and what, to the researchers, is a lack of interest by the Federal Government. The two major international firms are Marine Colloids Ltd and Genu Products, Canada Ltd. A third firm, Pierrefitte-Auby, were active buyers up to 1975. Two minor firms are P E I Seaweeds Ltd (owned by Stauffer Chemicals) and P E I Marine Plants Cooperative owned by the local fisherman.

In Maine and Massachusetts harvesting takes place between March and September but in Canada the season runs from May to September. This results in a new crop every 5–6 months. Since the weed grows at moderate depths most of the US harvesting is carried out by rakes 12–15 inches (30·5–38·1 cm) wide on a handle 15–20 ft (4·57–6·10 m) long and boat operated. Collection is usually made during ebb tide so that it is limited to about 4 hours per tide. Weed is also collected from the rocks by hand and if they are not stripped too clean at the first harvesting a second cutting may be secured before the end of the season. Hand-picked weed commands a high price for food and drug purposes whilst 'raked' weed is generally used for industrial purposes. A good collector is able to harvest about 400 lb (181·6 kg) of wet weed per tide.

After gathering, the wet seaweed is dried for storage so that grinding can take place. If there is blending with other carrageen producing algae this then takes place. The ground weed is soaked, washed and chemically treated to control the pH at an alkaline level. Where extraction is required this is performed by stewing in a water solution, normally in two stages at 15 psi and at a temperature of around 200°F using steam injection. The 'soup' so produced is filtered using a diatomaceous earth or else is centrifuged. The final stage is isolation of the carrageenan either by evaporation of the water or by precipitation using isopropyl alcohol.

Gigartina stellata can also be collected on the New England coast and it has a maximum biomass from 0–3 m above mean sea level. In New Hampshire growth is maximal during spring and summer with a maximum biomass in August and September. A study by Burns (1971) showed that harvesting in August was least detrimental to regrowth so long as holdfasts were not damaged. Careful to moderate harvesting enabled the plant biomass to be re-established by July of the following year. Even if no reproductive fronds appeared before harvesting, this was not unduly detrimental to the reproductive potential 12 months later. December harvesting led to recovery by August or September of the following year, though the biomass was not the equivalent.

Analyses of Rhodophyta from the Philippines showed that yields from species of *Halymenia* ranged from 40·33% to 51·30% which compared very favourably with 54·75% from *Eucheuma* species.

4.3.4 New Zealand

Early analyses in 1941 of *Gigartina* species in New Zealand indicated that *G. decipiens* seemed the most promising commercially. Analysis of the coarsely ground weed gave the following:

Moisture	14·3 %
Protein	11·1 %
Soluble matter in cold water	44·2 %
Soluble matter in hot water	63·0 %
Ash	15·4 %
Arsenic	4 ppm

It is superior to *Chondrus* in protein content but the gelling properties are not so good because it requires a 6% solution to set as firmly as a 4% moss solution. As a means of clarifying beer it is much superior to *Chondrus,* but it has the disadvantage of a rather higher arsenic content. At present the principal species used commercially for carrageenan is *G. angulata* from the shores of Stewart Island. Recently consideration has been given to the possibility of using other red algae containing carrageenans. One such that could turn out to be useful is the genus *Pachymenia* (see p. 274).

4.3.5 Mexico

Gigartina canaliculata is the principal source of carrageenan, and it has been exploited since 1966, mostly from Baja California and Magdalena Island. The extract contains about 47% lambda and 70–79% kappa carrageenan (see below) though the proportions vary depending on the ratio of gametophytes to sporophytes. In 1973, 526·5 tons of dry algae was obtained and exported to the USA and Japan as there was no local processing facilities available (Ortega, 1977). The output since then has been:

    1974   638·2 tons (648·2 tonnes) dry wt;
    1975   965 tons (980·2 tonnes)   dry wt;
    1976   768·5 tons (780·7 tonnes) dry wt;
    1977   569·3 tons (578·2 tonnes) dry wt (Ortega, personnal communication).

4.4 Composition

When freshly gathered the *Chondrus* plants contain a high proportion of water (80%), but when dry they are hard and horny, and the dried material contains 53–59% of two gelatinous galactan sulphates, plus 10–15% of salts. The carbohydrates and salts increase in quantity in spring and summer when metabolism is most active (Butler, 1931). As might be expected, a water extract of *Chondrus* therefore contains a number of substances, some of which can be extracted in the cold and some in hot solution (Haas, 1921). The proportion of gelling material in *Gigartina* is more variable and ranges from 49–58%, and upon its concentration depends the melting point temperature (Haas, 1921). The salts control the gelling process for which the weeds are used, and as they are readily washed out by fresh water, the plants must be protected from rain during the bleaching process and only sea water can be used for washing purposes.

The gelatinous substance in Irish moss was first extracted by Stanford and named carrageenin. About 47% is extracted by cold water (lambda) and up to 70–79% (kappa) in boiling water. Carrageenan, as now more correctly called, contains some 28% of galactans. The remainder of the dry material is protein (around 7%) and minerals (15%) including iodine (Dizerbo, 1964). Seasonal factors are not important, according to McCandless and Craigie (1974), in determining concentrations. Any seasonal variations are peculiar to the specific area. Likewise, the period of maximal values is related to area (Harvey and McLachlan, 1973). The addition of potassium salts during the extraction process increases the strength of the resultant gel. If sea water is used for extraction instead of fresh water the gel is also stronger (Kirby, 1963). Similarly a 2% hot water extract has a gel strength six times that of a 2% cold water extract. This is related to the existence of the various compounds in carrageenan (see below). Unbleached moss also yields a stronger jelly as no gel material has been washed out, but as consumers will probably continue to prefer the bleached material, whether because it looks cleaner and purer or not, bleaching will undoubtedly continue. In Great Britain the standard required is that 2·5 g of moss when heated with 100 ml of water will give a firm jelly on cooling. It must also be free of arsenic (less than ·01 grain per lb), sand, dirt and other algae.

An interesting feature of the chemical composition of Irish Moss is that Welsh material contains sito-sterols (Heilbron, 1942) whereas Canadian material contains cholesterol (Saito and Idler, 1966). This is an aspect that warrants further study. Speciation may, in fact, be taking place on both sides of the Atlantic (Harvey and McLachlan, 1973).

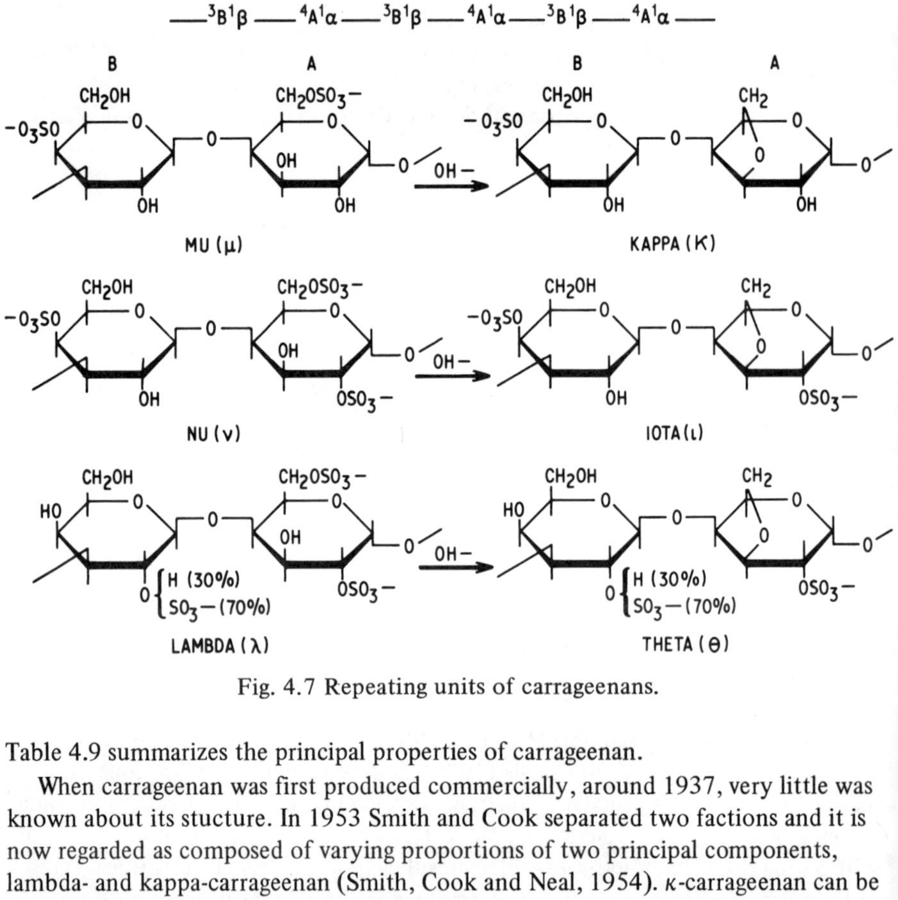

Fig. 4.7 Repeating units of carrageenans.

Table 4.9 summarizes the principal properties of carrageenan.

When carrageenan was first produced commercially, around 1937, very little was known about its stucture. In 1953 Smith and Cook separated two factions and it is now regarded as composed of varying proportions of two principal components, lambda- and kappa-carrageenan (Smith, Cook and Neal, 1954). κ-carrageenan can be separated from λ- because it is precipitated out by potassium ions whereas the other is not. Sexual plants of *Chondrus* possess κ-carrageenan and tetrasporic plants λ- (McCandless et al. 1973). Since then a number of other carrageenans have been isolated. An important one is iota-carrageenan, mainly derived from *Eucheuma* (see p. 127) but also found in *Chondrus**. Further work has shown that κ-carrageenan is derived from mu, iota from nu, and theta from lambda. The various members differ from each other in the amount of 3, 6 anhydro-D-galactose (3,6 AG) they contain

* Also reported by Mollion (1980) in *Anatheca montagnei* from Senegal.

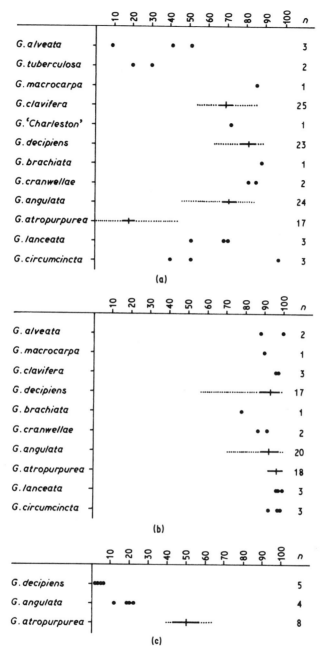

Fig. 4.8 (a) κ-carrageenan (% total gametophytic carrageenan) in species of *Gigartina*. (b) λ-carrageenan (% total sporophytic carrageenan) in species of *Gigartina*. (c) μ-carrageenan (% total gametophytic carrageenan) in species of *Gigartina* (after Parsons *et al.*, 1977).

Table 4.9 Selected properties of carrageenan (after Moirano, 1977)

	Kappa	Iota	Lambda
Solubility			
Hot water	Soluble above 70°C	Soluble above 70°C	Soluble
Cold water	Na^{2+} salt soluble. From limited to high swelling of K^+, Ca^{2+} and NH_4^+ salt	Na^+ salt soluble. Ca^{2+} salt gives thixotropic dispersions	All salts soluble
Hot milk	Soluble	Soluble	Soluble
Cold milk	Insoluble	Insoluble	Disperses with thickening
Cold milk (TSPP added)	Thickens or gels	Thickens or gels	Increased thickening or gelling
Concentrated sugar solutions	Soluble hot	Difficulty soluble	Soluble hot
Concentrated salt solutions	Insoluble cold and hot	Soluble hot	Soluble hot
Water miscible solvents	See text		
Organic solvents	Insoluble	Insoluble	Insoluble
Gelation			
Effect of cations	Gels most strongly with K^+	Gels most strongly with Ca^{2+}	Nongelling
Type of gel	Brittle with syneresis	Elastic with no syneresis	Nongelling
Locust bean gum effect	Synergistic	None	None
Stability			
Neutral and alkaline pH	Stable	Stable	Stable
Acid (pH 3–5)	Hydrolysis of solution, accelerated by heat. Gelled state stable.	Hydrolysis of solution, accelerated by heat. Gelled state stable.	Hydrolysis
Compatibility	Generally compatible with nonionics and anionics, but not with cationics.		

For effect of temperature upon viscosity see Fig. 6.4(b), p. 206.

Table 4.10 Carrageenans from red algal genera

Family	Genus	Species	Carrageenan
Furcellariaceae	*Furcellaria*	*F. fastigiata*	Kappa
Solieraceae	*Agardhiella*	*A. tenera*	Iota
	Eucheuma	*E. spinosum*	Iota
		E. cottonii	Kappa, lambda
	Anatheca	*A. montagnei*	Iota
Hypneaceae	*Hypnea*	*H. musciformis*	Kappa
		H. nidifica (Hawaii)	Kappa
		H. setosa (Hawaii)	Kappa
Gigartinaceae	*Chondrus*	*C. crispus*	Kappa, lambda, iota
		C. sp. (Hawaii)	Lambda
	Gigartina	*G. stellata*	Lambda, kappa, iota
		G. acicularis	Lambda, kappa
		G. pistillata	Lambda, kappa
	Iridaea	*I. radula*	Iridophycan, kappa, lambda
Phyllophoraceae	*Phyllophora*	*P. nervosa*	Phyllophoran
	Gymnogongrus	*G.* sp. (Hawaii)	Iota
Tichocarpaceae	*Tichocarpus*	*T. crinitus*	Lambda, kappa (Usov, 1977)

and the number and position of the ester sulphate group (Fig. 4.7). The B units represent the 1, 3 linked galactosides whereas the A units represent the 1,4 linked galactosides (Moirano, 1977; Luxton, 1977).

The reproductive state of the plants does not affect the total percentage of carrageenan that can be extracted, though it does affect the degree of sulphation. In both gametophyte and sporophytes there is SO_4^{2-} at C_6 of 4-linked units; however, in the sporophyte there is an additional SO_4^{2-} at C_2 whilst in the gametophyte any additional SO_4^{2-} is attached to C_4 (McCandless, Craigie and Walter (1973)). Subtidal populations have a higher percentage of carrageenan with a higher viscosity than do inter-tidal populations. Open coast populations have a higher percentage of carrageenan, with more κ-carrageenan and a greater gel strength than do estuarine populations (Harvey and McLachlan, 1973).

Carrageenan can thus be defined as 'that group of galactan polysaccharides extracted from red seaweeds of the Gigartinaceae, Solieraceae, Hypneaceae and Phyllophoraceae families that have an ester sulphate content of 18% or more and are alternately α-1, 3; β- 1,4 glycosidically linked' (Moirano, 1977). Table 4.10 sets out the types of carrageenan that have been recorded from various algae.

Recent studies (Pickmere, Parsons and Bailey, 1975; Parsons, Pickmere and Bailey, 1977) of twelve New Zealand species of *Gigartina* have shown that there is no marked variation in the total carrageenan content of either gametophytes or sporophytes from different localities, nor for κ-carrageenan in gametophytes or carrageenan in

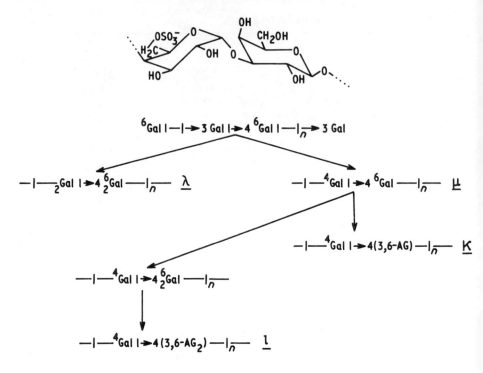

Fig. 4.9 Schematic pathway for carrageenan synthesis (after Yaphe, 1973).

sporophytes (see p. 125). In three species studied μ-carrageenan did vary between species (Fig. 4.8). So far as κ-carrageenan was concerned the gametophytes fell into four distinct groups (Fig. 4.7) so that species used can be important in any harvesting operations. A study of three species revealed no significant effects of season, maturity and stage of life cycle in the overall carrageenan level. Juvenile gametophytes did contain more μ-carrageenan than fertile plants and gametophytes of finely branched species (*G. angolata, G. decipiens*) contained more κ-carrageenan than a broad leaved one (*G. atropurpurea*). A more extensive survey showed that other broad-leaved species did contain as much κ-carrageenan as some branched species.

κ-carrageenan is now known to be a linear polysaccharide made up of alternating 1,3 linked D-galactose-4-sulphate and α- 1,4 linked 3,6 anhydro-D-galactose residues. Some of the 1,4 linked units may be 3,6 AG-2 sulphate, AG-6 sulphate or AG 2,6 disulphate (Anderson and Rees, 1963, Anderson *et al.*, 1968; Stancioff and Stanley, 1969a and b; Percival, 1969). An enzyme, 'de kinkase', is present in κ- bearing algae and this removes the suphate in the mu precursor resulting in ring closure to form a 3, 6 anhydrite (Fig. 4.10). Increasing the anhydride content from 28 to 35% results in increased gelling capacity. Stanley (1963) showed that this increase can be secured by the use of sufficiently alkaline extraction techniques.

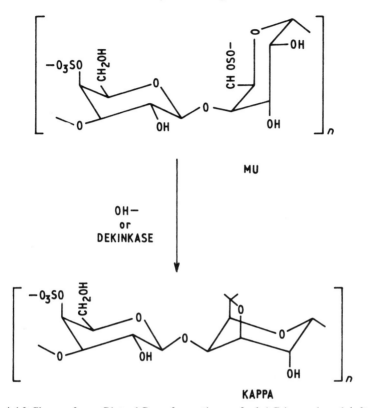

Fig. 4.10 Change from C1 to 1C conformation as 3, 6 AG is produced (after Moirano, 1977).

Lambda-carrageenan consists mainly of 1, 3-linked β-D-galactose 2-sulphate (about 70% contains the sulphate) and 1, 4-linked D-galactose, 2, 6 disulphate residues (Rees, 1963; Dolan and Rees, 1965). Theoretically there is no 3, 6 AG present though it is formed when the extraction is carried out under sufficiently alkaline conditions. When all the 6-sulphate is eliminated one is left with theta-carrageenan but this does not occur naturally. Some recent work, using antibodies produced in goats after injecting them with λ-carrageenan, indicated that there may be two classes, probably based on sulphation pattern (McCandless et al. 1979).

Iota-carrageenan, mainly from *Eucheuma* species, though it is also a minor component in *Chondrus* (Anderson et al. 1968), differs from κ-carrageenan in that initially all of the 3, 6 anhydride residues are sulphated. The presence of the extra sulphate is believed to determine its gelling properties, which differ markedly from those of kappa. The precursor, μ-carrageenan (Fig. 4.10) contains no anhydride (Stancioff and Stanley, 1969). In nature it is likely that only intermediates of both μ- and ι-carrageenan exist, and in any extract, as the ι- proportion increases so it becomes more sensitive to calcium than to potassium. μ-carrageenan, which is the

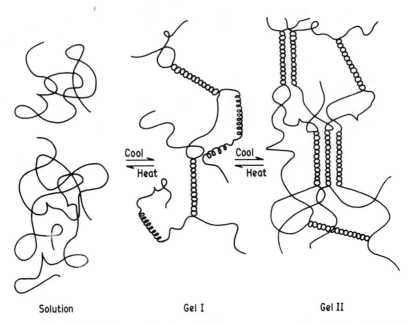

Fig. 4.11 Gelling mechanism of carrageenan according to Rees (1963).

precursor of both κ- and ι-carrageenan, consists of alternating 1, 3 linked galactopyranose 4-SO_4 and 4-linked galactopyranose 6-SO_4 units (Harvey and McLachlan, 1973).

κ-carrageenan occurs in nature in the cell wall of the algae as alkali and alkaline earth salts of carrageenic acid (the free galactan sulphuric acids, $ROSO_3H$) whilst λ-carrageenan is intercellular, mainly in the cortex, and μ is also intercellular in cortex and medulla. Biosynthesis of carrageenan involves the sulphation of linear galactans (Fig. 4.9), and in order for this to happen it seems that the alga *Chondrus* contains three sulphotransferases and a sulphohydrolase. Further work seems necessary as Craigie and Wong (1979) found this activity only in haploid plants.

Carrageenan is usually employed as either the sodium, potassium or calcium salt. Potassium carrageenate, as prepared commercially, is a mixture of λ- and κ-carrageenan which is soluble in hot water but only the κ-carrageenate is gel forming. Sodium carrageenate is soluble in cold water and does not gel. Because of its gelling properties κ-carrageenan is more effective as a stabilizer (see below) than λ-carrageenan. The proportion of these two compounds present in extractives from the various algal species is, therefore, highly important in determining their uses. λ-carrageenan, for example, is much richer in the product from *Gigartina acicularis* and *G. pistillata* than in the *Chondrus* product (Black *et al.* 1965).

Carrageenan gels are all thermally reversible. The formation of the gels is based upon the double helix structure (Fig. 4.11) which is lacking at high temperatures. On cooling a polymer network is formed with double helices forming the junction points which then undergo further aggregation. Sulphation at C_2 of the 1, 3 linked units acts as a wedging group and prevents the double helix from forming. This does not occur with sulphate at C_2 on 3, 6 AG units nor at C_4, on the 1,3 linked galactoside units as they project outwards. Sulphation at C_6 of the 1, 4 linked galactoside forms a kink in the chain which inhibits helix formation. If the sulphate is removed the chain is straightened and gel formation promoted. κ- and ι-carrageenans will not gel in the Na^+ form but will with K^+, Ca^{2+} or $NH4^+$. In the case of carrageenans, potassium ions produce the strongest gels, whereas with ι-carrageenan Ca^{2+} produces the strongest gel. Further details of the gelation process can be found in Moirano (1977) and Table 4.11 sets out the gelling conditions for carrageenan and allied compounds.

Table 4.11

Compound	Distilled H_2O	K salt solution	Ca salt solution	Na salt solution	Milk
Carrageenan	+	+++	+++	−	+++
Furcelleran	++	+++	++	+	++
Agar	+++	+++	+++	+++	+++
Alginate	−	−	+++	−	+++

In general, when carrageenan is added to flour, starches or albumins there is an increase in their gel strength and viscosity. Addition to gums and alginates gives a decrease in viscosity, whilst if added to agar it results in a lowered gel strength.

Blending with other compounds overcomes the cohesive, brittle characteristics of the substance. A very useful blend consists of 50% carrageenan, 33 1/3 % locust bean gum and 16 2/3 % of a potassium salt, usually KCl (Glicksman, 1962). Much of this is now used in canned dog and cat food, particularly in England (Stancioff, personal communication).

Since Irish Moss extractives normally pass unaltered through the alimentary canal they add no caloric value to foods. Swartz (1914) could only demonstrate 6% utilization in man though dogs could use 33% of the material.

Under usual conditions the extractives will withstand high heat and low temperatures provided the humidity is kept low and the pH is at 7·0 or higher. Once the compounds have been put completely into solution they will withstand violent agitation. An exception to this is the chocolate milk system. Oxidizing and reducing agents are destructive.

4.5 USES OF CARRAGEENAN

Hydrocolloids are generally employed for their physical functions in gelation, viscous behaviour, stabilization of emulsions, suspensions and foams and control of crystal growth. The viscosity depends very considerably on the method of preparation. High temperature is particularly adverse and the pH needs to be between 6 and 7. The viscosity of different concentrations of carrageenan and the other hydrocolloids are given in Table 4.12.

Table 4.12 (After Glicksman, 1962)

	Viscosities (CPS)			
% Sol.	Agar	Carrageenan	Furcelleran	Sodium alginate
1·0	4	57		214
1·5		5–800	850	1 102
2·0	25	397		3 760
2·5				8 300
3·0		4 411		29 400
4·0	400	25 356		

Marine Colloids Inc of the USA market a whole range of extractives with gel strengths ranging from slight to 900 g (1·5% in water, measured at 25°C with a 0·731 in cylindrical plunger). With the addition of potassium salts the gel strength is further increased. The refined extractives are used in a variety of preparations. They include hand lotions, mineral emulsions, chocolate milks, cream stabilizers, toothpaste, cough syrup, milk base puddings, ice cream stabilizers, etc. The extractive plus potassium salts is used for ice cream syrups and for tablet binding in pharmaceuticals. When locust bean gum is added the product is used for glazes on baked goods, thickening of fruit pie fillings, jellies, preserves, aspics and so on.

Bleached moss is principally used in the preparation of blancmanges and moulds and is often seen in 'health stores' as part of dietetic products.

As a means of suspending fine cocoa powder in milk carrageenan has to a large extent replaced algin (see p. 220). It has been used for feeding weak calves (one glass of jelly in milk per feed) with great success (Swartz, 1911) and it has improved the coats of red setter dogs (Kirby, 1953).

In the textile industry carrageenan is extensively used at a concentration of about 5% as a stiffening and binding material. It produces a soft finish and a surface to which printing will adhere. It is also used to stiffen and provide a gloss printing to leather goods. Leather manufacturers require a certain amount of Irish moss annually for smoothing leather and giving it a gloss and stiffness. The gelose is melted and brushed on to the leather, which is then polished with glass cylinders. Dried plants are also used extensively in shoe polishes, because the mucilage holds down and smooths out the tiny rough projections on the surface of the shoe leather. In Great Britain 100 tons (101·6 tonnes) or more are used in the manufacture of cold water or casein points in order to hold the film on the surface while the casein dries out. It can also be used to bind briquettes of vegetable charcoal powder.

Apart from toothpastes and hand lotions it is also used in the production of

Table 4.13 Algal gum applications (after Glicksman, 1962)

	Agar	Carrageenan	Furcellaran	Alginates
Dairy				
Ice cream stabilizer	x	x		x
Ice milk	x	x		x
Milk shake	x	x		x
Sherbets	x	x		x
Ice pops and water ices	x	x		x
Chocolate milk drink		x	x	x
Flavoured milk drink		x		x
Instant puddings		x		x
Cooked puddings	x	x	x	x
Eggnog mix		x		
Variegating syrups		x		
Cottage cheese		x		x
Neufchâtel-type cheese	x	x		x
Cream cheese	x	x		x
Cheese spread		x		x
Whipped cream		x		x
Yogurt	x	x		
Packageable milk/cream	x			
Beverages				
Soft drinks ± fruit pulp		x	x	x
Fruit juices		x	x	x
Beer foam stabilizer		x		x
Beer clarification		x		x
Fining wines etc.	x	x		x
Ageing of spirits		x		
Bakery				
Bread doughs	x	x		
Cake batters	x	x		
Fruit cakes		x		
Doughnuts		x		
Pie fillings	x	x		x
Fruit fillings		x		x
Bakery jellies	x	x	x	x
Boiled cream fillings		x		x
Doughnut glaze	x	x		x
Flat icings	x	x		x
Meringues	x	x		x
Cookies	x			
Batter and breading mixes		x		
Citrus oil emulsions		x		x
Cake fillings/toppings	x	x	x	x
Frozen pie fillings		x		

(*continued on the next page*)

Table 4.13 Algal gum applications (after Glicksman, 1962) (*continued*)

	Agar	Carrageenan	Furcellaran	Alginates
Confectionery				
Candy gels	x			x
Caramels, nougats		x		
Marshmallows	x			
Dressings, sauces				
French dressing		x		x
Salad dressing		x		x
Syrups, toppings		x		x
Relish		x		x
White sauces		x		x
Mustard, cocktail sauce		x		x
Catsup		x		x
Dietetic foods				
Starch free desserts	x	x		
Salad/French dressing		x		x
Jellies, jams		x		x
Syrups		x		x
Puddings		x		x
Sauces		x		x
Icings				x
Candies				x
Vegetable and health foods	x			
900 calorie foods		x		
Meat, fish				
Sausage casing	x			x
Fish preservation				x
Canned fish, meat etc.	x	x	x	x
Coated jellied meat		x		x
Antibiotic ice		x		
Sausage ingredient		x		
Preservative meat coat	x	x		x
Synthetic meat fibres		x		x
Miscellaneous				
Dessert water gels	x	x	x	x
Jams, preserves	x	x	x	x
Prepared cereals	x			
Processed baby food		x		
Soups	x	x		x
Coating fruits, vegs.		x		
Frozen foods		x		
Synthetic potato chips				x
Fountain toppings		x		x
Artificial cherries				x

shaving soaps and hair creams. For trade purposes carragheen is distributed in three grades known commercially as 'carragheen naturale', 'carragheen depuratum' and 'carragheen electum albissimum'. The last named is the purest and is used for the preparation of 'Decoctum chondri' which is the best-known pharmaceutical emulsifier. Small quantities of benzoic acid or sodium benzoate are added to this as a preservative. A discovery by Elsner, Broser and Burger (1937), which is important from the medical point of view, is that carrageenan, even in very great dilution, acts as an anti-coagulant for blood. Because of its mucus forming properties it was used in diseases of the lungs and also to correct the taste of bitter drugs. Irish moss has also been employed in irritations of the alimentary canal and in cases of diarrhoea and dysentery. In France and Great Britain it can, because of its mucous properties, be useful in control of stomach ulcers (Bhakuni and Silva, 1974). Carrageenan when used for its anti-ulcer effect does not have to be absorbed gastro-intestinally so that it must function on the surface (Anderson, 1969). Liver oil emulsions of carrageenan have been made whilst an interesting preparation is cotton-wool soaked in a carragheen decoction and dried. This, which is of French manufacture, can be used in place of linseed-meal poultices.

Dried *Chondrus* plants are still used as a fining agent for clarification of beers, coffee, honey and wine. Originally used as collected, it is now employed ground up as a powder. Too much cannot be added as it may remove matter in solution. Because it can contain a high arsenic content, Irish moss is diluted with tannic acid: about 10–15% is added and the moss is employed at the rate of 1·5–3·5 lb (0·67–1·57 kg) per 100 barrels of wort. A new important use is in connection with antibiotic ice used in fishing boats in order to preserve the fish. The antibiotic is CTC (chlortetracycline) which is far better distributed through the ice in the presence of carrageenan.

In recent years another new use is as an air freshener gel. The jelly is sometimes mixed with other gums and a perfume is gradually released as the gel dries. About one million kilograms are used in this way. In Europe there has been a large growth in the dessert field, especially in ready puddings. Apart from these there has been little change in uses (Stancioff, personal communication). Of the algal hydrocolloids there is little doubt that the carrageenans have by far the widest applications (Table 4.13).

4.6 FURCELLARAN

Since the last world war a new industry making use of the unattached red seaweed *Furcellaria fastigiata* forma *aegagropila* has been established in Denmark by the Litex Company. It was originally discovered by a barber who was trying to find a new permanent-wave liquid (Glicksman, 1962). After collection the weed is taken to the factory, washed and stored in alkaline solution for two to three weeks. It is then boiled with water to extract the furcellaran. After filtering or centrifuging a potassium salt is added and this promotes gelling. The gel is frozen and then thawed and pressed in order to remove impurities and water. This process is similar to that used in the preparation of agar (p. 165). The product is then bleached, dried, powdered and packaged. The material is soluble in water at 75°C and also in boiling milk. The gel is as strong as agar and reaches maximum strength at about pH 8 (Schachat and Glicksman, 1959). It reaches its maximum viscosity at 43°C (see Table 4.12). The principal product is a sulphated polysaccharide furcellaran, which is very closely allied to carrageenan (see Percival, 1969). The product is used for the same purposes, though to a lesser extent than the other two, e.g. jams, jellies, meat or fish preservation, milk puddings, food canning, icing bases, pharmaceuticals and toothpaste (Schachat and Glicksman, 1959) (see Table 4.13). Furcellaran, because of its mucous properties, has a use in the control of stomach ulcers. It is particularly valuable in the commercial manufacture of marmalades, jams and jellies because there is no prolonged boiling as when pectins are used. A typical procedure is to take 100 lb (45·3 kg) of the fruit pulp, add 100 lb (45·3 kg– of sugar and boil for 15 min. Furcellaran, at the rate of 1 oz (3·0 g) per quart, is dispersed in 10–15 gal (45·45–68·17 l) of hot water in order to dissolve it; this is then added to the fruit pulp and after mixing poured into the containers where it sets at 120–140°F (49–54·4°C).

Recent work indicates that furcellaran is a mixture of two components like carrageenan, one of which provides the gelling properties and the other the viscous properties. It is very close and almost identical with κ-carrageenan, having alternating 1, 3 linked D-galactose and 1, 4 linked 3, 6 anhydro-D-galactose residues with 10–16% sulphate present (Moirano, 1977) but differing from carrageenan in having rather fewer sulphate groups in the molecule.

The original harvest commenced in the Kattegatt (Table 4.14).

Table 4.14

Year	Raw weed	Furcellaran
1953	5 000	200
1956	10 000	500
1958	15 000	700
1960	–	704
1962	–	897
1964	–	769
1965	30 000	964

In the mid 1960s there was some over-harvesting and production declined. With more careful management 1000 tons of dry material were produced in 1975 and Moss (1979) believes the annual output could rise to 1500 tons.

In other countries the mean annual Canadian harvest between 1966 and 1976 was four million kg of wet weed. The present greatest production area appears to be the Russian Baltic Sea where the annual harvest of wet weed ranges from 27 000 to 40 000 tons (27 432 to 40 640 tonnes) of wet weed.

4.7 EUCHEUMAN

This is a carrageenan derived from species of *Eucheuma* and it is either κ- or ι-carrageenan. Originally first produced in the Malay Archipelago, where it was known as Macassar agar, Java agar, algal-algal or East Indian carragheen, it is now manufactured from wild and cultivated species of *Eucheuma* in the Pacific, especially in the Philippines. Doty (1973) records that between 3000–4000 dry tons (3048–4064 tonnes) reach the world's markets annually. The producing countries export it to Japan, China, USA, UK, France and Denmark. Apart from Indonesia and the Pacific, Tanzania has recently entered the production field. In this latter country the mean yields from the various species are as follows: *E. spinosum*, 72·8% ι-carrageenan dry weight; *E. striatum*, 69% κ-; *E. platycladum*, 65% ? ι-; *E. okamurai*, 58% κ-; *E. speciosum* var. *mauritianum*, 54% ? ι- (Nishigeni and Semesi, 1977). Western Australia also is reported to manufacture a carrageenan from *E. speciosum* as well as from *Hypnea musciformis*.

The species of *Eucheuma* fall into two groups depending on whether they yield κ- or ι-carrageenan (Table 4.15).

Table 4.15 Better known *Eucheuma* spp. containing carrageenan (after Doty, 1973)

κ-carrageenan	ι-carrageenan
(cottonii types)	(spinosum types)
E. cottonii (= *E. okamurae*)	*E. spinosum* (= *E. muricatum*, *E. denticulatum*)
E. striatum (= *E. nudum*, *E. edule*)	
E. procrusteanum	*E. isiforme* (= *E. acanthocladum*)
E. speciosum	*E. uncinatum* (= *E. johnstonii*)

The '*cottonii*' types are species of *Eucheuma* which do not possess a dense central axis of slender cells, whereas the '*spinosum*' types do possess a dense central axis. A study by Doty and Santos (1978) has shown that of six species studied there was no difference in the type of carrageenan as between cystocarpic and tetrasporic plants (Table 4.16). Extraction by lime produced the modified gel whilst extraction by sodium hydroxide yielded the unmodified gel. The gel strength of κ-carrageenan producers is, however, much greater when secured from cystocarpic thalli, so that those plants should be cloned and used in farms.

Table 4.16 Carrageenan properties from *Eucheuma* species (after Doty and Santos, 1978)

Species, voucher no.* and spore stage	Type of carragenan	CAY (%) Unmodified†	Modified†	Unmodified gel viscosity (cp)
Eucheuma arnoldii # 24531				
(a) cystocarpic	Iota	48·40	53·25	194
(b) tetrasporic	Iota	51·85	47·25	206
E. cottonii # 24975				
(a) cystocarpic	Kappa	61·74	58·45	34
(b) tetrasporic	Kappa	65·20	60·51	25
E. odontophorum # 31077				
(a) cystocarpic	Kappa	65·79	64·82	12
(b) tetrasporic	Kappa	63·67	65·52	85
E. platycladum # 402F				
(a) cystocarpic	Kappa	50·00	51·15	154
(b) tetrasporic	Kappa	50·90	47·35	14
E. procrusteanum				
# 31212 cystocarpic	Kappa	57·25	56·30	65
# 31311 tetrasporic	Kappa	61·40	57·10	62
E. sp. ined. # 24706				
(a) cystocarpic	Iota	50·10	54·85	256
(b) tetrasporic	Iota	58·00	60·93	131

* Specimen number in the Herb. Doty, University of California, Santa Barbara.
† See text for the difference between modified and unmodified gels.

Table 4.16 Carrageenan properties from *Eucheuma* species (after Doty and Santos, 1978) *(continued)*

Species, voucher no.* and spore stage	Type of carrageenan	Modified gel strength (g)	Sulphate [SO$_3$Na]$^-$ (%)		3, 6 AG (%)	
			Unmodified†	Modified†	Unmodified	Modified
Eucheuma arnoldii #24531						
(a) cystocarpic	Iota	280	35·58	32·53	20·38	21·50
(b) tetrasporic	Iota	280	35·29	31·59	20·89	23·22
E. cottonii #24975						
(a) cystocarpic	Kappa	1040	23·28	23·20	25·11	29·05
(b) tetrasporic	Kappa	700	25·02	22·05	25·84	28·70
E. odontophorum #31077						
(a) cystocarpic	Kappa	173	28·87	28·22	25·83	27·75
(b) tetrasporic	Kappa	140	28·49	27·64	23·57	27·17
E. platycladum #402F						
(a) cystocarpic	Kappa	860	26·82	24·79	21·53	24·62
(b) tetrasporic	Kappa	320	27·13	26·59	21·00	23·33
E. procrusteanum						
#31212 cystocarpic	Kappa	1110	25·57	23·44	23·26	25·12
#31311 tetrasporic	Kappa	680	24·85	24·18	21·83	24·74
E. sp. ined. #24706						
(a) cystocarpic	Iota	170	36·47	34·21	18·81	25·46
(b) tetrasporic	Iota	180	35·44	33·73	18·07	25·22

* Specimen number in the Herb. Doty, University of California, Santa Barbara.
† See text for the difference between modified and unmodified gels.

Studies of *Eucheuma uncinatum* from the Gulf of California show that the species grows best under a high light intensity and at a temperature of 20–28°C (Dawes, Stanley and Moon, 1977). The ι-carrageenan that it produces comprises 30–40% of the dry weight and there is a high sulphate content (32–34%). Comparable studies of three Florida species has shown that there are peak spring and low summer growth rates with maximum growth rate associated with low sea temperature, high nutrients and reduced light intensity. Growth was also naturally related to photosynthetic activity. Two species (*E. nudum* and Bahia Honda form) behaved as annuals, whereas *E. isiforme* was a perennial. In all three species vegetative reproduction by fragmentation or holdfast regeneration was most important (Dawes, Mathieson and Cheney, 1974). The maximal carrageenan content occurred between July and September but varied with the species (Dawes *et al.*, 1974).

At the present time the value of *Eucheuma* is such that it is being grown in the Philippines as a farm crop. In 1975 *Eucheuma* farms in the Philippines were either a family operation or an estate one. A farm of 1 hectare could be operated by 6–8 men. At that time there were about 1000 farms producing some 600 tons (609·6 tonnes) per month dry weight (2–3 tons (2·03 tonnes) per hectare or 30 tons (30·48 tonnes) per hectare per year) (Doty and Alvarez, 1973). Mobile teams of experienced workers help farmers to select suitable sites and then aid them (Fig. 4.12) with advice as to how production can be improved. Farms should not be in water deeper than 0·5 m and a limestone-rich substrate is best. Plants set out on the bottom are subject to grazing so that farms are best established as nets (Fig. 4.13) though recently the monoline system (Fig. 4.14) using very long lines has been found to be much better (Ricohermosa and Deveau, 1977). Growth rates on net or nylon farms range from 1·5–5·5% per day. Using the monoline system (the net system has almost disappeared), plants are tied on at a density of 50 per 10 m. The export of *Eucheuma* from the Philippines between 1966 and 1977 is given in Table 4.17 and the number of family farms involved in the first half of 1977 is provided in Table 4.18. At the present time it is extremely likely that farms will be established in Guam, Samoa, Fanning Island and Christmas Island. It is perhaps worth noting that in the Philippines a carrageenase has been extracted from a sea urchin (*Diadema setosum*) which feeds upon *Eucheuma* (Benitez and Macaranas, 1979). Outside of the Pacific an experimental farm at Djibouti (Somaliland) located 1 m below the surface, yielded a maximum growth rate in spring with minimal growth rate in the autumn. Using *E. spinosum* the yield of ι-carrageenan was 36·2% of the dry weight.

Efforts are currently being made to develop high-yielding, fast growing, pest resistant strains, especially of *E. spinosum*. There are two problems associated with the farming of *Eucheuma*. One is associated with senescence and the other with seasonality (Doty and Alvarez, 1974). The senescence problem does not yet appear to be understood, but may be related to some lethal imbalance between light, temperature, water quality and movement. Seasonality is more severe at high latitudes where the seasons themselves are more severe. Here again a new strain may be found which is not so subject to seasonal change.

Table 4.17 *Eucheuma* exportation from the Philippines, 1966–1977 in tonnes

Year	E. cottonii	E. spinosum	Total
1966	565	240	805
1967	430	245	676
1968	185	80	265
1969	306	122	428
1970	230	88	318
1971	195	145	340
1972	330	155	485
1973	751	214	965
1974	6286	304	6590
1975	2670	58	2728
1976	3277	253	3530
1977	1920	86	2006*
	2500	180	2680†

Source: Data collected by Marine Colloids (Phil.) Inc.
* Actual exportation, January–June, 1977.
† Estimated exportation, July–December, 1977.

Table 4.18 Participation of family farms in the production of *Eucheuma* in the Philippines from January to June, 1977

Months	E. spinosum		E. cottonii	
	No. of family farms	No. of plants (000)	No. of family farms	No. of plants (000)
January	294	1150	447	1 707
February	284	1166	436	1 958
March	351	1212	654	3 218
April	482	2071	733	4 218
May	461	1830	603	3 439
June	448	1924	755	4 068
Total	2320	9353	3628	18 608
Mean	386	1558	604	3 101

Source: Data collected by Marine Colloids (Phil.) Inc.

So far as returns are concerned, studies in the Philippines (Doty and Alvarez, loc. cit.). showed that in 1974 in the first 6 months 7624 man hours prepared and shipped 11 115 kg of weed (= 1·46 kg per man hour) for which the exporters were paying the farmers 3·05 pesos per kg, though the price does vary from collection point to collection point. The main use of the product is in food preparations, cosmetics and pharmacy.

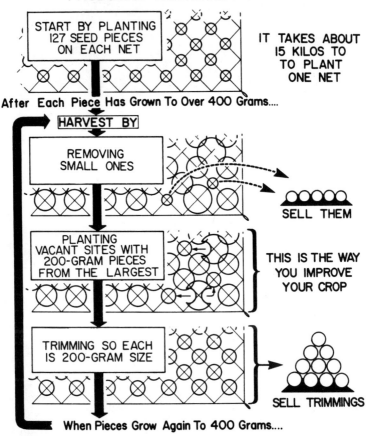

Fig. 4.12 Poster used in teaching the management of a *Eucheuma* farm. The English language posters are about 12 by 18 inches (30·4 by 45·7 cm). Pads of the same posters without lettering, except for the 'UH Research and Advisory Team' at the top, are to be used in the science and language classes in the schools for discussions of the biology and ecology involved and for translation of the English into the local language.

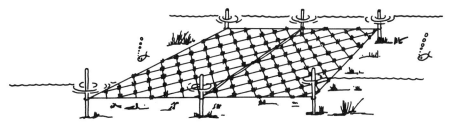

Fig. 4.13 The net system of planting *Eucheuma*. Each net is constructed of nylon monofilament, 80 lb* test for the meshwork, 100 lb test for the border. The meshwork is a square, 25 cm on a side and runs diagonally within the 2·5 x 5 m outline. There are 127 planting sites per net. One net requires about 13 kg of fresh seaweeds. (After Ricohermosa and Deveau).
* 1 lb = 0·454 kg.

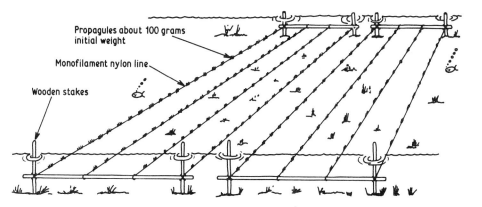

Fig. 4.14 The Monoline system of planting *Eucheuma*. The stakes are usually spaced 10 m along the line and 0·5 between lines. Distance of propagules along the nylon line is 0·2 m or planting density of 100 000 plants per hectare. One 10 m monoline could accomodate 50 plants and approximately 5 kg fresh weight. (After Ricohermosa and Deveau, 1977).

4.8 PHYLLOPHORAN

Species of *Phyllophora* (mainly *P. nervosa* and *P. rubens*) are another source for carrageenan, the principal production area being on the Black Sea (see p. 263) where some 150–200 tons (152·4–203·2 tonnes) dry weight are manufactured annually (Yamada, 1976) at Odessa. The *Phyllophora* plants are washed and then boiled for about seven hours in open vats at about 95°C. The product is allowed to set, cut into slabs and dried. It is regarded as a material with properties intermediate between those of agar and carrageenan (Tseng, 1944; Hoppe and Schmid, 1962).

It forms a valuable bacteriological agar and also yields a volatile oil (Geraniol, 1, 8 cineol; Güven et al. 1972). The quality and quantity of the product varies with season. Chemically it separates into two fractions, one soluble (40%) and the other insoluble (60%). The former consists of glucose and laevulose with 15·5% sulphate and the latter (which is the carrageenan) is galactose with 22% sulphate.

The addition of glycerine, electrolytes or sugar increases the strength of the gel and the temperature at which it sets.

4.9 HYPNEAN

This is derived from species of *Hypnea* and is essentially a κ-carrageenan (Table 4.10). *H. spicifera* gives an abundance of jelly, but if attempts are made to convert it to strips by freezing, it breaks down and becomes watery. In both Brazil and Venezuela, *Hypnea* is used as a source of carrageenan (Mollion, 1973). In Senegal the carrageenan extracted from *H. musciformis* decreases in the cold season (March) and increases in summer. The sulphate content bears an intense relationship to the 3, 6 anhydro-galactose content (Mollion, 1971).

Experiments have been carried out in North Carolina with the same species with a view to using it commercially since it is abundant along the Atlantic coast of North America. The plant is easily collected from shallow bays between mid May and mid July. In order to obtain an extractive that will gel, attention has to be given to the pH (adjusted to 6·0) together with the addition of 0·2–0·5% potassium chloride.

KCl is by far the best substance for control purposes. A 1% solution of hypnean plus 0·5% KCl yields a gel of greater strength than agar from either *Gelidium* or *Gracilaria* (see p. 188). The strength of the *Hypnea* gel can be further increased by adding more KCl up to a maximum of 1·5%. From these facts it would seem that this extractive, with the control that can be exerted over it, could be a very valuable substance.

4.10 IRIDOPHYCAN

This material, which is a mixture of κ and λ-carrageenan, is derived from species of *Iridaea* and *Iridophycus*. Comparable sulphated polysaccharides are also found in *I. capensis, Aeodes orbitosa, A. ulvoidea* and *Pachymenia carnosa* of South Africa (Allsobrook *et al.*, 1968). At the present time South Africa harvests up to 100 tons (101·6 tonnes) dry weight per year for use in beer refining. Iridophycan is also used elsewhere as a stabilizer in chocolate drinks, syrups and paint as well as for the sizing of paper and cloths. Chemically it consists of D-galactopyranose units with the SO_4 on the C_6 atom and forming about 8·7–10%.

4.11 FUNORAN

The Japanese have a seaweed glue called 'Funori' which is used as an adhesive and for sizing papers, fibre or cloth. It is also used, though not to the same extent, in China, where it is called 'Hailo'. It seems China now produces about 500 tons (508 tonnes) per year. It has been prepared in Japan since 1673. The word 'Funori' also refers to the seaweed from which the glue is made, although the word means 'material for stiffening fabrics'. The active principle has been given the name funoran, but this also is applied to the dried alga because the chemical compound is not extracted as is carrageenan and agar. Although funoran contains about the same amount of sulphate and 3, 6 anhydrogalactose as κ-carrageenan, it does not fall in the carrageenan group of polysaccharides since its 3, 6 anhydrogalactose is in the L-form, typical of agarose (Fig. 5.10), rather than the D-form typical of carrageenan. The proportions of SO_4 fraction to D-galactose to L-galactose are 8 to 12 to 1 (Hoppe *et al.*, 1974).

A detailed study of the chemistry of funoran by Hirase and Watanabe (1972) by fractionation showed that there were four fractions (I–IV), three of which contained varying proportions of D-galactose, 6, 0 methyl-D-galactose and 3, 6 anhydrogalactose. The fourth fraction also had 2, 0 methyl-3, 6 anhydrogalactose (Table 4.19).

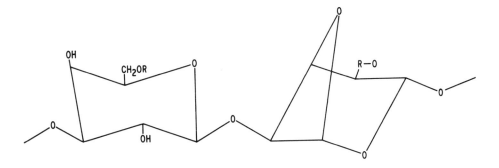

Table 4.19 Yields, compositions, and some properties of fractions I–IV of funoran (after Hirase and Watanabe, 1972)

Fraction	Relative yield (%)	Composition (molar proportion)				Sulphate	$[\alpha]_D^{28}$ (H$_2$O) (°)	Gelation
		Galactose	6-O-Methyl-galactose	3,6-Anhydro-galactose	2-O-Methyl-3,6 anhydrogalactose			
I	8	100	20	5	—	143	+6	None
II	3	100	20	5	—	173	+53	None
III	6	100	18	18	—	129	+14	None
IV	82	100	15	93	4	120	−22	—*

* The minimum concentration of solution which caused gelation was 2%.

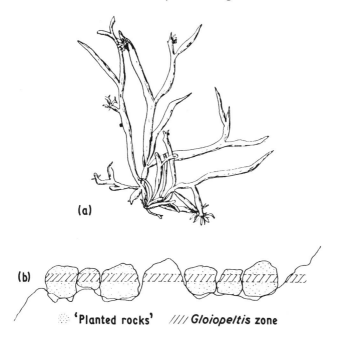

Fig. 4.15(a) *Gloiopeltis furcata* (XO7) (after Okamura, 1909); (b) *Gloiopeltis cultivation.*

The principal seaweed (Fig. 4.15) is the red alga *Gloiopeltis furcata* (funori or Fukoro-funori), but two other species are also used, *G. tenax* or 'Ma-funori', and *G. complanata* or 'Hana-funori'. *G. coliformis* and *G. intricata* are mentioned by some authors, but these are now regarded as forms of *G. furcata.* Hoffman (1939) gives 'Yanagi-funori' for *G. tenax* but this name is not employed by either Okamura (1909) or Takamatsu (1938). *Gymnogongrus pinnulatus* or 'Hira-kotiji', and species of *Iridaea* ('Ginnanso'), *Grateloupia, Chondrus* and *Ahnfeltia concinna* ('Saimi'), are also employed, but the glue is not of the same standard of excellence and the extractable compounds are not the same. The use of the subsidiary species varies in the different islands and prefectures: thus *Grateloupia filicina* (Mukade-nori) is used in Shikoku, Kiushui, Kii and Shima, *Chondrus elatus* (Naga-tsunomata) in Kazuoa, Shimosa, Idzu and Hitachi, whilst in the north of Hondo (the main island of Japan), and in the island of Hokkaido, *Iridaea* is the principal subsidiary alga. Other species of *Grateloupia* ('tamba-nori') are used in Owari, Mikawa, Shima, Kii and Echizen, and another species of *Chondrus* in Sado, Izumo and Iwami. The use of *Gloiopeltis tenax* (Ma-funori) is largely confined to Hizen and Isushima, whilst in Idzu and Chishima Hira-kotiji is frequently employed.

Gloiopeltis grows on rocks in all parts of the Japanese islands, but is most abundant in the warmer waters. There is no particular season for collection, and it is

Table 4.20 Algal harvest of Funori in Japan

Year	Harvest (tons* fresh wt)
1965	2117
1966	1655
1967	1224
1968	1472
1969	908
1970	1246
1971	699
1972	579
1973	1036
1974	959
1975	511
1976	673

* 1 ton = 1·016 tonnes.

gathered thoughout the year, long-handled hooks or rakes being used for the purpose, though in some places men dive for it. Generally, however, the seaweed is collected during the summer in the north and throughout the winter in the south. In certain places the Japanese even cultivate *Gloiopeltis* by elevating a flat shelf, which normally is not quite high enough above low-water mark, by means of large boulders (Fig. 4.15(b)). Spore suspensions are obtained and these are sprinkled on the rocks just before a flood tide in the spring and 2—29% of the spores become attached (Suto, 1949). The seaweed then colonizes the new boulders. The type of rock is also of some importance and quartzite appears to form the best substrate for funori. After collection the algae are sorted and dried and then sent to the manufacturing centres.

The preparation of Funori used to be carried out in about 100 factories, each employing between fifteen and twenty workers. The industry flourished most in southern Japan with Osaka as the main centre. At the factory the dried seaweed is first cleaned and then soaked in fresh water, or else softened by steaming, after which it is tightly packed in thin layers in large shallow trays or on thin mats: larger species, e.g. *Grateloupia, Iridaea*, have to be steamed and chopped. In order to prevent the sheets curling up during the drying process they are sprayed with water at short intervals. As a result of the watering and drying the alga becomes bleached, and when this had gone far enough it is completely dried and several sheets are packed together for market.

Funori is converted into a size by simply dissolving it in hot water. Apart from being used for glazing and stiffening fabrics, it is also employed for stiffening paper and threads, the cementing of walls and tiles and the decorating of porcelain. It also lowers the plasma cholesterol level when fed to rats and this could have pharmaceutical significance. When fractionated, one of the fractions has been found to possess anti-tumour activity (Hirase and Watanabe, 1977).

The alga gives a clear viscous solution in warm or hot water. The viscosity is dependent upon plant habit, species, temperature and the presence of salts. The price of funori natually varies with the quality. The production is not inconsiderable considering the size of the alga: in 1901 it was just over 1000 tons (1016 tonnes), but in 1936 only 714 tons (725 tonnes) were produced from 4595 tons (4669 tonnes) of wet weed. At present between 500 and 1000 tons (508–1016 tonnes) are gathered in Japan (Table 4.20) and rather over 500 tons (508 tonnes) in China.

5
Agar-agar

Two things are commonly called by the name of agar-agar. On the one hand it is the well-known name of a trade product in Europe, which is prepared from certain species of red algae, but on the other hand it is also used on occasion for drugs prepared from untreated red algae which have merely been dried. There is no doubt that much confusion exists in the use of the term and likewise there is also much erroneous information about its sources. The word is of Malayan origin and in that language it refers to red seaweeds of the genus *Eucheuma* (see p. 135). Hoffman (1939) states that the word agar in Malay referred to the red alga *Gracilaria lichenoides*, but this statement is not correct as *Eucheuma* is the genus most widely used in Malaya for making an agar-like material.

It is evident, though the material is really a phycocolloid (see p. 111), that the application of the word 'agar' requires to be standardized, and its use should be restricted to the dried extract and even here it should be used in the generic sense. It has been suggested by Tseng (1944a) that the term 'agarophyte' should be applied to the seaweeds used for its manufacture. The term agaroidophyte can be used for red seaweeds that yield substances that are agar-like chemically, but with different gelling and viscosity properties (see p. 129). The most important agarophyte genera are species of *Gelidium*, a few species of *Gracilaria*, *Pterocladia* spp., *Acanthopeltis japonica* and *Ahnfeltia plicata*. Species of *Eucheuma* are also used but the product (Eucheuman) is really intermediate between agar and carrageenan (see p. 136). Other red algae, such as *Gigartina* and *Hypnea*, also yield a compound more closely allied to carrageenan (p. 111). A critical definition of agar is not possible at present, because it is prepared from a number of different red algae and its composition varies. A tentative definition has been put forward by Tseng (1945a) as follows: 'the dried armorphous, gelatine-like, non-nitrogenous extract from *Gelidium* and other agarophytes, being the sulphuric acid ester of a linear galactan, insoluble in cold but soluble in hot water, a one per cent solution of which sets at 35–50°C to a firm gel, melting at 80–100°C, being composed of a neutral, partially methylated polysaccharide (agarose) and a sulphuric acid ester (agaropectin) of a linear galactan' (see p. 183). It is now (Levring, Hoppe and Schmid, 1969) suggested that a 1·5% solution should set between 32 and 39°C and not melt below 85°C. Agar differs from carrageenan (p. 182) chemically, but it can also be readily distinguished without elaborate chemical analysis by use of agarase from the bacterium *Pseudomonas atlanticum* which hydrolyses agar but not carrageenan (Yaphe, 1957). The agarases actually fall into two distinct categories depending upon whether they attack the α- or β-linkage of the agarose component (Young *et al.*, 1972).

5.1 SOURCES OF AGAR

The untreated algae are differentiated according to the country of origin e.g. Ceylon agar or Ceylon moss. This affords an example of the use of the word agar as applied to untreated red algae, a usage which should be abandoned. Ceylon moss refers to the dried red seaweed *Gracillaria lichenoides* obtained primarily in the island, though it also grows on the coasts bordering the Indian Ocean, where it is called Bengal isinglass. Chinese moss refers to another species of *Gracilaria, G. verrucosa*. In Japan agar-agar, made from *Gelidium* spp., goes under the name of 'Kanten', which means 'cold sky', and as such refers to the fact that the material used to be prepared on cold winter days, or else high up on the mountains where it is always cold.

Since the 1939–45 War a number of other countries have embarked upon the manufacture of agar, and various genera of red algae have been used, e.g. *Gelidium* (USA), *Pterocladia* (New Zealand), *Suhria* (South Africa), *Ahnfeltia* (Russia).

It will be realized, therefore, that agar is not derived from even a single algal genus (see Table 5.1), and it may be expected to be diverse in its characters and properties. It is very important, therefore, that in all investigations concerning this material, and in chemical analyses, the exact species and even variety should be clearly stated. There is little doubt that much of the older information about agar is rendered useless because no attention has been paid to this matter of proper identification.

Table 5.1 Rhodophytes used in the agar industry

Species	Country	Use
Acanthopeltis japonica	Japan Sea	Toriashi-agar
Ahnfeltia plicata	White Sea, Sakhalin	Base of Russian industry
Gelidiella acerosa	Japan	Agar and food
Gelidium corneum var. sesquipedale	Spain, Portugal, Morocco, California	
G. amansii	Japan, Korea, China	Tengusa
G. divaricatum	Japan, Korea, China	

(*continued on the next page*)

Table 5.1 Rhodophytes used in the agar industry (*continued*)

Species	Country	Use
G. japonicum	Japan, Korea, China	Onigusi
G. liatulum	Japan, Korea, China	Mixed with G. amansii as adulterants
G. pacificum	Japan, Korea, China	
G. subcostatum	Japan, Korea, China	Mixed with G. amansii as adulterants
G. subfastigiatum	Japan, Korea, China	Mixed with G. amansii as adulterants
G. vagum	Japan, Korea, China	Mixed with G. amansii as adulterants
G. cartilagineum	South Africa, Mexico, California	Base of USA agar industry
G. nudifrons	California	Used as adulterants
G. arborescens	California	Used as adulterants
G. densum		
G. lingulatum	Chile	
G. pristoides	South Africa	
G. spinulosum	Morocco	Principal species
Gracilaria verrucosa (confervoides)	Atlantic, N. America, California, S. America, India, Ceylon, Japan, Australia, S. Africa, China, Formosa, Philippines	Important in USA Japan and Australia
G. cornea		
G. multipartita	Atlantic, N. America	
G. lichenoides	Ceylon	Ceylon moss
Gelidiopsis rigida	Indonesia, Philippines	
Pterocladia pinnata (capillacea)	Japan, New Zealand USA	Base of New Zealand industry Important
P. lucida	Australia, New Zealand	
P. tenuis		
P. densa	Japan	
Suhria vittata	South Africa	Formerly important

5.2 AGAR INDUSTRY IN JAPAN

The world production of agar must be currently about 4540 metric tons annually, and of this Japan produces about 2040 metric tons (Moss, 1979). Half of this is manufactured industrially by some fifteen commercial firms, whilst the remainder is produced by about 400 small-scale operators. The principal centres of production are in Hokkaido and at Osaka, Kyoto, Hyogo and Nagano, together with the prefectures of Schizuoka, Miye and Dakayama (Fig. 5.1). To produce 2000 tonnes of agar about 11 000 tonnes of dry raw agarophytes are required (see also p. 159). 40 per cent of the raw material is collected around Japanese shores and the balance is imported from China, Chile, Formosa, USSR (Sakhalin), Australia, Mexico and South Africa. The value of the import figures in US dollars is set out in Table 5.2.

Table 5.2 (After Michanek, 1975)

Year	Raw agar algae value $
1965	1 600 000
1966	5 300 000
1967	7 100 000
1968	2 300 000
1969	2 500 000

The import figures above include *Gracilaria* and the relevant amounts of this alga imported are set out in Table 5.3.

Table 5.3 (After Yamada, 1976)

Country	Imports of *Gracilaria* (tons dry wt)		
	1972	1973	1974
Chile	3028	3000	3825
Argentina	2153	801	452
South Africa	499	649	245
Philippines	186	1317	634
Brazil	189	1281	2337
Taiwan	50	939	395
Other Asian areas	74	328	237
Other areas	128	89	214
Total	6307	8466	8340

Agar was originally produced in China and was introduced into Japan in 1662. According to Horiuchi (Tseng, 1944a), the following dates have all been suggested for the discovery of the present method of agar manufacture, involving freezing and thawing: 1647, 1655, 1658 and 1688. In these early days the produce had no connection with the word agar and as late as 1893 it was still called dried seaweed jelly, though it had been known to European scientists for some thirty years.

There are other names by which this important Japanese product is known, e.g. Oriental or Japanese isinglass, Japanese gelatine, gelose, Hai Thao or just Thao. It is not always possible to obtain a clear picture as to which algae are used in Japan for agar. This is partly because the different species have the same common name. It is

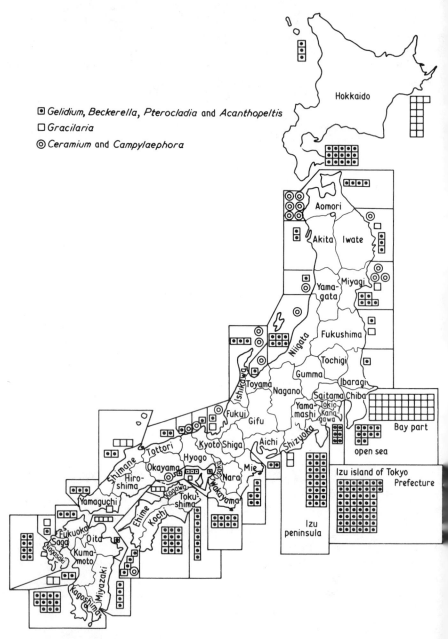

Fig. 5.1 Distribution of agarophytes (from Okazaki, 1971). One mark is equivalent to 20 tons (1 ton = 1·016 tonnes).

evident that *Gelidium amansii* is the most important species among all those used. Davidson (1906) uses the specific name *swansii* which must be a misprint for *amansii*. In Japan this particular species is called 'Tokoro-tengusa', or simply 'Tengusa', whilst 'Kinukusa' is applied to one form, f. *elatum*, of the species (Fig. 1.10). Tengusa, however, would seem rather to refer to the genera *Gelidium* and *Pterocladia* as a whole and not properly to this species: *Gelidium pusillum*, for example, is known as 'Hai-tengusa'. According to Takamatsu (1938) *G. amansii* (f. *typica*) is known in the middle part of Honshu as 'Makusa'. At the present time 22 species of *Gelidium* (see Table 5.1) are recognized in Japan, but the genus is really in need of further revision (Segi, 1966).

The economic importance of *Gelidium amansii* is such that considerable investigation of its biology has been undertaken. Katada (1949) has shown that the optimum temperature for the germination of spores is 25–26°C. Suto (1950) found that tetraspores were shed more abundantly at 20°C and carpospores at 24°C. Shedding occurs mainly in the afternoon and as many as 10^4-10^5 spores per 1 g frond are shed in a day. The spores become fixed to a substrate within 10 min. Growth is slow in summer and autumn, but then becomes rapid (Ueda and Katada, 1949). In the process of regeneration, grazing by gastropods can be a serious factor.

'Kinukusa' (*G. amansii* forma *elatum*) gives a very good yield because 25–30% of the dry weight is composed of the gelatinous material, although plants from another red genus, *Acanthopeltis japonica* ('Tori-ashi' or 'Yuikiri'), are of much the same value and yield a first-grade agar. *Gracilaria verrucosa* from southern Kyushu also yields a high quality gel (Ohta and Tanaka, 1964). There are a number of other species, some of which yield an inferior product so that they are used when mixed with a good-quality agar. These species, most of which are exported from Taiwan, include *Gelidium subcostatum* or 'Hira-Kusa', *G. japonicum* or 'Onigusa', *G. pacificum*, *Pterocladia pinnata* (= *P. capillacea*) and *Gracilaria verrucosa* ('Ogo-nori'). The last species yields an agar that is superior to that of other species (Funaki and Kojima, 1957). Despite this, *Gelidium amansii* and *G. japonicum* are regarded as the true Kanten weeds or Oyakusa.

When the manufacture was first commenced, agar was simply a mass of jelly obtained by boiling seaweed, but now it appears on the market in the form of bars (Kaku Kanten), strands (Ito Kanten), flakes (Fureiku-jo no Kanten) and powder (Funmatsu Kanten). The original preparation occurred by accident due to the fact that one day some jelly was put out of doors and solidified in various shapes. The story is that the Emperor was marooned in a snowstorm and took refuge in an hotel where some seaweed jelly was prepared for him. The remainder was thrown outside and froze during the night. The next morning the frozen jelly thawed and turned into a dry, papery translucent substance, which the innkeeper discovered could be reconverted back to jelly.

The algae are gathered in Japan from rocks between mid- and low-tide marks, with long-handled rakes in shallow waters, and by divers from the sublittoral. Women divers can collect down to a depth of 30 ft (9·15 m) with the use of goggles only.

The simplest method of harvesting is known as Taru-Ama. Collection is made from depths of less than 5–6 m and the weed is placed in a net suspended from a float. Ctiri-Ama involves diving to a depth of about 15 m using a sinker. The catch in this case is pulled up with the diver and put into a boat. Another method, not involving divers, is the use of a triangular dredge with bamboo teeth (Manga). A comparison of yields from Shirihama and Nishinahama using these three methods is set out in Table 5.

In Japan tetrahedral concrete blocks are widely used to provide an artificial substrate for the *Gelidium* plants and in the Izu peninsula the cost of such a bed was recouped in 4 years (Yamada, 1976). Manuring of the beds (8 parts NH_4^- to 4 parts PO_4^{3-}) has been tried by Yamada and co-workers (loc. cit), and it gave satisfactory results even when the cost of manure was involved. Manuring produced a slightly greater yield in agar and also a slightly greater jelly strength. As may be expected, the net increase in harvest value varied from locality to locality and also depended on whether the area had a thin or thick growth of *Gelidium*.

The best months for collection are July and August, though harvesting actually takes place continuously from May to October. The best weed comes from the deeper waters. After the weed has been collected it is dried on the shore and is partly bleached, and then sold to the factories. The rights to gather agarophytes are controlled by the Central Federation of Fisherman's Co-operative Association and they also are responsible for distributing the weed to the various processing factories. Originally it was stored until the arrival of the cold winter months, but nowadays with refrigeration this is not necessary.

Further treatment is as follows: it is first of all cleaned by beating and pounding, whilst larger lumps of foreign material are picked out by hand. Finally it is washed in fresh water, after which it is laid out to bleach on mats. This is done in warm weather, beginning in August, and the bleaching is much aided by dew. If the conditions are very favourable, twenty-four hours may be sufficient for the process, but it usually takes several days. The more modern method is to wash and stir the weed in autoclaves and then decolourize using charcoal. The importance of blanching for ordinary commercial use is exaggerated because it is not really necessary: it is, however, desirable for bacteriological purposes. As the drying and bleaching goes on the algae fuse into thick or thin sheets which can be loosely rolled. These sheets are then put into large wooden or iron vats together with some water, the exact quantity depending on the condition of the seaweed (Fig. 5.2). At this stage blending of the weeds takes place.

The final product is therefore not a pure kanten produced from a single species of *Gelidium*. Not only may a mixture of local algae be used but agarophytes imported from other countries are commonly added (p. 151). The amount of adulteration varies from 10–40%, the exact quantity depending on the purpose for which the agar is required. Marchand (1879) was able to identify thirteen different species of algae in one particular sample sent to Europe.

A mixture that is regarded as yielding a good quality agar is as follows:

	%
Gelidium amansii	45
Gelidium japonicum	10
Acanthopeltis	5
Campylaeophora	10
Gracilaria	15
Ceramium	5
Gelidium spp.	10
	100

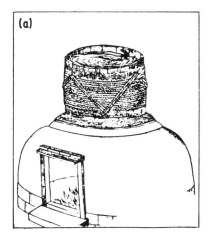

Fig. 5.2(a) Traditional furnace and tub for boiling *Gelidium*. (b) Press for straining crude seaweed jelly.

Table 5.4 Comparison of yields using different algal harvesting methods

	Giri: total per boat (tons wet wt) and per diver (kg wet wt)				Manga: total per boat (tons wet wt) and per diver (kg wet wt)				Taru-Ama: total per boat (tons wet wt) and per diver (kg wet wt)			
	Shirihama		Nishinahama		Shirihama		Nishinahama		Shirihama		Nishinahama	
Year	(tons)*	(kg)	(tons)	(kg)	(tons)	(kg)	(tons)	(kg)	(tons)	(kg)	(tons)	(kg)
1955	—	—	23.6	323.3	—	—	34.1	387.3	—	—	8.1	63.3
1956	143.9	45.2	24.6	332.2	89.1	60.9	36.1	425.5	18.8	14.9	7.5	65.2
1957	148.6	38.0	15.8	210.9	163.8	106.6	16.8	207.7	19.0	17.2	2.3	30.3
1958	150.1	52.8	28.8	339.0	96.3	76.2	35.8	350.9	17.3	14.5	10.5	40.2
1959	180.0	73.1	—	—	188.2	114.7	—	—	—	—	—	—
Average	—	52.3	—	301.3	—	89.6	—	345.4	—	15.5	—	49.8

* 1 ton = 1.016 tonnes.
This table effectively demonstrates that yields from different localities can vary greatly.

Fig. 5.3 Pouring liquid Kanten into cooling trays.

Sulphuric acid is first added in order to control the pH since this affects the quality of the agar. The 'hard' weeds (*Gelidium*) are boiled for 1 h 15 min and then the 'soft' weeds added and boiled for 15 min, after which the whole mixture is allowed to simmer for 12 h. The liquor is strained first through a coarse cloth and then through a fine linen cloth under pressure (Fig. 5.2), and the weed left is subjected to a second boiling of about 10 h. At the very end it is bleached by adding sodium peroxide. It is again filtered and the liquor is poured into wooden trays to cool by making use of rectangular wooden vessels (Fig. 5.3). The cold jellied material is known as 'Tokoro-ten'.

A freezing temperature is now essential for the next stage, but, before being subjected to the freezing process, it is cut up, by means of oblong iron frames with sharpened edges, into suitable sizes for further handling (Fig. 5.4); in this state it is known as agar bars, bar kanten or 'Kaku kanten'. Some of this is exported to Holland where it is largely used in the manufacture of beer. When the bar kanten

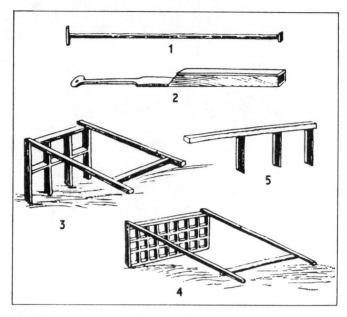

Fig. 5.4 Traditional articles used in cutting seaweed jelly into sticks and bars.

thaws after being frozen, the water flows away with the impurities, leaving behind relatively pure agar-agar, which is then dried. The freezing process, either in the open or in refrigerated rooms, take from one to three days and another three or four days are allowed for drying.

The kaku kanten is also put into cylinders and pressed out through a perforated base into thin fine threads 30–35 cm long, looking rather like macaroni. This material is known as 'Ito Kanten'. The kanten is finally parcelled up and made into bales weighing 133 lb (60·24 kg), but it is sometimes sold in a shredded or powdered form. The commercial product is pearly white, shining and semi-transparent, and is also tasteless and odourless. Examination and grading is carried out by the Japan Agar Distribution Control Co. If there is more than 3% crude protein it is rejected for export. The characteristics of the three export grades are shown in Table 5.5.

Davidson (1906) explains why, before refrigerated rooms became available, the site for the freezing of the tokoroten had to be carefully selected. The locality most suited to the operation was one bounded on the north-west by mountains or hills, and by meadows on the south-east. If the mountains are to the west the sun set sooner, congealing commenced early, and there was no rapid change to the required cold temperatures, and as a result the colour of the kanten was spoilt. When the mountains are in the north they intercept the cold north winds and cause them to deposit their rain, and then a cold dry wind blows over the meadows. The meadows were required for laying out the kanten, because no sand or dust was then likely to

Table 5.5

	Solidity	Crude protein (%)	Insoluble material (%)
1st grade	300	Less than 1·5	Less than 2
2nd Grade	200	Less than 2·0	Less than 3
3rd Grade	100	Less than 3·0	Less than 4

Moisture: less than 22% by weight.

Table 5.6

Year	Dried weed (tonnes)	Agar (approx. ·33 dry weed) (tonnes)
1952	10 130	2050
1963	13 476	2700
1964	15 135	3050
1965	15 150	3030
1966	16 804	3360
1967	21 300	—
1968	15 700	—
1970	12 314	—
1971	10 713*	—
1972	11 317*	—
1973	16 640	—
1974	13 414	—
1975	12 100†	—
1976	10 080†	2000 approx.

* Yamada, 1976.
† Shurtleff and Aoyagi, 1979.

become admixed during the freezing process.

Tondo (1923) has pointed out that the successful preparation of kanten demands great experience. 'It is necessary to take great care in the choice of materials and their mixing; the situation of the locality or how it was gathered, the nature of the ground, the kind of climate and the quality of the water are of the greatest importance'.

The output from Japan has been very considerable, ranging from 716 tonnes in 1945 to 2934 in 1940, the average annual output between 1926 and 1945 being 1885 tonnes. During this period there was a steady decline in the percentage exported (91% to 10%) but this has since risen to 27%.

Since the last world war the amount produced has greatly increased (Table 5.6).

In 1936, Horiuchi stated that there were 512 factories in operation, and the total value of the output was about £480 000. At today's (1977) prices Moss has estimated the value of the Japanese output at US $ 19 600 000 (£9 800 000). In 1967, 700 tons (711·2 tonnes) of agar were exported with a value of US $ 4 000 000. From the figures in Table 5.6, it appears that local consumption now uses some 4500 tons (4571 tonnes) of indigenous agarophytes so that the balance has to be imported (see Table 5.2, p. 151).

Japan not only leads the world in the production of agar, but also in its consumption. Thus in 1902, 164 tons (166·5 tonnes) or 16·4% of total production were consumed internally (Tseng, 1944a); in 1936, 886 tons (889·9 tonnes) or 31% of an increased production were consumed internally; in 1976, 1460 tons (1482 tonnes) of product or 73% were consumed internally, showing that internal consumption has markedly increased.

Japan also manufactures agar from *Ahnfeltia plicata* which is imported from Sakhalin (see p. 270). A recent newer method of preparation increased the yield of this product to 16·8% and the gel strength to 650 g cm^{-2} (Kojima et al., 1960). The product is known as 'Itani' agar and the ash contains 5–18% of sulphate depending on the method of preparation (Agar ash from *Gelidium* or *Gracilaria* contains over 50% (Tagawa et al., 1960)).

Although the process of agar manufacture came to Japan from China, the industry in the parent country declined considerably, and in later years China was importing substantial quantities from Japan. In 1906, 73% of the total Japanese export went to China and Hong Kong, but in 1938, because of the Sino-Japanese War, the imports had been reduced to about 49 tons (49·8 tonnes). Prior to 1937 China did possess a small agar industry of her own with three factories at Ningpo, Tsingtao and Chefoo. It has been estimated by Tseng (1944a) that their annual production was about 75 000 lb (34 050 kg). More recently a new process has been used that eliminates the freezing process. The resulting agar is also said to have a higher gel content.

5.3 AGAR INDUSTRY IN AMERICA

In 1920 it was decided to ascertain whether agar-agar could not be produced in America, because it could be important as a measure of national security. A number of the Rhodophyta growing on the Californian coast were regarded as probably suitable, and an industry was established in California.

The man who laid the foundation of the now flourishing US industry was a Japanese, Chokichi Matsuoka. He built the first factory in 1920 at Tropica (now Glendale). Tseng (1944b) has pointed out that Matsuoka in his first patent emphasized the use of a red alga, *Gloiopeltis,* and only casually mentioned *Gelidium* as a substitute. Since the former does not grow to any extent in California, it is evident that Matsuoka employed *Gelidium*, but presumably in order to mislead possible competitors he purposely emphasized *Gloiopeltis* in the patent specification.

According to Johnstone and Feeney (1944), the first concern discontinued operations in 1934, but prior to that in 1923 it was purchased by a company headed by John Becker, who modernized the whole process in the course of the next few years. He took out a number of patents covering a special combined congealer and sizer, a de-waterer for flaked agar and a special dehydrator. This process is still employed today in the United States. In the early years the output varied considerably as may be seen from Table 5.11.

The new company was unable to weather the last depression and finally closed in 1933. In that year a former employee of the company established the United States Agar Company in National City, California and between 1934–8 reprocessed Japanese agar, since this material was so cheap. This small company became the nucleus of the present American Agar and Chemical Company of San Diego, a firm which now obtains most of its raw material from Mexico and Laguna, San Pedro, and Redondo in Southern California, and is one of the world's largest producers of agar. Because supplies of suitable weed are increasingly insufficient to meet the demand, efforts are being made to find alternative sources in other parts of the world*. In 1942, when the USA issued an agar freezing order, the factory had a monthly output of about 4000 lb (1812 kg), and in 1945 there were eight additional factories producing, but none of these now exist. Tseng (1944a) estimated that the expected annual output in California would be about 100 tons (101·6 tonnes) and the figures for 1976 (Moss, 1979) confirm that this amount is produced. South of the border (Tseng, 1945b) at least two factories have been established in Mexico in the last 20 years, Agar-Mex at Ensenada being one of them.

Although after the war (1974) re-importation of agar from Japan took place, scientists and industrialists on the Atlantic seaboard of the United States had also become interested in the production of agar. Investigations by Humm (1942) showed that *Gracilaria verrucosa* was present in commercial quantities near Beaufort, North Carolina. In 1944 three factories were in existence on the Atlantic coast and experiments were in progress with a view to utilizing yet another red alga, *Hypnea musciformis* (De Loach *et al.*, 1945). The work was later extended to Florida (Humm, 1947) where two species, *Gracilaria blodgettii* and *G. foliifera*, were found to be abundant, but these two species apparently yielded an inferior product which can best be termed an agaroid. One can, however, obtain a good agar gel from *G. foliifera* (Mitsubishi and Hayashi, 1972) which shows that the extraction method is important.

On the Pacific coast the principal species used is *Gelidium cartilagineum* (recorded as *G. corneum* in error by Tressler (1923) and Hoffmann (1939)), which in extreme cases may grow to 4 ft (122 cm) tall, but is usually 3 ft (91·5 cm). It is regarded as harvestable when it reaches 18 in (45·7 cm). It is found in commercially harvestable quantities in Baja California: here it occurs in small concentrations where the tidal current is strong. About 80% of the agar manufacture now comes from this species. A further 10% is made from *G. sequipedale* and the balance from other species, e.g. *G. nudifrons, G. arborescens, G. densum* and *Pterocladia pinnata.* This last genus was apparently used by Matsuoka at Glendale (Gardner, 1927). There is no dearth of red algae of the agarophyte class in California; species which could be used, if localities are found where they grow in commercially harvestable quantities, are *G. pyramidale* (previously referred to as *G. amansii* but now shown to be *Pterocladia pinnata*)

* At present New Zealand is one of these sources.

G. pulchrum (= *G. australe*), and *Gracilaria verrucosa*. The local weed is not sufficient for the full needs of the American Agar Co. and dry weed is now bought from South America, South Africa, Portugal, North Africa, and New Zealand.

The importance of *Gelidium cartilagineum* in the production of agar has naturally directed the attention of scientists to problems connected with its survival, and also to a study of seasonal changes in the amount of gelling material. A prime necessity is to discover whether the methods and times of harvesting will have any detrimental effect upon the future supplies of the seaweed. In this connection Johnstone and Feeney (1944) have shown that there is no seasonal periodicity in reproduction, and that therefore time of harvesting is not of particular importance. It has also been found that regeneration of cut plants and maximum vegetative activity both occur chiefly in spring and autumn.

Baja California is particularly rich in stands of *Gelidium cartilagineum* and eight companies have concessions along the coast, the most important being Agar-Mex SA with 27 areas and Gel-Mex SA with 35 (Fig. 5.5). Because of the importance of the industry the National Institute of Fish Biology has a research programme directed at the ecology of this species and of *Macrocystis* (p. 6). The objective is to determine the optimum tonnage which can be removed without depleting the beds and also to see whether areas of standard and minimal production can be improved to bring them up to areas of high production. The annual production from this region for 1955–76 is shown in Fig. 5.6, the greatest amounts each year being collected from May to August (Table 5.7). This production has increased still further because in 1974 2451 tons (2490 tonnes) dry weight were collected, principally from areas shown in Fig. 5.5. There are still greater quantities that could be exploited (see p. 273).

There are several extraction processes used in Mexico and one of them is set out below (Ortega, 1976).

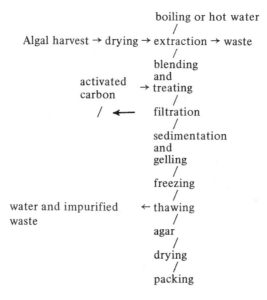

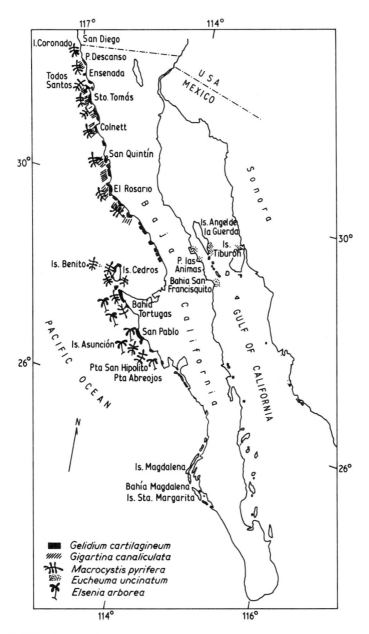

Fig. 5.5 Mexican marine algae in commercial quantities (after Guzmán del Próo, 1969).

Table 5.7 (After Guzmán del Próo, 1967)

Month	Gel-Mex. (tons)	Coop. PNA (tons)	Coop. Ensenada	Agar-Mex (tons)	Total (tons)
January	14	2·5	2·7	10·3	29·5
February	16·8	4·7	2·4	11·0	34·9
March	22·65	5·05	6·9	5·8	40·4
April	21·5	–	2·1	29·3	52·9
May	61·1	–	8·2	52·9	122·2
June	61·4	–	18·6	70·75	150·75
July	64·5	13·45	41·2	56·3	175·45
August	64·5	17·25	71·1	73·3	226·15
September	31·4	–	9·9	48·1	89·4
October	44·6	3·75	1·6	37·4	87·35
November	41·1	1·70	7·75	41·7	92·25
December	27·5	–	5·0	47·2	79·7
	471·05	48·4	177·45	484·05	1190·95

Table 5.8

Month	Weight of dry agar (g)	
	Boiled 2 h	Boiled 4 h
January	3·89	7·11
February	1·98	6·86
March	3·54	7·91
April	6·26	7·37
May	6·88	9·94
June	7·40	11·30
July	5·17	7·86
August	4·26	5·82
September	5·79	8·23
October	3·24	6·97
November	5·38	8·13
December	5·34	7·15

Cooper and Johnstone (1944) studied the seasonal production of the agar material in *Gelidium cartilagineum*, and they found that there is a peak production in June with a general summer maximum (May–July). Their figures also show that the length of time of boiling in the production process may affect to a considerable extent the quantity of material extracted (Table 5.8).

The peak production is associated with the period of maximum photosynthesis, and it has therefore been suggested that the price should be determined by the season. In view, however, of the wide fluctuations, especially after four hours' boiling, this would hardly seem feasible.

In Baja California, harvesting is carried out by divers who collect the plant by hand. The best harvesting period is from August to November after reproduction has occurred and the agar content is close to maximal. Beds are harvested bi-annually (Barilotti and Silverthorne, 1972). Each boat has a crew of three—a diver, boat operator, and life-line tender. The alga is put into a rope basket which is hauled to the surface when full (it then contains 60–70 (27–31·5 kg) of seaweed). In a single working day a diver can harvest up to 1·5 tons (1·52 tonnes) of fresh *Gelidium*, but quantities much less than this are more normal. Diving can only be carried out when the water is clear and there is no strong ground swell. Because of this most of the collecting is restricted to the period between May and November. Some companies furnish the boat and facilities and pay the harvesters per fresh ton ($85 in 1945), whilst in other cases the harvesters provide the boat and equipment. In the latter case they also dry and bale the weed. Under these conditions the harvesters received from $350 to $500 per ton (1945) of the sun-dried seaweed.

The preparation of agar from these seaweeds follows much the same sequence as the modern Japanese process (see p. 157). If species of *Gigartina* are used, a preliminary soaking is necessary first of all, followed by a boiling in a solution containing 2% chloride of lime. Unless this is carried out the resulting jelly is not firm enough for commercial purposes. This is because the product is not a true agar. Matsuoka not only introduced artificial freezing, but he also substituted bleaching by chlorine for sun bleaching. The value of this method of bleaching is open to some doubt because of possible effects upon the gel strength.

The *Gelidium* is first washed and soaked for 12–14 h, after which it is transferred to special pressure cookers, where it is cooked for 6 h at a pressure of 15 lb in^{-2} (6·79 kg per 6·45 cm^2) in a dilute agar solution from the third and final cook. It then receives two more cookings before being discarded. The extract is clarified and filtered, and then poured into open tanks where it gels after 24 h. The gel is chopped up and put in cans in the freezing rooms at 14°F (-10°C) for about two days. On removal it is thawed and placed in the de-waterer, which removes the washing water with the soluble impurities.

The purified agar flakes, which now contain about 90% moisture, are dried by hot air until they contain about 35% moisture. After this they are bleached in 1% sodium hypochlorite solution at room temperature. The excess bleaching reagent is reduced by sodium sulphite, after which the agar is removed, washed, and finally dried until only about 20% moisture remains.

In the original Becker process there was a special combined congealer and sizer, which was used to cool and gel the solution. This has now been abandoned in favour of the older method of cooling and gelling in open tubs. Also in the original process activated carbon was used to decolourize the agar. Unless great care was taken, particles of carbon passed into the final product, and so the function of the carbon has been replaced by the bleaching process.

Work on agar seaweeds (Tshudy and Sargent, 1943) has shown that the pH of the extracting solution is of very considerable importance, but the effect varies for the

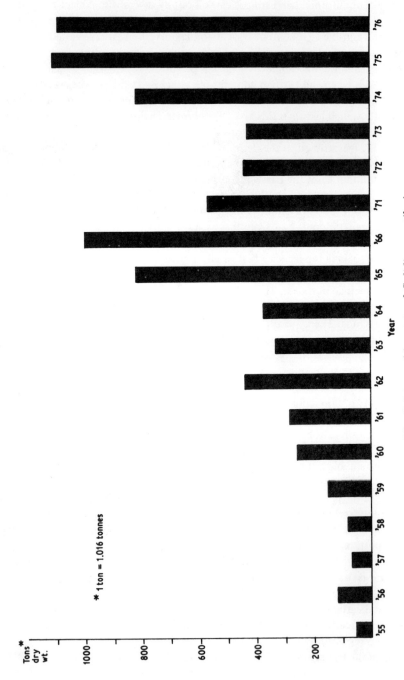

Fig. 5.6 Annual harvests of *Gelidium cartilagineum*.

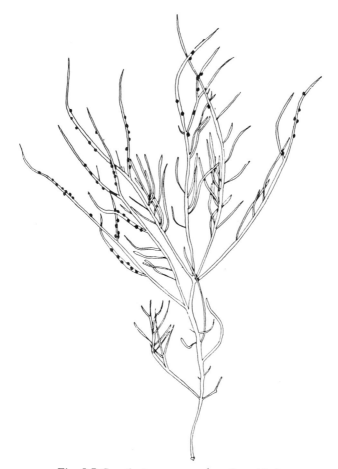

Fig. 5.7 *Gracilaria verrucosa* (*confervoides*).

different species. Thus the yield from *Gelidium cartilagineum* and *Pterocladia* spp. is maximum around pH 6–8 and falls off with increasing alkalinity. In the case of *Endocladia muricata* the reverse holds true, the maximum yield being secured at pH 12. Attempts to extract an agar-like substance from *Gigartina canaliculata* and *G. serrata* were only successful when they adopted the treatment used by Kizevetter (1937) to secure agar from *Ahnfeltia plicata* on the maritime coast of Russia and from *Phyllophora rubens* in the Black Sea.

On the east coast of America an agar industry using *Gracilaria* had made rapid headway early on. In the first nine months of 1944, 25 000 lb (11·35 tonnes) were produced, whilst prior to that period, from August 1943 to January 1944, approximately 50 000 lb (22·7 tonnes) of agar had been produced (Humm, 1944). Collection commences about 1 August and continues until mid November, when it ceases, so that the remaining *Gracilaria* can serve as a 'stock' from which the succeeding year's crop is developed. Since then attempts have been made to transplant *Gracilaria* from North Carolina to Miami, in order to start the industry in Florida (see p. 250).

The plant grows in shallow waters, which are more or less protected against wave action, and it exists in two forms, a free-living sterile phase and an attached fruiting phase (Fig. 5.7). The former increases solely by vegetative means and the latter by the normal sexual and asexual processes. It is possible that these two phases are completely distinct, in the same way as the free-living marsh fucoids of European salt-marshes are distinct from the attached parent species. It has been found that low temperatures and insufficient light may retard or even inhibit growth (Causey et al., 1945). Thus the most rapid development takes place at sea temperatures between 25 and 28°C, whilst the light intensities that are found at depths of 2–4 ft (0·6–1·2 m) permit of the most rapid growth (see also p. 252).

Because of its habitat the *Gracilaria* is readily harvested by forking the seaweed into small boats, by accumulating it in 'stop nets' placed at right angles to the tidal current, or by trawling in deeper waters. After collection it is air-dried, the process taking three to four days. Three tons (3·05 tonnes) of fresh *Gelidium* yield one ton (1·02 tonnes) of dry weed, but according to De Loach et al. (1945) it requires 10 tons (10·16 tonnes) of wet *Gracilaria* to give 1 ton (1·02 tonnes) of dry weed. Tseng (1945b) on the other hand reports seven tons of wet weed as giving one ton of dry weed. In Australia (p. 175) 7–8 tons (7·11–8·13 tonnes) are required to give one ton (1·016 tonnes) of dry weed. Seven tons (7·11 tonnes) is therefore probably the minimum, and in general it is likely to be rather more.

Extraction must take place under steam pressure or by boiling, and it would seem that although the extraction is easier than with *Gelidium* the amount extracted is not so great. Not only does the agar content vary with the species and season, but also with the method of extraction. Thus material collected between June and early August can be extracted satisfactorily without attention to the acidity (pH) of the extracting solution. For material collected later in the year the pH of the extracting solution must be adjusted to a point below 6·5 (i.e. it must be distinctly acid) as may be seen from Table 5.9 (Samples 5–8). Michanck (1975) reports that two factories in North Carolina use two different processes, one involving boiling followed by freezing and the other pressure cooking without freezing. These two different processes yield agar gelling at 38°C or at 55°C.

The total agar content of North Carolina *Gracilaria* lies between 55 and 65% on a dry weight basis, and although 25–35% can be obtained in the laboratory, such yields are not likely with commercial equipment. One factor which tends to bring about this result is the presence in the late autumn of an adulterant, the red alga *Laurencia poitei*. An extract of this alga, when heated, brings about partial hydrolysis of the *Gracilaria* agar, and this introduced filtration difficulties, so that the full yield is not obtained. The gel strength of this *Gracilaria* agar compares favourably with that of *Gelidium* agar and it can be increased 5–15% by the addition of potassium chloride.

Table 5.9 Variation in yield of *Gracilaria verrucosa* (*confervoides*)

Sample	Date of collection	Yield %	pH
1	20 June	21	7·3
2	3 July	37	7·3
3	1 August	40	7·3
4	15 August	11	7·3
5	31 August	1·3	7·3
6	31 August	30	0·1% acetic added
7	4 October	1·6	7·3
8	4 October	38	0·1 % acetic added

Monthly yield of *Gracilaria*

Month	No. of samples	Yield % dry wt
July	3	38
August	6	34
September	9	32
October	10	29
November	4	18

Studies by Hoyle (1978a and b) on two species of *Gracilaria* from Hawaii, *G. bursapastoris* and *G. coronopifolia*, have shown that there is no significant difference in the yield of agar or in its gel strength from the gametophytes and tetrasporophytes (Table 5.10).

Table 5.10

Agar yield from two species of *Gracilaria* (% dry weight)			
Species	Male plants	Female plants	Sporophytes
G. bursapastoris	17·6−19·7	17·3−19·8	16·3−18·5
G. coronopifolia	24·8−36·70	25·5−33·2	24·8−32

The yield from *G. coronopifolia* can be up to twice that of *G. bursapastoris* and the gel strength up to four times greater. The agar content is apparently inversely proportional to the total nitrogen content of the thallus.

By way of contrast, cystocarpic plants of *Gracilaria verrucosa* in Chile give higher agar yields (21−26%), but lower gel strength (111−520 g cm^{-2}) than do tetrasporic plants (17−20% yield, 260−692 g cm^{-2}) (Kim and Henriques, 1979). Extraction in Chile, where there is one factory with an annual capacity of 300 tons (304·8 tonnes), involves treatment with 2·5% NaOH for 90 min, washing, neutralizing with

Table 5.11 American agar production

Year	Quantity (lb)*	Year	Quantity (lb)
1923	7 755	1934	1 802
1924	7 281	1935	8 061
1925	117 773	1936	—
1926	29 877	1937	21 208
1927	—	1938	7 170
1928	22 797	1939	8 098
1929	5 140	1940	24 000
1930	44 895	1941	52 000
1931	28 395	1942	110 054
1932	10 009	1943	165 954
1933	41 557	1944	113 762
		(Jan.–Aug. only)	

* 1 lb = 0·454 kg)

Table 5.12 Agar consumption in the USA

	1956 (lb)*	1966 (lb)
Microbiology	180 000	300 000
Baked goods	200 000	200 000
Confectionery	80 000	100 000
Meat and poultry	60 000	100 000
Desserts and beverages	50 000	100 000
Laxatives and health foods	50 000	50 000
Pet foods	—	50 000
Moulages, including dental	40 000	30 000
Pharmaceuticals	10 000	20 000
Miscellaneous	20 000	50 000
Total	690 000	1 050 000

* 1 lb = 0·454 kg

0·3 N H_2SO_4 for 15 min, washing, extracting at 100°C for 90 min and then separating out by freezing.

Humm (1962) has reported on gel strengths of agars from *Gelidium* and *Gracilaria* and hypnean from *Hypnea* all collected from Chesapeake Bay.

The amount produced annually by the American agar industry is given in Table 5.11 (Chase 1942; Tseng, 1945b). From these figures it will be observed that there was a considerable drop in production in 1934 after the principal agar company closed down. A statement by Ferguson Wood (1942) that just prior to the Second World War (1938)

America was producing 6 000 000 lb (2722·2 tonnes) annually, cannot be regarded as correct. It is possible that this figure should have been 600 000 lb (272·2 tonnes) or else it included production from the Irish moss industry (Section 4.3.3).

Unfortunately the seaweed industrialists are extremely secretive about amounts of weed used and output of finished products so that later figures are not available. Some clue to the output is provided by Table 5.1 showing the estimated consumption of agar in the USA. The latest figures give the production for 1976 as 100 tons (2 240 000 lb) (101·6 tonnes).

In 1966, therefore, about 525 metric tons were produced, and on a basis of Japanese data it is probable that some 2500 metric tons of dried weed were used.

In British Columbia, Saunders and Lindsay (1979) have estimated that *Gracilaria* could reach a density of 75 tons (76·2 tonnes) per hectare but no industry appears to have commenced.

5.4 CENTRAL AND SOUTH AMERICA

Agarophytes (*Gelidium, Gelidiella, Pterocladia, Gracilaria, Hypnea*) occur in Venezuela and there may be sufficient to justify exporting to the USA for processing (Diaz-Piferrer, 1967).

The distribution, occurrence and end product of a number of potential agarophytes from Cuba has been published by Diaz-Piferrer (1961). Of 50 species investigated some 54% produced a compound that could be classed as an agar and 46% gave rise to an agaroid. A study by the same author (Diaz-Piferrer and Perez, 1964) of the same or similar red algae of Puerto Rico has shown that, of those investigated, 22 species produce an agar and 20 species an agaroid of good quality (Table 5.13). In neither country has an agar industry been established as yet.

It will be observed that several species yield a product that behaves very differently depending on whether the plants are collected in Cuba or Puerto Rico, even after some allowance has been made for a 0·5% difference in concentration, e.g. *Gracilaria caudata, G. foliifera, G. verrucosa* and *Hypnea musciformis.*

In Brazil the annual output of agar, mainly from *Gracilaria,* is 20–30 tons (20·3–30·5 tonnes) (Yamada, 1976) though 144 tons (146·2 tonnes) were manufactured in 1973 (Michanek, 1974). The Argentinian output (mainly from *Gracilaria verrucosa*) is 500–550 tons (508–558·8 tonnes) and the Chilean (mainly from *Gelidium ungulatum*) 80–200 tons (81·3–203·2 tonnes) (Yamamada, 1976; Moss, 1979).

5.5 AGAR INDUSTRY IN RUSSIA

Before the 1939 war the USSR began a search for seaweeds suitable for agar manufacture. The most suitable algae were *Phyllophora, Iridaea* and *Ahnfeltia plicata.*

Table 5.13 Characteristics of Cuban and Puerto Rican Agarophytes

Species	1 % gel	Cuba		Puerto Rico	
		Melting point °C 2 % sol.	Gelling point °C 2 % sol.	Melting point °C 1·5 % sol.	Gelling point °C 1·5% sol.
Brongniartella mucronata	Liquid	41	21	–	–
Bryothamnion triquetrum	Solid	68	37	68·5–81	40–44
Bryothamnion seaforthii	Liquid	71	32	–	–
Digenea simplex	Solid	79–86	32–37	75	34–36
Enantiocladia duperreyi	Liquid	49	29	–	–
Gelidiella acerosa	Liquid	77	31	68·5–87	37·5–43·5
Gelidium corneum	Solid	78	36	–	–
Gracilaria caudata	Liquid	70	31	38	27
Gracilaria cervicornis	Liquid	82	35	–	–
Gracilaria cornea	Liquid	52	33	–	–
Gracilaria crassissima	Liquid	78	34	78·5–80	35–42
Gracilaria cylindrica	Liquid	79	36	–	–
Gracilaria damaecornis	Liquid	81	37	–	–
Gracilaria ferox	Solid	73	39	61·5	37
Gracilaria foliifera	Liquid	62	27	87	47
Gracilaria mamillaris	Solid	77	39	62	38
Gracilaria verrucosa	Liquid	78	37	53–54	28–30
Gracilariopsis sjoestedtii	Liquid	79	36	–	–
Gymnogongrus tenuis	Liquid	75	41	–	–
Hypnea cervicornis	Liquid	44	27	–	–
Hypnea musciformis	Liquid	46–48	22–28	56–68	42–49

It is now known that the first yields phyllophoran and the second iridophycin or iridophycan, and only the third a true agar (see p. 111). Three factories were built, at Archangel, Vladivostock and Vladimir, and in 1936 they produced 55 tons (55·85 tonnes) of agar from *Ahnfeltia*, mostly harvested in Sakhalin (Kizevetter, 1937) With *Ahnfeltia*, 8% of lime is added in the digestion process in order to keep the pH alkaline. The quality of the product varies with the digestion time and best results have been obtained with 6 h for the first, followed by subsequent treatments of 4 h and 2 h. During filtering the temperature has to be kept above 80°C. Activated carbon is added as a clarification agent and the liquor is then set in moulds and put through a normal agar freezing process.

5.6 AGAR INDUSTRY IN GREAT BRITAIN

When Japan entered the 1939 War the principal source of agar supplies immediately became cut off. Since agar is of very considerable importance in bacteriological work (see p. 190) every country conserved its stocks with great care, and research was immediately undertaken in order to provide agar from local seaweeds. In Great Britain it was found that species of *Gelidium* are not present in commercially usable quantities. In Ireland, agar had been manufactured from *Gelidium pulchellum* and *G. latifolium*, since these two species occur in sufficient quantity to be harvested (Yarham, 1944). In England, it was found that a suitable agar substitute could be produced from *Chondrus crispus* (Irish moss) and *Gigartina stellata.* Both these species occur in considerable abundance, and a survey of the available supplies was carried out. Since 95% of the weed harvested is *Gigartina,* much attention has been given to it. It has been recommended, after studies of the life cycle, that there be only one harvest per annum and that the weed be hand-picked so as to leave the hold-fast from which regeneration will take place (Marshall, Newton and Orr, 1945; Newton, 1949). Gel strength reaches its maximum at the time of reproduction. Production was commenced in 1943 when 10 cwt (0·508 tonnes) was manufactured. The agar-like compound differs from that of the Japanese in having a higher ash content and a lower melting point, but on the other hand it dissolves more readily and yields a clearer, highly viscous gel. As in the case of *Gracilaria* and *Hypnea* agar, the melting and gelling temperatures are both affected by the presence of electrolytes, particularly potassium chloride. A better product was obtained by heating with potassium hydroxide and then neutralizing with hydrochloric acid. The amount of added hydroxide should not exceed 15%. Small quantities of a very good agar were also manufactured from *Ahnfeltia,* but this alga is not very abundant in Great Britain and so could not be employed on a commercial scale. At the present time agar is not manufactured any more from local algae in Great Britain.

5.7 EUROPEAN CONTINENT

Several European countries now manufacture part or all of their agar supplies. In Italy, Caassini-Lokeur and Bruni (1970) examined the best conditions for agar extraction from *Gracilaria* species and recently Yamada (1976) reported that 30—50 tons (30·5—50·8 tonnes) are produced annually. In Portugal there are four factories using *Gelidium sesquipedale* or *G. corneum* as the raw material and two more in the Azores using *Pterocladia pinnata (capillacea).* The raw weed is collected by hand or by divers and Michanek (1974) estimates there is sufficient to yield 1300 tons (1321 tonnes) agar annually, but at present only 300—400 tons (304·8—406·4 tonnes) are produced (Yamada, 1976). Neighbouring Spain has an annual output of 700—800 tons (711·2—812·8 tonnes) (Moss, 1979) and France 50—100 tons (50·8—101·6 tonnes) (Yamada, 1976; Moss, 1979).

In Spain, where *Gelidium corneum* occurs in some size and quantity around the Galician coast, collection has been regulated since 1942 by the Department of Fisheries which requires the alga to be collected by hand. In the factories the dried and bleached weed is extracted in large metal boilers with false bottoms. Autoclaves can be used but the agar is not so good and the yield is not so high (32% as against

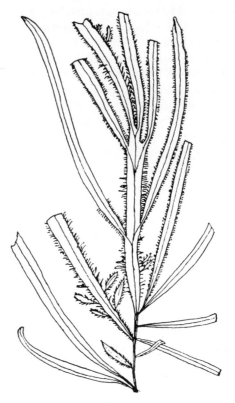

Fig. 5.8 *Suhria vittata.*

50% by boiling). Since natural freezing is not possible and artificial freezing is expensive, the agar is purified by washing with cold water. In concentrations of 3–4% the gel is said to be as good as Japanese agar.

On the opposite shores of the Mediterranean in Morocco a very good agar is produced from *Gelidium spinulosum* (Nielsen *et al.*, 1979).

5.8 AGAR INDUSTRY IN SOUTH AFRICA

In South Africa a number of seaweeds would appear to be suitable for the manufacture of agar-agar, though the quality of the product varies with the species. The most satisfactory species appears to be *Suhria vittata* (red ribbons) (Fig. 5.8). The gelatinous material is apparently very easy to extract from the species. It yields a very good agar, but it is not easy to harvest since it occurs mainly as an epiphyte on the large brown seaweed, *Ecklonia maxima*. The chemistry of the extractive has yet to be determined or publicized in order to show whether it is a true agar or only an agar-like compound.* The species that is commonly used, because it occurs in quantity, is *Gracilaria verrucosa* (Fig. 5.7, p. 167), which grows below low-tide mark on the Atlantic side of the Cape. The best plants are found where there is sheltered water and the bottom is sandy or sand with rocks. Big meadows occur in at least two localities (Hont Bay, Langebaau) and much of the weed is cast ashore after the winter storms (see p. 264). The agar produced from this seaweed serves extremely well in the manufacture of sweets. *Gelidium cartilagineum*, or 'red lace', grows in bunches hanging down from the rocks near low-tide mark. This gives a good yield of agar, but only when it is extracted under pressure. A related species *G. pristoides*, or 'brown sea parsley', is also common and abundant from mid- to low-tide marks. This species provides as good a source of agar as *Suhria* and is just as easily processed.

5.9 AGAR INDUSTRY IN AUSTRALIA

In the last century, a company was formed at Dongarra in Western Australia for the purpose of making agar from the red seaweed *Eucheuma speciosum*. Now, however, agar is made from *Gracilaria verrucosa*, which was very abundant in Botany Bay (recorded in error by Ferguson Wood (1942) respectively as *E. spinosum* and *Gracilaria furcellata*). It has been estimated that in New South Wales, where the species grows down to depths of 20 fathoms (36·6 m) especially in estuaries, along a coastline of 850 miles (1368 km), there is enough of this red seaweed to produce 100 tons (101·6 tonnes) of agar annually. This would seem to be sufficient for Australian home needs as only 71 tons (72·1 tonnes) were consumed in 1938. In 1943 it is reported that 200–250 tons (203–254 tonnes) of dried weed were harvested. In 1945, 20 tons (20·3 tonnes) of agar were produced. The latest figures (Yamada, 1976) give an annual output of 80–100 tons (81·2–101·6 tonnes). Apparently Australia intends to continue the industry, because attempts have been made to cultivate the *Gracilaria*. Ferguson Wood (1942) reports that there are great variations in the quantities and type of *Gracilaria* available and so a periodical examination of each estuary will be necessary. Even if the *Gracilaria* is not harvested, it still disappears from the beds at the end of the season. This is exactly comparable with its behaviour in North Carolina, and, so long as harvesting stops in sufficient time to leave a 'stock' for the next year, there should not be any great decrease in the source of supplies. It has been suggested that fences should be built along the shore near low-water mark in order to catch the weed coming ashore when the plants break off, and so prevent it being buried by the sand.

After the weed has been collected it is laid out on wire-netting racks to dry. Drying is important because extracts of wet weed do not gel. Sometimes it is dried by artificial heat but in such cases the product is not so good. When dry it is treated by boiling with steam in open vats, using mechanical stirrers in order to avoid 'cold spots', and

---

* It does contain agarose (Levring, Hoppe and Schmid, 1969).

great care has to be taken to control the acidity of the solution since below pH 5 the yield falls off rapidly. It has also been found that the use of iron and copper vessels tends to discolour the agar. The acidity of the solution is controlled by using sodium acid phosphate and an extraction time of 2–4 h. After filtering, re-digesting and re-filtering the liquor is clarified by activated carbon. It is then subjected to freezing, thawing and dewatering. The most satisfactory drying process, evolved by the Karna Vita Company, is to wash in alcohol and dry under infra-red lamps. It does not provide such a good agar as the Japanese product but for meat canning it is regarded as superior. The arsenic content is low and it can be used for most bacteriological work. Bleaching does not seem to produce a better product and in fact it reduces the yield. It requires 7 tons (7·1 tonnes) of wet drained *Gracilaria* to produce one ton of dry dark weed, but 8 tons (8·1 tonnes) of wet weed are necessary in order to produce one ton of bleached weed (Ferguson-Wood, 1942). One ton of agar can then be produced by boiling 3 tons (3·05 tonnes) of the dark dried seaweed and the product is said to be not inferior to imported Japanese agar. Some idea of the labour involved can be gauged from the fact that it is estimated that twenty men can harvest, dry and bleach 8 tons (8·1 tonnes) of wet weed per day.

5.10 AGAR INDUSTRY IN NEW ZEALAND

Moore (1944) and Kirby (1953) have provided accounts of the agar industry which arose in New Zealand as a result of the war. Two species of the genus *Pterocladia* are principally employed, as they are found in commercial quantities in the North Island (especially in the Bay of Plenty and East Cape area), and around Kaikoura. Investigations have shown that renewal of these algae after harvesting is satisfactory so they can be collected annually, e.g. *P. pinnata* (= *P. capillacea*) regains its full length eight months after being cut. Reproduction of the species appears to occur throughout the year, though the amount of fertility differs greatly in the various habitats. Since continued supplies of the weed are important, further research into the methods and effects of harvesting are desirable. Recently a first study has been made of the physiology of *P. pinnata* (Lee, 1967). The weed is collected by Maoris, dried and cleaned and then sold by them to the sole manufacturing firm, the Davis Gelatine Co. of Christchurch. The larger weed, *P. lucida,* dominates in the material and the agar produced from it is of very good quality, both in colour and gel strength, and is therefore highly suitable for culture media and meat canning. A more recent study by Luxton (1977) shows that the species could possibly be cultivated.

Since both species of *Pterocladia* grow in the lower part of the littoral the time available for collecting per tide is not very great. In rich areas a single collector may be able to obtain from 50 to 75 lb (22·7 to 34·1 kg) per tide but generally it will be le After collecting, the weed is washed with fresh water and laid out to dry or hung from fences for the same purpose. After drying, the plants, especially those of *P. lucida,* have to be cleaned of epiphytic corallines. At the factory the dried plants are washed again and then digested several times under pressure. The liquor is subsequently treated with a bleaching agent, filtered and allowed to set. It is purified by freezing and thawing after which the agar is dried and ground into a powder. The agar is of a very high quality and it is reported (Kirby, 1953) that 1 part by weight is equivalent to 1·6 parts of Japanese agar and in some commercial uses it may be equivalent to 4 parts of Japanese agar. Setting and melting points as compared with some agars are shown in Table 5.14. There is a seasonal variation in gel strength as

Table 5.14

Origin	Setting point	Melting point
Japanese agar	92°F (33·3°C)	193–194°F (89·4–90°C)
USA agar	96°F (35·5°C)	195–196°F (90·5–91·1°C)
New Zealand agar	95°F (35°C)	199–201°F (92·7–93·9°C)

well as a slight variation dependent upon locality (Fig. 5.9; Luxton 1978).

During the first year of operation 60 tons (61 tonnes) of weed were collected. This yielded 15 tons (15·2 tonnes) of agar, which was more than sufficient to cover the annual requirements of the country. In 1945, 105 tons (106·6 tonnes) of weed were collected and in 1946, 110 tons (111·8 tonnes) of weed. Recent production is rather less (Table 5.15).

Table 5.15

Agar production in New Zealand

		Weed used	Agar produced	Imported weed (tonnes)
1950–53 annually		108 tons*	26 tons	
1954–58 annually		114·5	24·6	
1959–63 annually		109	22·3	
1964–68 annually		126	24·3	
1970	annually	125 tonnes	20 tonnes	20 from Mexico
1971	annually	–	22	
1972	annually	110	18	
1973	annually	100	13	
1974	annually	75	–	64 from Australia, Taiwan, Philippines
1975	annually	80	–	
1976	annually	110	–	

* 1 ton = 1·016 tonnes.

Currently work is being undertaken to cultivate *Gracilaria secundata* var. *pseudoflagellifera* as an agar source (see p. 252).

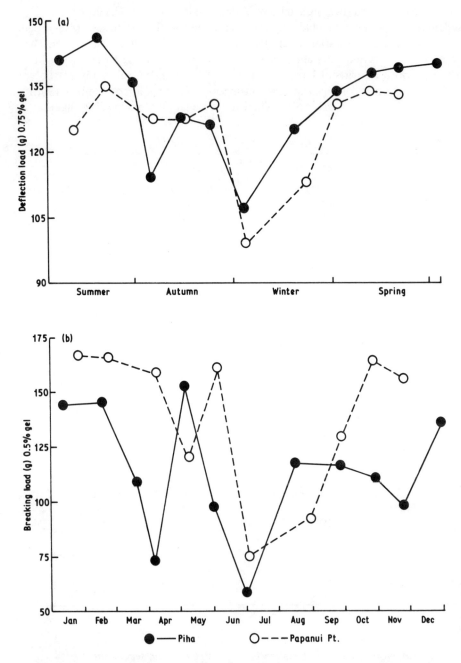

Fig. 5.9 Seasonal variation in gel strength of agar from *Pterocladia lucida* (after Luxton, 1977).

5.11 OTHER COUNTRIES

An agar industry, as distinct from the production of dried weed, commenced just prior to the last war in Indonesia. Tseng (1944a) reported that two factories existed in 1938 but their output was unknown. Spoon (1951) stated they still existed in 1950 when *Eucheuma gelatinosa* and *Gloiopeltis coliformis* were the main algal resources. Yamada (1976) reported an annual output of 50—60 tons (50·8—61 tonnes). In the Philippines, *Gracilaria verrucosa* is a source of agar that is mainly used as food (Sulit, Salcedo and Pangani Bau, 1956). Some 30—40 tons (30·5—40·6 tonnes) are produced annually (Yamada, 1976). Analyses by Anglo *et al.* (1973) have shown that *Gracilaria eucheumoides* also gives a high agar yield (58·15%).

On the Gold Coast of Africa, *Gracilaria henriquesiana* yields a very good agar. A 2% solution sets at 46—47°C and has a gel strength of 63 g cm^{-2} (Lawson, 1954).

In Ghana the maximum yield is made during the dry season from October to April, leaching apparently occurring during the wet season. Maximum harvests were obtained between January and June and minimum harvests in August and September. Temperatures at which gelling and melting of a 2·5% solution took place have also been investigated though these can vary with the season of collection (Table 5.16) (John and Azare, 1975).

Table 5.16

Species	Gelling temperature (°C)	Melting temperature (°C)	Product
Gracilaria dentata	36·5	79·2	Agar
Gigartina acicularis	30	43	Carrageenan
Hypnea musciformis	30	51	Hypnean
Hypnea flagelliformis	30·5	50·6	Hypnean

In North Africa *Pterocladia pinnata* (*capillacea*) produces an agar, the yield being 25% of the dry weight (Mohamed and Halim, 1952). However, the yield is seasonal, being maximal in August and minimal in February (Abdel-Fattah *et al.,* 1972). The current Moroccan output is stated (Yamada, 1976) as ranging from 200—250 tons (203·2—254 tonnes) annually.

In 1941 investigations were carried out in India with a view to exploring the use of agar from *Gracilaria lichenoides* (Bose *et al.* 1943). The weed was dried and bleached and the agar extracted by successive hot water extractions, the first being 4 h at 95—100°C. The extract was frozen and then thawed and filtered and the solution poured into glazed trays and allowed to gel. In this process the alga yielded about 43% of the dried weed. This agar proved to be too high in starch content and had to be purified by steeping in 0·2% acetic acid solution for about 1 h at 20—24°C. It then underwent several washing with fresh, cold water (Kirby, 1953). Another method of extraction evolved at Coimbatore used about 20% of the dried weed. This agar

Table 5.17

Species	Yield (%)
Gracilaria verrucosa	25–35
Gracilaria multipartita	17
Gracilaria edulis	20–25
Gracilaria corticata	14·5–25·5
Gelidium amansii	23

Table 5.18

Country	Agarophyte	Crop (tons*)	Harvest method	Agar output	First year of production
Japan	*Gelidium*	3000–3300	P, D	2000–2300	1944
	Gracilaria	1000–1500	P	2000–2300	
RO Korea	*Gelidium*	2000–2500	P, D	530–650	1924
Korea DPR	*Gelidium*	80–100	P	30–40	1960
	Ahnfeltia	500–700	P, G	30–40	
China	*Gelidium*	100–150	?	40–50	1937?
	Gracilaria	100–150	?	40–50	
Taiwan	*Gelidium*	50–60	P	30–50	1950
	Gracilaria	400–500	C	30–50	
Philippines	*Gelidiella*	200–300	P	30–40	1964
	Gracilaria	1000–1500	P	30–40	
Indonesia	*Gelidium*	400–500	P	50–60	1939
	Gracilaria	300–400	P	50–60	
India	*Gelidiella*	400–500	P	30 40	1968
	Gracilaria	400–500	P	30–40	
Australia	*Gracilaria*	500–600 †	P	80–100	1943
New Zealand	*Pterocladia*	50–100	P	40–50	1943
USA	*Gelidium*	150–200	?	100–150	1920
Mexico	*Gelidium*	1000–1500	D	20–30	1950
Brazil	*Gracilaria*	2000–3000	G	20–30	?
Argentina	*Gracilaria*	4000–5000	G	500–550	1963
Chile	*Gelidium*	100–150	P	80–100	1950
	Gracilaria	4000–4500	G	80–100	

P, picking in intertidal zone; G, from drift; C, cultivation; D, diving.

† This figure is probably too high.

* 1 ton = 1·016 tonnes. (*continued on the next page*)

Table 5.18 (*continued*)

Country	Agarophyte	Crop (tons*)	Harvest method	Agar output	First year of production
USSR	*Ahnfeltia*	2000–2500	G, P	150–200	1940
Spain	*Gelidium*	1850–5500	G, P	300–800	1944
Portugal	*Gelidium*	2500–4300	G, P	300–700	1951
	Pterocladia	2500–4300	G, P	300–700	1951
France	*Gelidium*	300–400	G	40–50	?
Italy	*Gracilaria*	100–200	?	30–50	?
Morocco	*Pterocladia*	1000–1850	G, P	200–300	1953
South Africa	*Gelidium*	2000–2500	G	150–200	1943
	Suhria	2000–2500	G	150–200	1943

P, picking in intertidal zone; G, from drift; C, cultivation; D, diving.
* 1 ton = 1·016 tonnes.

proved very acceptable and compared very favourably with Japanese agar. The gel strength ranged from 52–98 g cm^{-2} compared with 75 g for Japanese agar. Agar obtained from *Gelidium micropterum* is even firmer, 84–190 g cm^{-2} (Kappanna and Rao, 1963). Agar has also been manufactured from *Gelidiella acerosa* (Rao, 1968). Apart from local manufacture, over 10 000 kg are imported annually, nearly all of it coming from Japan (Govt India Statistics, 1967–68).

Local manufacture has been declining from 198 tons (201·2 tonnes) in 1967 to 30–40 tons (30·5–40·6 tonnes) in 1976 (Yamada, 1976), but is sufficient to meet internal needs – bacteriological agar and edible (Gopal, personal communication). The percentage of yield from different species is et out in Table 5.17.

In the case of *Gracilaria corticata* the yield is maximal in June–July and again in November–March. The minimum gel strength (17 g cm^{-1}) occurs in agar made in July and the maximum (27 g cm^{-1}) in agar made in June–November (Oza, 1978).

The current yearly crop and agar output of a number of countries is set out in Table 5.18 (from Yamada, 1976).

At present, whilst weed may be abundant, in some cases at least it is not easy to collect and in most countries, other than Japan and Baja California, the labour costs render it difficult to produce a product that will compete economically with the Japanese and American agar.

5.12 CHEMISTRY OF AGAR AND AGAROPHYTES

(See Selby and Selby, 1959; Jones and Peat, 1942; Bodard and Christiaen, 1978). The gelatinizing material in all the various red seaweeds that have been used in the manufacture of agar or agariferous materials is often referred to as gelose. Sauvageau (1918) has pointed out that the gelose-producing algae can be divided into three groups on the basis of the setting power of the gel.

(a) *Gelidium* type: decoction sets firm if dilute (= true agar).
(b) *Eucheuma* and *Hypnea* type: decoction sets firm if in medium concentration or if electrolytes added (= agaroids).
(c) *Chondrus* type: decoction only sets firm if concentrated (= carrageenans).

The second and third types have already been considered (p. 109). The differences between the three types have been further emphasized by Holmes (1907), and in the following table from this worker it will be observed that the melting point of the jelly from the different sources varies. This fact is of considerable importance in bacteriological work. The figures can also vary depending upon season and method of extraction.

Table 5.19

	Parts required to gelatinize 1000 parts of water	Melting point of jelly (°C)
Gelidium corneum	8	90
Gelose (*Gelidium* extract)	4	90
Chondrus	30	80
Carrageenan (*Chondrus* extract)	30	70
Isinglass (country not stated)	32	70
Eucheuma spinosum	60	90
Gracilaria henriquesiana	–	87

Analyses of commercial Japanese agar show that it contains on an average 16–20% water, 2·3–5·9% protein, 0·3–0·55% fat, 67·85–76·15% carbohydrate, 0·8–2·1% fibre, and 3·4–3·6% ash. Matsui (1916) gives some interesting figures for the composition of three different algae used in Japan for the manufacture of agar, and it will be observed that their composition is somewhat different from that of the commercial agar (Table 5.20). The variations in chemical composition reflect the different properties exhibited by the various species, which depend not only upon the area but also the season of collection.

The 'gelling' material of Japanese agar has been subject to considerable study. At one time it was thought to be α-galactose united with sulphuric acid in the form of an ester. There is now little doubt that it is composed of galactose residues united together in chains, but the early views have had to be modified.

Table 5.20 Composition of three algae used in the manufacture of agar (Japan), compared with that of commercial agar

	Main source	Subsidiary sources		
% of dry material	*Gelidium amansii* (Tengusa)	*Campylaeophora hypneoides* (Yego-nori)	*Gracilaria* sp. (ogo-nori)	Commercial agar
Ash	4·23–6·16	3·04	3·54–6·71	3·4–3·6
Nitrogen	2·01–2·97	2·19	0·69–1·44	2·3–5·9
Fibre	17·89–14·71	12·25	4·32	0·8–2·1
Galactans	23·7	24·88	22·14–22·70	0·8–2·1
Pentosans	3·23–4·87	2·13	1·94–2·18	–

Among the early workers Barry, Dillon and McGettrich (1942) considered that the galactose residues could not exist in the form of an open chain because the agar molecule is very resistant to oxidation. Jones and Peat (1942) argued that in each chain there were nine D-galactopyranose units, combined by a 1, 3 glycosidic linkage, the chain being terminated by an L-galactopyranose. It has since been confirmed that both these galactopyranoses are in fact present. Early workers (Butler, 1934; Haas and Hill, 1921; Hoffmann and Gortner, 1925; Takahashi and Shirigama, 1934) had also demonstrated the presence of an ethereal sulphate which was the calcium salt of an acid with the formula $R-R_1(OSO_2O)_2$ Ca. Jones and Peat (1942) considered that the sulphate radical was attached at carbon atom 6 of the L-galactopyranose unit. They further suggested that the SO_4^{2-} radical fulfilled the same function in the biological synthesis of agar as does the PO_4^{2-} radical in the synthesis of starch from glucose in higher plants. However, their formula required a sulphate content of 1·8%, whereas analyses generally showed a content ranging from 0·5–1·5%.

This situation has become clarified by recent work (Araki, 1966) which had shown in 1956 that agar from *G.amansii* comprises two polysaccharides, agarose and agaropectin. Stoloff (1943) had secured evidence that indicated *Gracilaria* agar is also a complex of two or more gelling materials. Separation of the two agar polysaccharides can be effected by acetylating the agar with acetic anhydride and pyridine. The resulting acetate is fractionated with chloroform into soluble agarose acetate and insoluble agaropectin acetate. Both acetates can subsequently be deacetylated with alcoholic alkali (Araki, 1937). The sulphuric acid component is found in the agaropectin moiety so that agarose is a neutral polysaccharide. The agarose was initially regarded as consisting of alternate residues of 1, 3 β-D galactopyranose and 3, 6 anhydro α-L galactopyranose connected through the 1, 4 positions.

The evidence for the above structure and its formula $(C_{12}H_{14}O_5(OH)_4)n$ was obtained by other means as well as by the separation procedure above. One of these is the use of methylation. In 1937, Araki and Percival and Somerville (1937) obtained 2, 4, 6 tri-*O*-methyl-D-galactoside, and in 1940, Araki isolated 2-*O*-methyl-3, 6 anhydro-L-galactose dimethylacetal. The existence of these two compounds showed that the D-galactose residues were connected through the 1, 3 positions and 3, 6 anhydro-L-

$^3_{d \cdot G}{}^1$ —————— $^4{}_{l \cdot \alpha G}{}^1$ —————— $^3_{dG}{}^1$ —————— $^4{}_{l \cdot \alpha G}{}^{1*}$ ——

* dG = β-D-galactopyranose   lαG = 3, 6 anhydro-α-L-galactopyranose.

galactose residues were united through the 1, 4 postions. Yet another line of evidence was mercaptolysis (Araki and Hirase, 1953) which yielded both galactose residues. The last evidence was concerned with the arrangement of the components in the agarose molecule and this was arrived at by subjecting the agar to hydrolysis. When agar from *G. amansii* is subjected to partial hydrolysis by dilute sulphuric acid it yields a disaccharide named agarobiose (Araki, 1944). If agar from *G. amansii* is hydrolysed by enzymes from agar-digesting bacteria two compounds, neo-agarobiose and neo-agarotetraose are obtained (Araki and Arai, 1956, 1957) and these two results provided strong evidence for the structure of agarose as given in the formula above, where the two residues (1, 3 linked and 1, 4 linked) are repeated alternately in the molecule. The same structure was proposed by O'Neill and Stewart (1956) for unfractionated agar from *Gelidium cartilageneum* and likewise for *Gracilaria verrucosa* agar by Clingman, Nunn and Stephen (1957).

When work was extended to agars from other sources all were found to contain agarose and agaropectin but in different proportions (Table 5.21). The agarose component contained in all cases only traces of sulphate and it was therefore regarded as neutral. When the component sugars were estimated the proportions of β-D-galactopyranose residues that became 6-O-methylated varied from species to species of agarophyte (Table 5.21). Thus in *G. amansii* the methylated content is one fifieth of the total.

It is low also in *Pterocladia tenuis* and *Campylaeophora hypneoides*. In *Ceramium boydenii*, on the other hand, there are two methylated residues in every five. A high value is also found in agar from *Gracilaria verrucosa*. However, the sum of the amounts of the methylated and unmethylated residue always constitutes 51–53% of the polysaccharide (A + B, in Table 5.21). The evidence indicated then that some residues are 6-O-methylated and the nature of the bonds involved was arrived at by partial methanolysis of agarose from *Ceramium boydenii* (Araki, 1966). The latest structural formulae for the three agaroses and the sulphated galactan is given in Fig. 5.10 (Luxton, 1978).

Agarose is the component responsible for the gelling properties of agar whilst agaropectin provides the viscous component, the viscosity varying depending upon species of alga, method of production and sulphated content. Increase in the SO_4^{2-} content reduces the gelling capacity. Removal of the SO_4^{2-} molecules is regularly carried out in the production of Ion, Noble and Reifagars.

Table 5.21 Percentage composition of agarose (after Araki, 1966)

Species	% Agarose	Galactose	6-O-methyl D-galactose	3, 6 anhydro-L galactose	L-galactose	A + B	A + B / C
Gelidium amansii	61	51·0	1·4	44·1	1·9	52·4	1·05
Gelidium subcostatum	89	45·1	7·3	43·8	1·8	52·4	1·05
Gelidium japonicum	69	50·8	1·6	44·0	1·9	52·4	1·05
Pterocladia tenuis	85	51·7	0·8	44·2	1·4	52·5	1·06
Acanthopeltis japonica	28	49·4	3·2	44·5	2·0	52·6	1·04
Campylaeophora hypneoides	55	50·0	0·8	43·7	4·0	50·8	1·04
Gracilaria verrucosa	61	36·3	16·3	44·0	2·1	52·6	1·04
Ceramiun boydenii	82	31·9	20·8	44·1	1·1	52·7	1·03

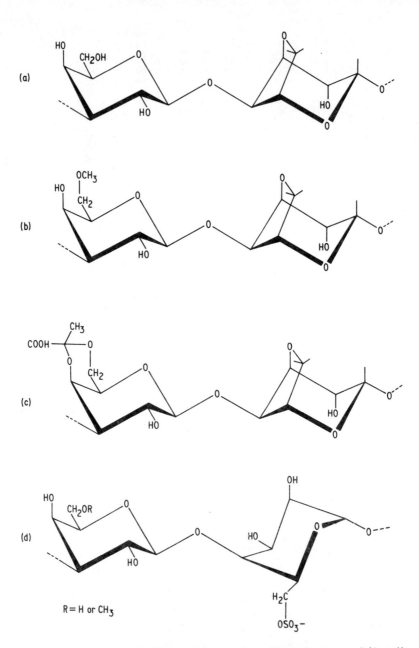

Fig. 5.10 Agar structure: (a) Neutral agarose; (1 → 3) D-galactose and (1 → 4) anhydro-L-galactose; (b) Methylated agarose, (1 → 3) 6-*O*-methyl-D-galactose and (1 → 4) anhydro-L-galactose; (c) Pyruvated agarose, (1 → 3) 4, 6 *O*-(1-carboxyethylidene)-D-galactose and (1 → 4) anhydro-L-galactose; (d) Sulphated galactan, (1 → 3) D-galactose and (1 → 4) L-galactose-6-sulphate (after Luxton, 1977).

Table 5.22

	Agarose	Agaropectin
Ash	0·06—0·20 %	5·1—9·9 %
SO_4^{2-}	0·02—0·04 %	3·7—9·7 %
Uronic acid	Nil	+ +
Pyruvic acid	Nil	Nil (6 spp.)
		1·3 in *G. amansii* and *G. subcostatum*

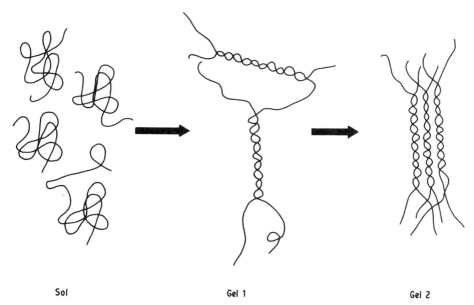

    Sol   Gel 1   Gel 2

Fig. 5.11 Schematic model of the mechanism of gel formation (after Yaphe and Duckworth, 1972).

The structure of the agaropectin component has still to be completely elucidated. In the case of *G. amansii* it appears to be a polymer of agarose with acid residues. In particular it contains the ester sulphate of the earlier workers, but it also contains D-glucuronic acid and pyruvic acid. The most recent suggestion (Percival, 1964) is that it consists of a mixture of 1, 3 β-D-galactose with a 1, 4 linkage to 3, 6 anhydro-L-galactose plus a small proportion of sulphate, D-glucuronic acid and 1% pyruvic acid*. The difference in composition between agarose and agaropectin for the 8 species in Table 5.21 is shown in Table 5.22.

Agarobiose seems to be the common basic structure of other red algal compounds, e.g. funoran (p. 143), porphyran (the polysaccharide of the genus *Porphyra* — see

---

* *G. amansii* and *G. subcostatum* only (Percival, 1969).

p. 109). The latter compound can be extracted by hot water but so far no use has been found for it. The actual chemical composition seems to vary (Levring, Hoppe and Schmid, 1969). The disaccharide, carrabiose, derived from κ-carrageenan (see p. 126) is structurally similar to agarobiose (Araki, 1966).

According to Tsuchiya and Hong (1966) the gel strength of an agar is related to the percentage of agarose present in relation to agaropectin. In *Gelidium* agar the proportion is 1 to 5 to 1 whilst in some *Gracilaria* agars it is 20 to 1 and this accounts for the greater gel strength and value of such *Gracilaria* agars (see below). It is considered that the agarose molecules are present as double helices and that, in gelling, these aggregate (Fig. 5.11) (Yaphe and Duckworth, 1972). It is the acidic nature of agaropectin that probably gave rise to the earlier concept of agarinic acid (Tseng, 1945a) and the suggestion that in nature it existed as the calcium salt.

The existence of the two major components, agarose and agaropectin, has made a considerable difference to an industry that earns substantial financial rewards. Two methods now exist whereby agarose can be separated out commercially by itself, and this substance, because of its gelling properties, is obviously as important, if not more so, as the parent agar material. One method of separation, the Hjerten method, secures the precipitation of the sulphated agaropectin by means of quaternary ammonium salts (Hjerten, 1961). The second method makes use of propylene glycol in order to separate the two compounds.

It has already been made clear that the introduction of ions into various agar-like materials has an effect upon the tensile strength of the gel. The following lyotropic series of cations has been established (Pavlov and Engel'shtein, 1936), the effect being, in descending order: caesium, rubidium, ammonium, potassium, sodium, lithium, calcium, barium, strontium. The corresponding anion series is: nitrate, bromide, sulphate, chloride, iodide and acetate. The strength of true agars is not affected by addition of either anions or cations.

In India the gel strength of agar from *Gracilaria* spp. and *Gelidiella acerosa* can be increased by the method of extraction. Pillay (1977) found that boiling the weeds first with 0·1% KOH solution and discarding the liquid, followed by another boiling in 20–40 times the weight of water containing 2 ml l^{-1} acetic acid for 1–4 h represented the initial steps. These steps are followed by $CaCO_3$ precipitation at pH 6·0–6·5 with subsequent gelling and freezing between 12–20°C. After thawing the product is decolourised by bleaching powder solution, washed and dried. The *Gracilaria* agar has a gel strength of 125 g cm^{-2} at 1·5% whilst *Gelidiella* agar has a strength of 300 g cm^{-2} at 1·5%. In both cases there is a 45–50% yield on the air-dried weeds.

There are several interesting physical properties of agars which vary, depending on the source of the agar. One of these is hysteresis or the lagging in response to changed conditions. Thus in *Gelidium* agar there is a range of 40°C over which the solution may exist either as a sol or gel depending on whether it is being cooled or heated. In the case of *Hypnea* carrageenan, hysteresis is also affected by the presence of electrolytes (see p. 142).

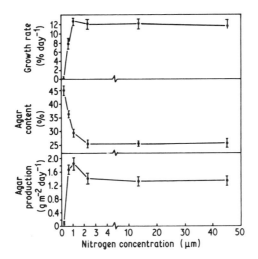

Fig. 5.12 Effects of nitrogen enrichment on the growth rate, agar content and agar production in *Gracilaria foliifera*. Standard error indicated about the mean of two acid and two alkaline agar extractions (after De Boer, 1979).

All gels have a tendency to shrink with the elimination of tiny droplets of water. The droplets contain a higher concentration of solutes than were present in the original sol. This phenomenon is known as syneresis and it varies inversely with concentration of agar in the gel. The degree of inhibition of water by dehydrated agar is primarily determined by the moisture content and nature and concentration of the solutes. The acidity of the gel is also of importance and both cations and anions exert an effect in order of a lytropic series. For *Gelidium* agar this series appears to be the reverse of that for *Hypnea* agaroid.

In the light of the information that has been presented above, it is justifiable to conclude that the agar produced from red seaweeds is composed not only of galactose but also of closely related substances with similar physical characteristics. At present, however, the information available about the chemical composition of agar made from species in the different countries is limited, but it is clearly not a uniform commercial product. Although commercial Japanese agar contains 67–76% galactose, the raw algae do not contain so much because it becomes concentrated during manufacture. Thus *Gelidium cartilagineum* contains 40–45% galactose (gelose), *G. amansii* 25–35%, *G. pulchrum* (= *G. australe*) 32–37% and *Gracilaria lichenoides* 70%.

Not only, therefore, are there differences in chemical composition as between the parent materials in the different algal species, but there are also differences in the composition and physical properties of the manufactured product. It is most desirable that a refined and purified product should be secured which could be standardized, especially in respect of the temperature of transition from liquid to gel, the gel strength, elasticity, viscosity, transparency, ash content and content of impurities.

Such a product, it has been suggested (Humm and Wolff, 1945), might well be marketed as 'bacteriological agar'. Such standardization would be of great value to agar users who are at present compelled to accept the standards of the different producers.

The agar with the strongest gel strength is derived from *Gracilaria verrucosa* of North Carolina. If this is recorded as 100 the relative strengths of other comparable agar gels are as follows (Lee and Stoloff, 1946): Californian *Gelidium* agar 69, South African *Gracilaria* 52, Japanese *Gelidium* agar 42, Californian *Gracilaria* agar 30, Australian *Gracilaria* agar 26. It is probable that these figures would vary from sample to sample so that they merely indicate a general trend. Six species of *Gracilaria* were studied by Hoag and Yaphe (1968) who found that only one, *G. debilis,* yielded a gel of high strength. This was associated with a decreased SO_4^{2-} content and an increased 3, 6 anhydrogalactose. Treatment with alkali reduces the SO_4^{2-} content, as Tagawa and Kojima (1972) found with *G. verrucosa*. They used 0·5–1·0 N Na OH for 1–5 h at 60–90°C. Pillay (1977) boiled *Gracilaria* with 0·1% KOH and after discarding the extract boiled the residue with 2 ml l^{-1} acetic acid in 20 times the weight of water for 1 h, neutralized with $CaCO_3$ to pH 6·0 and K_2CO_3 to pH 6.5. This was followed by filtration, gelling and freezing and gave a yield of 40–50% of the dry weight with an average gel strength of 125 g cm^{-2}.

Experimental work with nitrogen, mainly in relation to cultivation, has shown that in *G. foliifera* increased nitrogen enrichment decreased agar content from 45% to 25% with a maximum productivity at 1·0 μm N (De Boer, 1979). Below 0·5 μm N there was also a decrease in agar production (Fig. 5.12).

Apart from investigations into agarinic acid and the sugars of hydrolysis, it has also been suggested (Robbins, 1939) that commercial agar contains a growth substance akin to plant auxins. Although there is some evidence to support this view, the present author feels that its unqualified acceptance would not be justified in the absence of additional information.

5.13 USES OF AGAR

The uses of agar are manifold (Morel, 1950), but probably its most important use is in bacteriological and fungal culture work, because after nutrient materials have been added even a dilute solution sets to a firm jelly upon which the bacteria or fungi can grow. Although this is its most important use, the actual quantities of agar employed form only a small portion of the total produced. The melting point of the agar is of the greatest importance, and for this reason the material produced from some of the red seaweeds is unsuited for this type of work. Bacteriological agar remains liquid when cooled to 42°C and hence organisms can be thoroughly distributed within it at a temperature which will not hurt them. It also remains a firm gel at 37°C, which is the temperature commonly employed for incubating bacterial and fungal cultures.

Another reason why it is so valuable for bacterial cultures is that it resists liquefaction, e.g. many bacteria convert solid media such as gelatine into a liquid solution. There are, however, some agar-digesting bacteria, the best known probably being *Vibrio agar-liquefaciens*. Recently twenty such bacteria have been isolated from marine habitats (Humm, 1945), and in similar places three agar-digesting actinomycetes have been discovered (Humm and Shepard, 1946) (see p. 149 for agarases).

The first use of agar for culture purposes dates back to 1881 when Frau Fanni Hesse suggested it to her husband, though it was probably not known in its present-day form, but was supplied as the dried seaweed (Hitchins and Leikind, 1939). The first published note of its use, however, was by Robert Koch in 1882.

Different nutrient materials are added to the agar, depending upon the type of organism intended to be grown. Thus, there is malt agar, potato agar, heart extract agar and so on. Some of the bacterial organisms which are grown on agar are highly coloured, and in certain cases the colour of the colony can be altered by changing the type of agar nutrient. Only the best and purest agar can be used in much of this work, and so there is a considerable demand for it.

Not more than 5% of ash is allowed by the British Pharmacopoeia and the USA Pharmacopoeia requires not more than 1% of foreign organic matter and not more than 1% of acid insoluble ash. In a comparison of agars from different countries Forsdike (1950) considered that products from New Zealand, South Africa, and Australia (to which the USA product should also be added) were true agars, but that the British product was different and should be marketed under some other name. It was, in fact, more closely allied to carrageenan (see p. 173).

In 1956, 10 000 lb (4500 kg) of agar were used for pharmaceutical purposes in the USA and this had doubled to 20 000 lb (9080 kg) in 1966 so that by 1976 it could well have been approaching 40 000 lb (18 160 kg). Thus alkyl agaroses are employed in hydrophobic chromatography and amino acids can be attached to the agarose molecule (Sigma pamphlet).

Apart from its value for the culture of micro-organisms, agar has a variety of other uses. A rather interesting use in a number of countries is concerned with the transport of preserved cooked fish, which is protected from breakage by being embedded in the firm jelly. In 1958 the Japanese started using agar for canning tuna fish. It also prevents the constituents of certain fish, e.g. herrings, from blackening (detinning) the contents of the can and so rendering it unsaleable. The greater part of the agar required by Australia is used for this purpose, whilst New Zealand uses most of its production for canning sheep's tongues. A somewhat similar use is seen in hot countries, where cooking with some agar provides a temporary method of preservation for easily spoiled foodstuffs.

Another industrial use of agar is in the sizing of fabrics, but as it is much in demand the finest grades of agar are only used for the valuable silks, whilst poorer grades can be used for nansooks, muslin, voiles and tulles. The best agar has to be used for silks because it is most important that the sheen should not be destroyed. Japanese agar is stated to be definitely superior as a sizing material to products obtained from *Chondrus* or *Gigartina*. Poorer qualities of agar are used as a coating in paper manufacture, in making waterproof paper and cloth, as a glue and as a cleaning medium for liquids.

Another use is as a lubricant in the hot drawing of tungsten wire for electrical lamps. For this purpose a suspension of powdered graphite in agar gel is used. A very promising use of agar is seen in the photographic industry for making plates and films. Early attempts were not satisfactory because of difficulties encountered in adding certain chemicals, which make the emulsion sensitive. These difficulties have now been overcome, and the agar is regarded as superior to gelatine because the film need be only one eighth the thickness of a gelatine film; it is also soluble in water, does not melt in tropical heat and is cheaper. Agar is also employed to a considerable extent in the finishing processes of leather manufacture, in order to impart a gloss and stiffness. It is sometimes a constituent of high grade adhesives, and as such is used in the manufacture of plywood. In cooking it is invaluable for thickening soups and sauces, whilst considerable quantities are used for making fruit jellies because it is more economical than gelatine and sets readily. Its use in food products is primarily associated with its tolerance of high temperatures. It is widely employed in both Europe and America as a thickening agent in the manufacture of ice-creams, malted milks, jelly, candies and pastries. In the preparation of sherbets, ice-creams and cheeses its function is mainly that of a stabilizer and to give smoothness. It has, however, a low whipping capacity so that gums have to be added, and has now largely been replaced for this purpose by sodium alginate. It has also been used in the manufacture of cream cheeses (Dahlberg, 1927; Marquardt, 1903) as a means of improving the texture of the cream, and in making icings, custards and mayonnaise (Hart, 1937). In icings the hydrocolloid ties up the free water and prevents the icing sticking to paper wrapping. In doughnut glazes the same property stops the glaze from cracking. A product known as 'imperishable milk' has also been marketed in the USA in which the cream is separated from the solids and then emulsified with agar. This process removed most of the material upon which bacteria grow and so kept the milk fresh.

The principal potential food material in agar is the carbohydrate or galactose, but Swartz (1911) has shown that very little of this in ordinary agar is digestible, although, if the galactose is first hydrolysed, up to 50% may subsequently become digestible. As a result agar is not used to modify the nutritive value of the foods with which it is incorporated, but instead to form a jelly which will preserve them against the action of the air, or else to stiffen liquid products which are more suitable for consumption in solid form. It is also extremely useful as roughage. Agar is sometimes used in place of pectin for making jellies, jams, marmalade and preserves. For a complete summary of the uses of agar as a natural gum see Table 4.13 on p. 131.

The Japanese use less agar for food, compared with their other algal products, but it is used extensively as such in China and Indo-China. In both the latter countries it is sold to poor coolies by travelling mobile canteens. The proprietors of these vehicles offer badly cooked rice or bowls of rice flour, small fishes grilled or fried, some vegetable floating in soya bean sauce, and at the finish a faintly scented agar jelly. It is very doubtful whether the coolie obtains any energy from the jelly, but it serves as roughage, and agar also has a very valuable laxative action. It is sometimes served as a substitute for the expensive 'birds nest'; whilst in Japan it is often cooked with rice to form 'agar rice'.

An important use in western countries is in connection with the brewing of beer and the manufacture of wines and coffee, where the agar is used as a clarifying agent. It also finds a use as a pill and suppository excipient, as a base in shoe-stains, shaving soaps, cosmetics and hand lotions, and is sometimes employed as a stabilizer in chocolate drinks. There is perhaps one use which may be important in war-time; this is in connection with wound dressings because agar contains a principle that stops blood clotting and thus enables wounds to be properly cleansed. One of the very early uses was by the Chinese and Japanese for gastrointestinal disorders. In 1955 the USA Dispensatory listed it as a laxative (Schwimmer and Schwimmer, 1955), but this use has probably ceased. Under the trade name of 'Agarol', or associated with other compounds, agar is valuable as a lubricant in cases of constipation. If necessary it can be made into chocolate-coated pills or a mineral oil emulsion. In dentistry as the material 'Dentocoll' it is employed for making impressions of gums for manufacturing false plates though it has now largely been replaced by alginates (see p. 220). Another use is as an activator in nicotine pesticide sprays that are used by gardeners. It is not completely satisfactory in this case, because there are difficulties in dispersing it through the spray liquid. It is also employed in the making of moulds required by those who model in plaster of Paris. An extension of this usage is as a mould for the casting of artificial legs. Other uses are as a raw material in making linoleum, artificial leathers and silks; as an insulating material against sound and heat; as an ingredient in water-base paints and in the manufacture of storage batteries for submarines. Agarose *per se* did not originally have much value but recently it has become increasingly used as a reagent in molecular sieve chromatography where it has some advantages over other compounds. No doubt other comparable uses will emerge in the future.

No really suitable substitute for agar has so far been produced chemically, and agar manufacture from Rhodophytes seems likely to continue for many years to come. It is evident that production is much greater now than it used to be and that the two major producing countries, Japan and the USA, are having to import supplies of raw weed from other parts of the world or to establish branch subsidiaries in countries with particularly rich supplies. At present there are two options open to mankind. Since 1920 the use of agar has fluctuated and the point has now been reached where normally there would be a decline in production (Moss, 1979). The alternative is a major programme involving cultivation of agar-producing seaweeds. The supply of natural and cultivated dried agarophytes could well become a valuable source of overseas income to some of the smaller independent island countries in the West Indies, Africa and in the Pacific. It is evident that increased supplies of suitable agarophytes is becoming an urgent need.

6
Algin and Alginates

Algin was first discovered by Stanford in the early 1880's and there is little doubt that the event was the beginning of a new era in the use of seaweeds. Stanford, when he first prepared this substance, did not succeed in obtaining it in the pure state, and because of the impurities he described it as a nitrogen-containing compound. It was later properly prepared by Krefting (1896), who thought he had a new substance which he called 'tangsaure', or seaweed acid. In the seaweeds the algin is present as a mixed salt of sodium and/or potassium, calcium and magnesium. The exact composition varies with algal species. Since Stanford discovered algin, the name has been applied to a number of substances, e.g. alginic acid, all alginates, sodium alginate, etc., derived from alginic acid. It has been suggested (Tseng, 1945a) that it is desirable to reserve the name algin for the soluble sodium salt of alginic acid, and it will be so used here. In recent years the amount of information has been accumulating at a very high rate. It now appears that algin production may not be restricted to marine algae as Larsen and Howe (1972) have found it in *Azotobacter vinelandii.*

6.1 CHEMISTRY OF ALGINIC ACID

Alginic acid is a complex organic compound originally thought to be composed of polymers of D-mannuronic acid, first isolated by Cretcher and Nelson in 1930. In 1933 Schoffel and Link obtained both α and β-D-mannuronic acids from it and it was given the chemical formula of $(C_6H_8O_6)n$ according to Marsh and Wood (1942), Lunde *et al.* (1938), Hirst (1939) and Speakman (1944), or $(C_6H_{10}O_7)n$ according to Dillon (1938), where n is regarded as being some number between 80 and 83.

Speakman and Chamberlain (1944) devoted considerable space to a discussion of the correct formula. They favoured the pyranose ring $(C_6H_8O_6)n$, and this appeared to have been confirmed (Astbury, 1945) by means of X-ray analysis.

Up to 1955 this formula was generally accepted and alginic acid was regarded as β-D-mannuronic acid with the residues united by a 1,4 linkage (Hirst, Jones and Jones, 1939; Lucas and Steward, 1940). In 1955 Fischer and Dorfel, using paper chromatography, demonstrated that in algin obtained from *Laminaria* 1 guluronic acid formed from 30–70%, whilst in alginic acid obtained by other workers from *Macrocystis* it formed from 20–40%. It is, therefore, now quite evident (Vincent, 1960; Frei and Preston, 1962; Hirst, Percival and Wold, 1964) that alginic acid contains two monomers, poly-D-mannuronic acid and poly-L-guluronic acid. These can be separated electrophoretically as well as by paper chromatography, but they are bound in the same polymeric chain. Haug and his co-workers in Norway have shown that the alginate molecule contains portions with separate homo-polymeric blocks of the two monomers, and also at the same time other portions with predominantly alternating mannuronic and guluronic acid residues (Haug, Larsen and Smidsrød, 1967 and 1968). It appears that, in nature, crystalline alginate is composed essentially of salts of the guluronic acid component.

The molecular weight of alginic acid varies with its mode of preparation and also the source. In the case of sodium alginate it varies from 35 000 to 1·5 million (Cook and Smith, 1954; Smidsrød and Haug, 1968). The theoretical equivalent weight of alginic acid is 176 and values close to this have been obtained, though other workers have got values close to 194. As Haug (1964) points out, this discrepancy could be accounted for by the presence of one molecule of water of hydration bound to each uronic acid residue. Such an interpretation would also explain an unknown element that is present even in 'pure' algin obtained from *Laminaria digitata*. Both uronic acids consist of chains of ring units linked into the chain molecule by 1, 4 linkages.

Poly-D-mannuronic acid (M)

β–1,4 guluronic acid (G)

The chain arrangement is of the greatest importance, because it provides strength for the molecule and is also responsible for an important property, e.g. the power of forming fibres. After partial acid hydrolysis three fractions separate out and their relative proportions determine ion building and gel strength (Grasdalen *et al.*, 1979). The three components (Haug *et al.*, 1966, 1967a and b) are formed by separate chains of the two monomers and a chain with alternating components. (Table 6.1).

–M–M–M–M–M–   –G–G–G–G–G–   –M–G–M–G–M–G–

The alginic acids from *Macrocystis* and *Ascophyllum* are very comparable, but differ markedly from the algin of *Laminaria hyperborea* which is very high in the poly-guluronic component (Table 6.1). The alginic acid from *Azotobacter vinelandii* differs from all others in possessing a very high alternating component (Penman and Sanderson, 1972).

Table 6.1 Proportions of three components in different alginic acids

Source	Polymannuronan (%)	Polyguluronan (%)	Alternating (%)
Macrocystis pyrifera	40·6	17·7	41·7
Ascophyllum nodosum	38·4	20·7	41·0
Laminaria hyperborea	12·7	60·5	26·8
Azotobacter vinelandii	17·8	0·5	81·7

One peculiar feature is that algin extracts in the presence of $ZnSO_4$ and suitable foci produce characteristic capillaries, the shape and degree of branching varying with the algal source (Baardseth and Grenager, 1962).

6.2 OCCURRENCE

Alginic acid is found in all the larger brown seaweeds as well as some of the smaller ones. There has been considerable debate as to the actual site of algin in the algae. Until recently the intercellular mucilage has been regarded as the principal site, but there is some clear evidence that it occurs also in the cell walls (Frei and Preston, 1962). There is also reason to believe that it occurs in such sites as the salts of divalent metals (Bird and Haas, 1931; Dillon and McGuiness, 1931), but in particular as calcium alginate.

In the rockweeds a substance called fucin has been reported in the past and Kylin considered that fucin was a separate compound from algin. In the light of the recent investigations on the chemistry of algin it is now thought there is no distinction between fucin and algin. Numerous analyses have been carried out on various species in order to determine the amount of potential algin available (Table 6.2).

Whilst there is substantial agreement among the different workers, it is clear that source and habitat can affect the result as well as the part of the plant used. Moss (1948) has shown that open sea plants of *Fucus vesiculosus* are richer in algin than loch plants and Black (1950) has demonstrated the same for *Laminaria saccharina*. In *Laminaria* stipes are richer in algin than fronds, but in *Durvillea* and *Ecklonia radiata* the reverse is the case. Togasawa and Miue (1954) have shown that there is greater quantity of alginic acid in the blades than in the sporophylls. From the figures currently available it seems that *Laminaria, Ecklonia, Durvillea* and *Ascophyllum* are potentially the richest source of supply.

Table 6.2 Algin present as % dry weight

Species	Lunde (1937, 1938)	Black (1950)	Kirby (1952)	Munda (1973)	Hoppe-Schmid (1962)	Nielsen et al., (1977)
Laminaria digitata				35		
frond	15–40	14·5–26·5	25–44		15–40	20–35
stipe		27–33	35–47			30–45
Laminaria saccharina						
frond	15–35	13–21†	17–22		12–23	
		11–21*	25–31			
stipe		20·5–25*				
Laminaria hyperborea						35
frond	15·5–24·5	14–21·5	11·3–24		14–24	10–25
stipe	17–33	8·5–19	21·27			20–30
	25–38	18·5–23·5				
Alaria esculenta (Ireland) (Munda, 1963)	30–35		20–42		30–35	
Ascophyllum	18·8–23		22–30		20–30	
					24–29·5 (Norway)	
Fucus serratus	18–28		20–29			
Fucus vesiculosus	18–28		16–21			29
Fucus ceranoides					21·05–29·48	
Fucus virsoides (Michanek, 1975)						20
Himanthalia	38		22–29			

* Open sea plants
† Loch plants

(continued on the next page)

Table 6.2 Algin present as % dry weight (*continued*)

Species	Lunde (1937, 1938)	Black (1950)	Kirby (1952)	Munda (1973)	Hoppe-Schmid (1962)	Nielsen et al., (1977)
Halidrys siliquosa			16–17			
Desmarestia						20
Ecklonia maxima					29·6–40	
Ecklonia radiata	(Stewart et al., 1961)					
frond	17–24·5					
lamina	19·8–30·2					
Macrocystis					13–24	
Nereocystis					14–20	
Pelagophycus					15–17	
Egregia					18–20	
Durvillea antarctica					33	
					(Moss and Naylor, 1954)	
disc					20·6–26	
stipe					36–39·8	
frond					32–43·5	
Durvillea willana						
stipe					37–40·5	
frond					36·9–44·8	

(*continued on the next page*)

Table 6.2 Algin present as % dry weight (continued)

Species	Lunde (1937, 1938)	Black (1950)	Kirby (1952)	Munda (1973)	Hoppe-Schmid (1962)	Nielsen et al., (1977)
Sargassum longifolium (Ligthelm et al., 1953)					17.0	
Sargassum sp. (Davis, 1950)	13					
Sargassum sp. (wightii, tenerrimum) (Pillai ans Varier, 1953)					29.8–34.6	
Myriocystum (Channubothla et al., 1979)						15.15–34.5
Stoechospermum marginatum (Kalimuthu et al., 1977)						17.75–23.75
Turbinaria decurrens (Kaliaperumal et al., 1976)						16.3–26.3
Turbinaria spp. (Pillay, personal communication)						25–37

* Open sea plants.
† Loch plants.

Table 6.3 % Ascophyllan or fucose present in algin

Species		(%)
Ascophyllum (Ascophyllan)		10–25
Pelvetia canaliculata	(Fucose)	18–24
Fucus vesiculosus	(Fucose)	12·4–15·2
Fucus spiralis	(Fucose)	15·6–23·2
Fucus serratus	(Fucose)	9·6–11·6
Halidrys	(Fucose)	12·4–14
Himanthalia	(Fucose)	7·2–8·0

Recently alginate-modifying enzymes have been reported from *Pelvetia canaliculata* (Madgwick et al., 1978) and other brown algae should obviously also be examined. There are some earlier studies by Shiraiisa et al. (1975) concerning alginate lyase activities in extracts from brown algae.

Algin from *Ascophyllum* and *Laminaria hyperborea* is not pure as it contains a fucose-containing polysaccharide which ranges from 1·8–6·2% of the dry weight or 10–25% of the algin fraction. This substance, which has been called ascophyllan when derived from *Ascophyllum*, contains sulphate groups attached to some of the residues. Chemically ascophyllan consists of a chain of glucuronic acid residues with side chains of a high molecular weight, fucoxylan, and also a polypeptide. It can be separated from algin by means of acid precipitation (Larsen, Hans and Painter, 1962). The percentage of ascophyllan present in algin derived from rockweeds is set in Table 6.3. In view of the high proportion of alginic acid recorded from *Durvillea* (Table 6.2), it seems important that the species should be examined for the presence of any fucose.

Another feature of the rockweed alginates that requires further study is the differential solubility at low pH. This may well be due to ascophyllan but further investigations are necessary before this could be accepted. It is equally possible

Table 6.4

Species	Average % hydrolysable fraction	Average % resistant fraction
Laminaria digitata	30	65·5
Laminaria hyperborea	20·5	68·5
Ascophyllum	56·6	41

that it is related to a difference in the sequence of the uronic acids along the alginate chain (Myklestad and Haug, 1962) (Table 6.4).

The brown rockweeds to a large extent, and the oarweeds to a lesser extent, also contain phenolic compounds, of which fucosan is an example. They are significant in that the amount present exerts an effect upon the quality of the alginate that is

Table 6.5 Ratio α-D-mannuronic to L-guluronic acid in alginates from various algae

Species	Haug (1964)	Fischer and Dorfell (1955)
Laminaria digitata	1·15—2·35	3·1
Laminaria hyperborea	0·4—1·65	1·6
Laminaria saccharina	1·25—1·35	—
Alaria esculenta	1·2—1·7	—
Ascophyllum	1·55—2·25	2·6
Fucus vesiculosus	0·9—1·2	1·3
Fucus serratus	1·15	2·7
Macrocystis pyrifera	1·56	—
Azotobacter vinelandii (bacterium)	1·50*	—

* Penman and Sanderson, 1972.

obtained. It is for this reason that alginates from oarweeds tend to be better than alginates from fucoids. Ecological parameters can also affect the alginic acid content of fucoids (Berard-Theriault and Cardinal, 1973).

Since the discovery of the existence of the two uronic acids in the algin molecule some work has been carried out to determine the relative proportions of the two components. The relative proportions in various algal extracts is set out in Table 6.5. These variations exert an effect upon the properties and qualities of the various extracts obtained. It has also been found that in *L. digitata* new fronds are much richer in D-mannuronic acid than old fronds so that proportions can change with age. In *L. hyperborea* the stipes are richer in L-guluronic acid than the fronds and even in the stipes there is a higher proportion in the outer cortex than in the medulla. Extracts from the tips of *Ascophyllum* are much higher in mannuronic acid than extracts from older parts of the thallus (Haug *et al.*, 1968). In the Japanese algae *Eisenia bicyclis* and *Ishige okamurai,* the proportions of the two acids vary in different parts of the frond. The synthesis of alginic acid in the latter alga has been studied by Abe *et al.,* 1973.

Because of all the potential sources of variation, samples for analysis should consist of plenty of plants. Of the three main methods of extraction for analysis, two of them, by involving preheat treatment with formalin or extraction at a pH between 6 and 7, provide for removal of phenols and this is important.

Studies by Ricard (1931) and Lunde (1937a) showed that in *Laminaria digitata* there is a definite seasonal fluctuation with a maximum content between February and April (Fig. 6.1). A more recent study by Cardinal and Breton-Provencher (1977) has given a very different result with a maximum in August. In *L. longicruris* they found a maximum in September (Fig. 6.2). In view of the discrepancy further studies would seem essential. Black (1948) recorded similar seasonal fluctuations in *L. hyperborea* with a maximum in February. Comparing different organs he found that the fluctuations were primarily in the frond and that the amount in the stipes

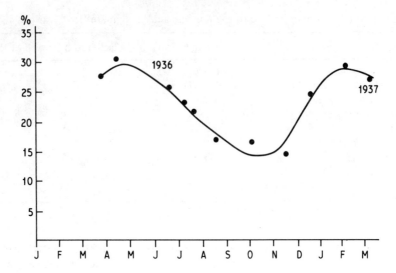

Fig. 6.1 Seasonal variation of algin in *Laminaria digitata*.

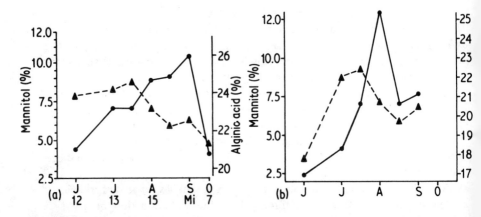

Fig. 6.2(a) Seasonal variations in the percentage of mannitol (▲—▲) and alginic acid (●—●) of *Laminaria longicruris* (dry weight) (b) seasonal variations in the percentage of mannitol (▲—▲) and alginic acid (●—●) of *Laminaria digitata* (dry weight) (after Cardinal and Breton-Provencher, 1972).

varied very little. In *Ecklonia radiata* maximum amounts occur in the autumn (April) and minimum values are found in September (early spring) (Stewart *et al.*, 1961). In Canada the algin content of the rockweeds (*F. vesiculosus*, *F. evanescens*, *Ascophyllum*) reaches a maximum in the winter, whereas in Norway it appears to be somewhat lower in winter in *Ascophyllum* (Jensen, 1966). In India the algin content reaches its maximum in *Sargassum wightii* and *Turbinaria conoides* during peak growth between October and December (Rao, 1968), but there is no such variation in *T. ornata* (Rao and Kalimuthu, 1972). It is evident that there is scope for further studies in the metabolism of the different algae in order to determine whether there is a general pattern or not. In the New Zealand *Durvillea antarctica*, South (1979) found that 7-month-old plants had an alginate content 5–6% higher than older plants and that it was higher than in either *D. willana* or the Australian *D. potatorum*.

Table 6.6

Organ	Locality	Algin % fresh wt
Leaf	Pacific Grove (north)	1·8
Leaf	San Diego (south)	2·7
Stipe	Pacific Grove	1·7
Stipe	San Diego	2·3

There is also some evidence from the USA Pacific coast of variation with latitude. The rather inadequate data for *Macrocystis* from Hoagland (1916), presented in Table 6.6, show that plants from the south contain considerably more algin than those from farther north. This result is probably related to higher metabolic activity in the warmer southern waters. It is comparable to a somewhat similar phenomenon mentioned previously (p. 22) for iodine, potash and nitrogen.

Genetic variation within a species does not appear to be significant. In the case of *Laminaria longicruris*, Chapman and Doyle (1979) reported that there was only a low additive genetic component. Hybrids between *Macrocystis* and *Pelagophycus* and *Macrocystis* and *Nereocystis* have been produced (Sanbonsuga and Neushul, 1979), but there is no data on the algin content of such hybrids.

6.3 WEEDS AND THEIR COLLECTION

Alginic acid or its salts were used to some extent in the 1914–18 war, and in 1915 a factory was erected at Sydney in British Columbia for the manufacture of potash and algin. This factory used 30–40 tons (30·5–40·6 tonnes) of wet weed daily, but, in spite of legislation to encourage the industry, it does not appear to have flourished. During the same war alginates were also prepared in Tasmania from *Macrocystis* (see p. 207) and in New South Wales from *Ecklonia*, though for what purpose is not stated. This manufacture ceased after the war. It was suggested by Isaac (1942) that the large beds of *Laminaria pallida* and *Ecklonia maxima* (= *E. buccinalis*) on the west coast of South Africa might prove a profitable source for both potash and algin.

On the Pacific coast of the USA, *Macrocystis* and to a lesser extent *Nereocystis* and *Pelagophycus* form the raw material, and these are cut by mechanical harvesters. Extraction was originally commenced by the Hercules Powder Co. in the 1914–18 period. Later the Kelco Company commenced operations in 1929 and it is now the world's major producer. In Mexico some 37 000 tons (37 580 tonnes) of *Macrocystis* were harvested in 1974, and there could be an annual exploitable harvest of around 65 000 tons (66 030 tonnes) (Michaneck, 1975). Recent harvests are 27 480 tons (27 920 tonnes) in 1975, 41 570 tons (42 220 tonnes) in 1976 and 41 746 tons (42 390 tonnes) in 1977 (Ortega, personnal communication). On the US east coast rockweed and *Laminaria* were used and in 1961 Scotia Marine Products (now a subsidiary of Kelco), established a factory making use of *Ascophyllum*. The latest yearly output figures for this region are 1200 tons (1219 tonnes) of algin (Moss, 1979) derived from about 7·0 million kg of wet weed collected primarily from S.W. Nova Scotia (Pringle, personal communication).

Alginate Industries of the United Kingdom use *Laminaria hyperborea* with lesser quantities of other *Laminaria* species, together with *Ascophyllum*. They are now importing *Macrocystis* from Tasmania (see p. 205). *L. digitata* is the raw material from Morocco and France, and the same species together with *Ascophyllum* is used in Norway. In the latter country *L. digitata* grows in shallow water and can therefore be harvested from small boats at low tide. *L. hyperborea* from deeper water presents difficulties as it is not readily amenable to mechanical harvesting. *Ascophyllum* can be collected from small boats or directly from the shore at low tide. At present algin production in Norway annually requires 7000–8000 tons (7112–8128 tonnes) of dry *Ascophyllum* and 15 000 tons (15 240 tonnes) of dry *Laminaria* (Jensen, personal communication). In Europe both cast and cut weed are used in either the fresh or dry state. Supplies of cast weed are naturally very dependent upon weather and in England at the beginning of the present century it was proposed to preserve kelp by means of heavy gas oils, but now sulphur dioxide is used. Algin from preserved algae is more viscous than that from fresh material (Dillon, 1938).

In Japan manufacture was commenced in 1940 and by 1962 there were four factories producing some 1200 tons (1219 tonnes) of alginates. At present about 1425 tons (1448 tonnes) are produced from 33 000 tons (33 528 tonnes) of weed, 3000 tons (3048 tonnes) being imported (Michanek, 1975). Species of *Laminaria, Eisenia bicyclis, Ecklonia cava, Sargassum horneri* and *S. natans* are the principal raw materials, the most widely used being *Eisenia* and *Ecklonia*. *Sargassum* spp. could also be used in Brazil where Filho and de Paula (1979) estimate that some 18 000 kg of algin could be manufactured annually. The two potential algae are *Laminaria*

Table 6.7 (After Filho and Quege, 1978)

Year	Kombu	Medicinal	Algin
1974	61·48	–	197·85
1975	21·40	4·21	175·50
1976	21·60	1·5	196·41
1977	8·02	9·82	131·78

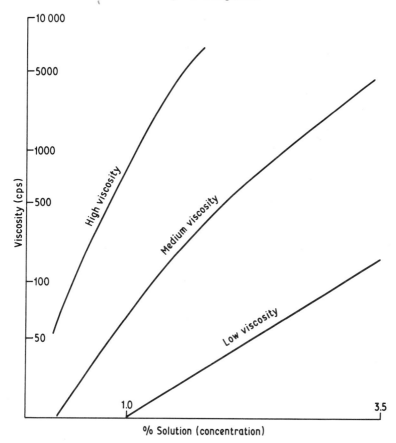

Fig. 6.3 Effect of concentration on viscosity (after Cottrell and Kovacs).

abyssalis and *L. brasiliensis,* but more work is needed on harvesting effects and growth of biomass. A study should also be made of any ecological consequences of harvesting. The present importations into Brazil of brown algae and their uses are shown in Table 6.7. Chile has been experimenting with the production of alginates on a small scale using *Durvillea.* Also in the southern hemisphere a factory was established in Tasmania, using *Macrocystis,* in 1964 (Pownall, 1964) but it has since closed. More recently a new company, Kelp Industries Pty, has been formed, but this is purely for the collection and export of dried *Macrocystis* to Alginate Industries of Great Britain. Export commenced in 1975 and there has been a steady increase since then:

1975	38 tonnes
1976	775 tonnes
1977	2442 tonnes
1978	3000 tonnes (target)

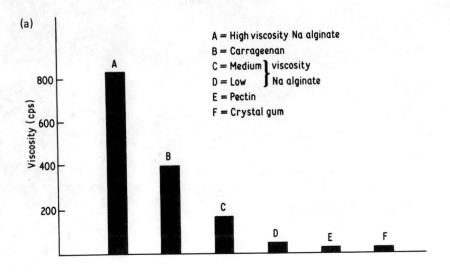

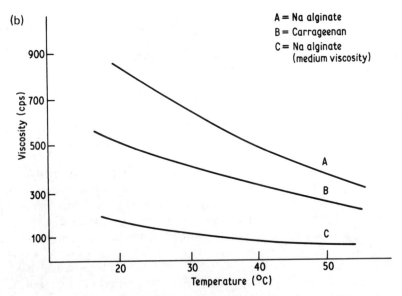

Fig. 6.4 (a) Comparison of viscosities of various solutions. (b) Effect of temperature on viscosity of algin. (After Cottrell and Kovacs).

Small-scale manufacture is also taking place in the USSR, and is probably expanding, though information is difficult to obtain.

In Europe, apart from Norway, economically usable supplies of *Laminaria* and *Ascophyllum* are almost fully exploited and this is holding back the industry. A study by Tolstikova (1977) on plant age and regeneration of *Fucus vesiculosus* and *Ascophyllum* may help in improving resource management. In the meantime the two major USA producers are capturing the main market (SSRA Rept. for 1967).

One of the principal problems facing all European concerns is the collection of the raw material. Alginate production requires large quantities of weed and, therefore, mechanical harvesting is a necessity. Unfortunately the plants are not large like the giant Pacific kelps so that harvesters of the American pattern cannot be readily employed. It was reported (SSRA Rept.) in 1967 that new harvesters have been developed in both the USA and Norway. In Russia also various types of mechanical harvesting have been proposed. One involves a boat with knives near the bottom and a moving hay conveyor. Another suggestion involved maceration in the growing zone and hoovering up the cut material. A third prototype used a trawl on wheels that travelled over the bottom (Transl. All Union Conf. Algae, 1962). In Great Britain about 4% of the wet weight or 30% of the dry weed is the normal yield. On the Pacific coast of the USA and in Tasmania the yield is probably higher and around 35% of the dry weight. In 1976 the estimated total world production of alginates was 19 200 tons (19 507 tonnes), the principal producers being 5500 tons (5588 tonnes) from the USA, 6500 tons (6604 tonnes) from the UK and 1425 tons (1447 tonnes) from Japan, 1200 tons (1219 tonnes) from France (mainly Brittany) and 3000 tons (3048 tonnes) from Norway. Chile and India each produce about 200 tons (203·2 tonnes), though for the latter country this is perhaps an under estimate (Pillay, personal communication). The Indian production is mainly manufactured from *Sargassum* spp. but *Turbinaria* spp. are now being considered. Up to 19 000 tons (19 304 tonnes) of wet weed are available from the Gulf of Kutch alone (Michanek, 1975). The total annual production of 19 200 tons (19 507 tonnes) for the world represents a marked increase in recent years as McNealy and Pettitt estimated the annual yield in 1973 as 10 000 tons (10 160 tonnes).

6.4 PRODUCTION

One of the simpler ways of obtaining alginic acid is to macerate the seaweed with dilute hydrochloric acid in order to remove any soluble mineral salts. Gloess (1932) regarded the material left after this first treatment with acid as 'coarse algin'. The alginic acid can then be extracted by using a solution of sodium carbonate, during which treatment the tissues swell up and lose their shape. The resulting solution, which is viscous, is filtered, and the algin is precipitated from it by treatment with more acid (cf. Table 4.12, p. 130) for viscosities of sodium alginate solutions). The crude material is filtered off and washed, whilst from the solution Glauber's salt can be obtained. Dillon (1938) has described another easy method of preparation as follows. The weed is first extracted for a short time with boiling water in order to remove fucoidan (see p. 232). It is then soaked in dilute hydrochloric acid for a day, after which the alginic acid is removed with ammonia. Wet weed was commonly used for these extractions, but Gloess (1919) considered that maximum extraction was achieved if the weed was dry rather than wet. Factories, on the other hand, want to use fresh weed in order to obviate the drying costs, which may outweigh the advantages of using dried weed. The preparation of the pure material is said to be difficult (see Maass, 1959, for references to patents) and firms do not reveal their techniques. Barry and Dillon (1936) did describe a fairly elaborate technique by means of which pure alginic acid can be obtained.

The very first method of production described above is the basis of the lixiviation process, which was proposed by Stanford (1884) at the end of the last century, in order to enable the Scottish kelp workers to compete with the production of potash from mineral resources (see p. 18). The principal stages in the lixiviation process are set out in the scheme below.

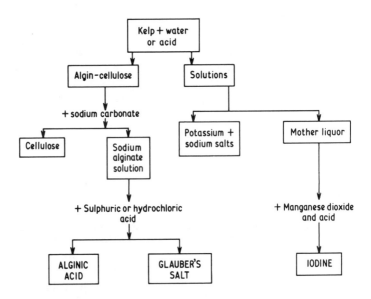

No published details of the commercial processes used in England are available, but in America there are two basic processes: Green's cold process (as used by the Kelco Co.) and the Le Gloahec-Herter process, as used by the Algin Corporation of America.

In the first, or cold process, so called because it is conducted at the relatively low temperature of 50°F (10°C), fresh kelp is first leached for several hours with 0·33% hydrochloric acid. The liquid is removed and after being chopped and shredded the leached kelp is digested with soda ash solution (40–50 lb (18–22·9 kg) per ton of

Algin and Alginates

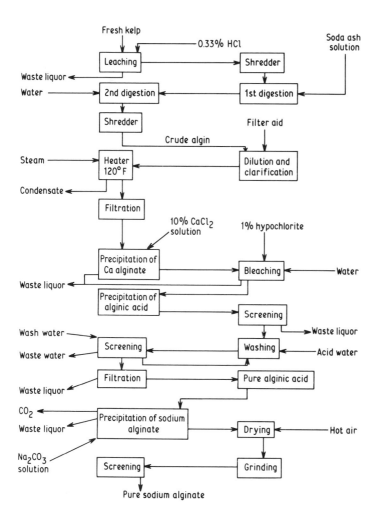

fresh kelp) at a pH of about 10: this first digestion occupies about 30 min and it is then repeated. The crude pulp is shredded again and six volumes of treated water added at a pH of 9·6–11. At this point the fibrous material can be dried and sold as crude sodium alginate.

To obtain a purer product the liquor is filtered, using filter aids and presses, and the temperature may be raised to 120°F (48·9°C) to assist the process. The filtered liquor is then added slowly to a 10–11% calcium chloride solution (about 100 lb (45·4 kg) $CaCl_2$ to 8 tons (8·1 tonnes) of alginate liquor), under constant agitation, and when the agitation is stopped calcium alginate slowly rises to the top. The remaining liquor is drained off and more water and a bleacher (sodium hypochlorite) added to the precipitate. Too much bleacher must not be added as it exerts an adverse effect on the final product.

The bleached precipitate is separated off and added to a 5% hydrochloric acid solution (1 part alginate solution to 42 parts acid). This converts the calcium alginate into fibrous alginic acid, which is purified of calcium salts by being passed several times through screens, agitation with dilute acid taking place between each screening. The alginic acid so produced can be filtered and stored or else converted into a salt by treatment with an appropriate carbonate, oxide or hydroxide. The process is generalized in the scheme (Tseng, 1945c) on p. 209.

In the Le Gloahec-Herter process the raw kelp can be either fresh or dried. Recent work (Whyte et al., 1977) shows that *Macrocystis* dries 1·3 times more rapidly than *Nereocystis* and further, that higher temperatures (up to 300°C) can safely be used with *Macrocystis*. To one part of kelp, three parts of 0·8–1% calcium chloride solution are added, the solution being either hot or cold. The function of the calcium salt is to remove laminaran, mannitol and other salts. These salts and the calcium are then removed by washing with softened water, after which it may receive an additional treatment with 5% hydrochloric acid in order to dissolve any residual alkaline earth salts. It is washed once more with softened water, and is then digested with 4% soda ash solution in the proportions of two volumes solution to one volume of kelp. Lixiviation is continued for about two hours at 104°F (40°C) and the kelp is macerated at the same time until it is reduced to a paste.

The resulting paste is diluted with water in the ratio 3 to 7, and after being beaten into a homogeneous suspension it is vigorously aerated, but if ozone or hydrogen peroxide are used it is merely agitated mechanically. The liquor is then passed continuously at high speed through a centrifuge, where it is charged with air bubbles, and then led into a clarifying tank. Here after 6–10 h the cellulose particles agglomerate to form a floating cake and the liquor is removed.

The coloured liquor is decolourized by the addition of an absorbent jelly, usually made of hydrated alumina, gelatinous silica and aluminium alginate. The proportions are usually 20–25 parts of jelly to 100 parts of dry alginous material. The jelly is removed by centrifuging and can be reclaimed by various methods.

The alginic acid is now precipitated by running it into a mixing baffle, where it meets a stream of strong hydrochloric acid, arranged in such a way that the precipitate runs at once into another tank. All through this process the pH of the solution is maintained at 2·8–3·2. The precipitated alginic acid is placed in baskets and drained, after which it is purified by solvents such as alcohol and dried.

Decolourization can also be achieved by the use of formaldehyde, alcohol, tannic acid or other protein coagulants. If formalin is used, it is added before digestion with soda ash solution, and after the mixture has stood for one hour the kelp is taken out and kept in a store for 15–21 days. When the soda ash solution is now added, only the alginous matter is dissolved, and the pigments remain behind, 'fixed' by the protein–cellulose mixture. The flow-sheet for this process is set out in the scheme (Tseng, 1945c) on p. 211.

A factory in Baja California produces alginates based upon the flow sheet (Ortega, 1977) on p. 212.

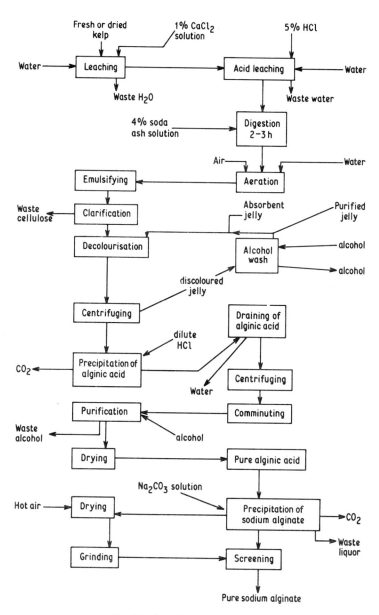

Le Gloahec-Herter process

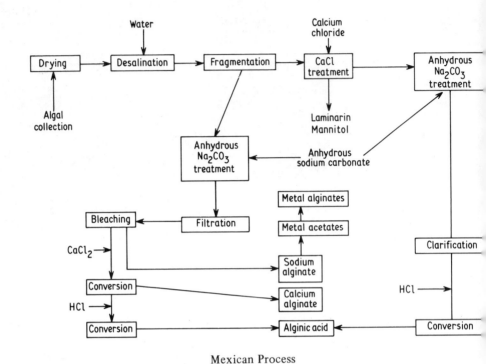

Mexican Process

6.5 PROPERTIES

Originally algin was not considered to be of sufficient value alone, and the process by which it was obtained was suggested as a commercial necessity, in order to enable the Scottish kelp trade to compete with mineral imports from Chile. Now, however, alginic acid is of greater importance than the kelp salts, even when used agriculturally. From the technical point of view the importance of alginic acid lies in the properties of its salts, although the acid itself is of interest because of its capacity to absorb 200–300 times its weight of water. Since deterioration of algin and its salts sets in above 32°C, they should be stored in a cool place. The sodium, potassium, magnesium and propylene glycol alginates all dissolve readily in water to give a solution that is extremely thick and viscous. The solution properties of sodium alginate from *Eisenia bicyclis* have been studied by Saito *et al.* (1977). Typical properties of food grade sodium alginates are as follows (Cottrell and Kovacs, 1977):

Moisture content	13 %
Ash	23 %
Specific gravity	1·59 %

Colour	Ivory
Bulk density	54·6 lb ft^{-3} (874 kg m^{-3})
Ashing temperature	480°C
Heat of combustion	2·5 cal g^{-1} (10·47 > g^{-1})

Because of the nature of alginic acid, the viscosity of the product varies from species to species of alga and from sample to sample (Vincent, Goring and Young, 1955; Levring, Hoppe and Schmid, 1969). The viscosity of the solutions depends on four major factors:

(1) increasing the polymerization grade increases the viscosity;
(2) increasing the concentration increases the viscosity (Fig. 6.3);
(3) increasing the temperature decreases the viscosity (Fig. 6.4);
(4) solution in hard water or the addition of electrolytes affects viscosity

(Schulzen, 1962) when the effect is related not only to the Ca^{2+} ions and electrolytes but also to the relative proportions of the two uronic acids.

The viscosity is greatly affected by the addition of small amounts of NaCl, Na_2SO_4, Na_2CO_3 and sodium ammonium salts. The viscosities are constant between pH 5 and pH 10. Below 4·5 the viscosity increases and precipitation occurs below 3·0. Above pH 6·5, propylene glycol alginate rapidly degrades (Cottrell and Kovacs, 1977). The various alkali solutions, which are tasteless, odourless and almost colourless, do not coagulate on heating nor do they set to a firm jelly when they cool. A comparison of the viscosity of sodium alginate with carrageenan and pectins is provided in Fig. 6.4. The ion exchange properties of sodium alginate depend on the relative proportions of the two uronic acids. The addition of divalent ions leads to gel formation (see below), though gelation also depends on the concentration of sodium ions (Haug, 1964).

An important property is the capacity of alginic acid to react with polyvalent metal ions to form gels or solutions with a high viscosity. An important divalent ion that is commonly used for this purpose is calcium. With calcium, adjacent polymeric chains become linked, the poly-guluronic first and then poly-mannuronic, and as the calcium concentration is increased the algin becomes more and more viscous, until it is finally precipitated (Fig. 6.5). The gel, when formed, consists of 99·0–99·5% water and 0·5–1·0% Ca alginate. In the initial linkage of the polyguluronic segments an aggregation is formed with interstices into which the calcium ions fit (egg box model) (Fig. 6.6). In algae such as *Laminaria hyperborea* with a high percentage of poly-guluronic acid, rigid, brittle gels are formed. Use is made of gelation under varying conditions in a variety of commercial ways, including canning, production of fruit or vegetable flavoured particles, film making etc. (Andrew and MacLeod, 1970)

Further use is made of this property in the manufacture of imitation low calorie spaghetti, fabricated cherries, meatballs and potato chips etc.

The heavy metals (e.g. mercury, beryllium, copper, cobalt) also yield salts of

Calcium alginate

alginic acid, but these are not soluble in water. They do, however, form a plastic material that can be moulded, when it is moist, and which sets hard on drying. This is an extremely useful property with great potential. Another property of considerable significance from the industrial view point is the ease with which the soluble alkali salts can be converted to the insoluble heavy metal salts. Algin solutions derived from *Laminaria* and *Macrocystis* are stable at pH 5—10 but algin derived from *Ascophyllum* will degrade unless it has been purified of the reducing phenolic compounds etc. The rate of degradation depends mainly upon the concentration of the reducing compounds. In general it may be said that the quality of alginates derived from the rockweeds (*Fucus, Ascophyllum*) depends on three factors:

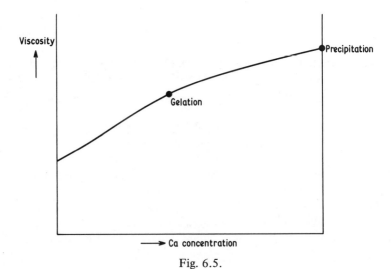

Fig. 6.5.

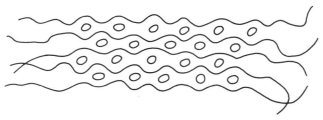

Fig. 6.6 Egg box model (algin).

(1) Uronic acid composition, (2) amount of ascophyllan or fucoidan present, (3) amount of phenolic compounds present. The numerous patents that have been granted for processes concerned with alginic acid indicate rapid progress, not only on the technical side but also in the knowledge of its chemical properties.

6.6 USES

The soluble salts, especially the sodium salt, of alginic acid are used in the textile industry because they form an excellent dressing and polishing material. The soluble alkali salts can also be used as a thickening material for colours that are employed in printing fabrics, and as a hardener and adhesive for joining threads in weaving. They are used alone or mixed with starch or tragacanth gum. Deschiens (1926) has also described how they have been used for impregnating manure sacks. The cementing and filling property is further utilized for glazing and sizing in the paper and cardboard industries. Another use for the soluble salts is as an adhesive in the manufacture of briquettes, especially those made from brown coal or lignite, and for this purpose they are said to be markedly superior to other substances. In the manufacture of the briquettes a 2·5% solution is used as the binding medium. Other possible uses are dependent upon the emulsifying power of the salts, e.g. in casein emulsion paints. This property should open up further uses for water-insoluble substances such as tar products, petrol, oil and disinfectants. Because of its suspending properties sodium alginate is used in drugs, e.g. penicillin, in car polishes, in paints, cosmetics, insecticides and pharmaceutical preparations (Maass, 1959, 1962). In the paint industry its chief functions are in (1) suspension of the pigment, (2) stabilization of the emulsion, (3) a better flow to the paint, (4) providing a continuous film to the painted surface, (5) giving a better coverage. Paints mixed with algin are also less affected by wide variations in temperatures (Maass, 1962).

When the soluble salts have been converted into insoluble salts, the latter can be used for the production of waterproof cloth such as tents and wagon covers. Ammoniated aluminium alginate is employed in this connection because it becomes insoluble after drying. Because of their pliability whilst moist, the insoluble salts can also be employed in the preparation of plastics, vulcanite fibre, linoleum and imitation leather. Another use has been found for them in the clarification of sugar solutions and mineral waters. Copper and mercury alginates are said to be valuable components of paints for use under the sea because of their insoluble character. This latter application, however, is one that needs confirmation and further study. All the insoluble heavy metal salts can be dissolved in ammonia and when the solution is evaporated a waterproof film is left that can be used as a varnish. Thus Gloess (1932) says that ammoniacal copper alginate has been employed successfully in France for impregnating and preserving wood.

One of the increasingly important uses of algin, especially in the United States, is as a stabilizer to give smooth body and texture to ice-cream and in preventing coarse crystals being formed. It is also employed as a suspending agent in milk shakes. In fact about half the total output in the USA is applied to these purposes. When used as a stabilizer it must be in alkaline solution because in acid solution it gels (Steiner and McNeely, 1954). It is even better for this purpose if used as propylene glycol alginate. It is also used as a stabilizer on icings, sherbets and light beers, where it is replacing agar, whilst it is also put into cream cheeses and into whipping creams for decorating fancy cakes. In the confectionery trade it is used as a filler for candy bars and also in salad dressings. Its value in these uses rests with its property of dissolving without heating. In Norway it is much used in the deep-freezing of fish.

Another very important use of algin is as a latex creaming and stabilizing agent in the production of rubber from natural sources. For this process it has been found that ammonium alginate is the best salt.

The insoluble calcium salt readily gives gels and films. If a solution to which a small proportion of tannic acid has been mixed is violently agitated, the mixture emulsifies and can be poured upon a glass or polished surface to give a thin transparent film resembling cellophane (Pehorey, 1937). The film is cheap, almost noninflammable, and does not become quite so brittle in light as does cellophane. When algin is mixed with resins or lacquers it yields a rubber or gutta-percha like product, for which no doubt a number of uses will be found. For this purpose the weed is treated with sodium carbonate, formalin and tan bark to which either rubber, glue or resin is added. The product is said to be a very good substitute for panelling board or linoleum. Calcium alginate with another alkali is marketed as 'Tragaya' which is a substitute for gum tragacanth.

A modification of the alginic acid manufacture has been used on a small scale in Galway (Dillon, 1938), in order to produce a hard horny mass which could be used as a wall or ceiling board. In this process the weed is allowed to rot until the algin becomes degraded or else disappears entirely. Acid is then added and the remaining alginates are precipitated leaving a solution of inorganic salts. The precipitated organic matter is finally boiled with alkalis in order to yield the hard horny mass.

A more important potential usefulness for algin can be visualized in the production of an artificial fibre. If the purified alkaline extract is forced through a fine aperture it forms a viscous thread, which can then be spun in a bath containing a mixture of furfurol, caustic soda, formalin and other substances. This process was originally patented by Sarason in 1912 and later by Godha using the whole weed, and a full account is given by Speakman and Chamberlain (1944), Speakman (1945) and Kirby (1953). Marsh and Wood (1942) give a brief description of the various methods that have been patented for obtaining silk threads from algin. One method involves the extrusion of a 7% solution into a bath containing 10% chloride of lime. The Japanese, using alginic acid prepared from a species of *Sargassum,* obtained a 3% solution of the ammonium salt, which is forced through holes and coagulated in a bath containing 10% sulphuric acid. The fibres are subsequently immersed in a 10% solution of aluminium sulphate, and are then given a final bath in a weak solution of lead acetate.

Speakman and Chamberlain (1944) studies the potentialities of alginic fibres in some detail and arrived at the following conclusion: 'It seems clear that alginic acid fulfils the main requirements of a substance intended for use in the manufacture of fibres. It consists of chain molecules of high molecular weight possessing reactive side chains, and abundant supplies are readily available'. Supplies may indeed be abundant but experience shows that they are not so easily harvestable (see p. 207).

The experiments showed that at present the best solution for the production of an alginic yarn is a 7·5–8% solution of air-dried sodium alginate, of such a grade that the viscosity of a 1% solution at 20°C is approximately 40 centipoises. For use with this solution there should be a coagulating bath of normal sulphuric acid saturated with sodium sulphate. This gives a thread of sodium alginate, but by passing the thread into a normal solution of calcium chloride in 0·02 N hydrochloric acid a yarn of calcium alginate is secured. Speakman and Chamberlain found that the threads of yarn produced in this way tended to adhere, but they overcame this difficulty with the calcium thread by using 2·5% olive oil emulsified by the neutral detergent Lissapol C; for alginic acid threads other emulsifying agents had to be used.

It has been found that the strength and extensibility of calcium alginate rayons increase with increasing viscosity of the spinning solution. The 'handle' of the final product is also improved by increasing the concentration of the spinning solution. The best results were obtained when a liquid was used that contained 7·5–8·0 g of sodium alginate per 100 g of solution. Improvements in the yarn have also been effected by improved washing and handling techniques, because these are apparently very important items. As originally produced the tenacity of the calcium alginate yarn was low, 1·23 g per denier* as against 1·3–1·8 for cellulose and viscose threads. This weakness was due to degradation resulting from the carrying over of acid and to mechanical damage. The former was overcome by washing and the latter by winding on to the reel direct from the drier. The tenacity then increased to 1·8–2·1 g per denier. The breaking load of calcium alginate increases with increasing metal content, but density also increases so that there is an optimum metal content for maximum tenacity at about 10 g atoms per 100 g of alginic acid (Fig. 6.7).

Unfortunately threads prepared from alginic acid, sodium alginate and calcium

---

* Denier = weight in g of 9000 metres.

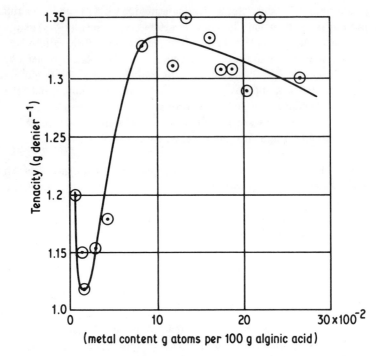

Fig. 6.7 Optimum tenacity of calcium alginate in relation to metal content.

alginate are readily dissolved by soap and soda. Sodium alginate is even soluble in water. This fact was largely responsible for the abandonment of much calcium or beryllium alginate fibre produced during the 1939 war for camouflage netting (Tseng, 1945c). In spite of precautions the fibre contained sodium alginate as an impurity and the material, therefore, gradually rotted in a wet climate. Attempts have naturally been made to obtain a resistant fibre, and it has been found that chromium alginate is resistant but is too highly coloured to form a satisfactory general solution. Calcium alginate fibres can be made insoluble by means of a resin treatment (B.P. 575 611).

The degradation of ordinary alginic acid threads is relatively rapid, and the tenacity falls from 1·06 g per denier after 28 days to 0·36 g per denier after 480 days. Calcium alginate on the other hand can be stored for 30 weeks without any loss of strength. The introduction of chromium or beryllium stabilizes the yarn and also makes it alkali-resistant, though better yarns are obtained by introducing these metals into alginic acid direct. Yarns containing 1·29% chromium and 9·04% Ca or 2·89% Be and 5·2% Ca are stable, and suffer less than 5% loss in tenacity after 30 min treatment with soap and soda at 25 or 45°C. The tenacity of beryllium alginate even exceeds 2 g per denier. A yarn resistant to alkalis can also be obtained by cross-linking with formaldehyde (Speakman, 1945). The discovery that beryllium alginate is resistant to soap and soda was important because threads prepared from this salt then assumed particular importance. One method of preparing them is as follows. Filaments of sodium alginate are obtained first of all, and they are then passed into an emulsion of sulphuric acid and olive oil, after which they go into a bath of beryllium acetate kept at the temperature of the boiling point of water. In this bath the sodium is replaced by the beryllium.

Speakman and Chamberlain (1944) have improved on this technique. They found that calcium alginate can be woven or knitted, and then converted to an alkali-resistant rayon when in fabric form by treatment with the basic acetates of either beryllium or chromium. It would seem therefore that any large-scale production of a seaweed rayon will depend upon the preparation of a stock material of calcium alginate, which can be converted to a resistant salt later. This method of manufacture is also determined to some extent by the fact that alkali-resistant rayons are difficult to prepare by direct spinning, and also the extensibility of chromium and beryllium alginate is too low to permit of their successful use in weaving and knitting.

The artificial silk obtained by these various methods is of very good quality, and it should ultimately be able to compete with artificial silk made from other sources. At present, however, artificial fibres can be produced at much less cost, and no seaweed fibre is, so far, alkali resistant. The best, chromium, beryllium or cadmium alginates, are all rather harsh to the touch. One asset of these threads is that the fibre does not burn so that woven material would be non-inflammable. A further improvement can be foreseen because certain salts of alginic acid are coloured, and hence cloth made of such threads would not require to be dyed; furthermore the colour would be fast and would not come out in the wash. As examples of these coloured salts there are copper and nickel alginates which are green, whilst cobalt alginate is red and chromium alginate is blue. Apart from this, the non-coloured salts also have a marked affinity for basic dyes.

During the 1939 war, experiments with alginate rayons resulted in the production of new fabrics and new effects. Some of the latter have undergone some exploitation and there should be a great future for the industry. Two examples may be quoted: if cotton and calcium alginate yarns are woven so as to remove the normal twist in cotton, the calcium alginate can subsequently be removed by solution in an alkali, thus leaving an untwisted cotton fabric. The 'disappearing fibre' technique, as it is called, has also been used effectively with mohair and calcium alginate.

The recent production of organic derivatives, e.g. the propylene glycol ester, has greatly widened the uses to which alginate can be put, especially as a natural gum. A dramatic application is in the quick freezing of fatty fish where the alginate coat prevents rancidity. It is particularly useful for coating mackerel and herring fillets. A 5% solution of algin is used in jam manufacture in order to assist the setting.

Other uses for alginates have been considered, though there is no evidence they have been employed. For example, Gloess (1919) suggested that brick and cement buildings could be coated with a 1 or 2% solution of the soluble alginates, in order to make them weatherproof. Such solutions have in fact been used in the building

Table 6.8 Applications of Kelco algin products (from Kelco)

Food applications

Property	Product	Performance	Property	Product	Performance
Water-holding	Frozen foods	Maintains texture during freeze-thaw cycle.	Emulsifying	Salad dressings	Emulsifies and stabilizes various types.
	Pastry fillings	Produces smooth, soft texture and body.		Meat and flavor sauces	Emulsifies oil and suspends solids.
	Syrups	Suspends solids, controls pouring consistency.			
	Bakery icings	Counteracts stickiness and cracking.	Stabilizing	Beer	Maintains beer foam under adverse conditions.
	Dry mixes	Quickly absorbs water or milk in reconstitution.		Fruit juice	Stabilizes pulp and concentrates and finished drinks.
	Meringues	Stabilizes meringue bodies.		Fountain syrups, toppings	Suspends solids, produces uniform body.
	Frozen desserts	Provides heat-shock protection, improved flavour release, and superior meltdown.		Whipped toppings	Aids in developing overrun, stabilizes fat dispersion, and prevents freeze-thaw breakdown.
	Relish	Stabilizes brine, allowing uniform filling.		Sauces and gravies	Thickens and stabilizes for a broad range of applications

(continued on the next page)

Table 6.8 Applications of Kelco algin products (from Kelco) (continued)

Food applications

Property	Product	Performance	Property	Product	Performance
Gelling	Instant puddings	Produces firm pudding with excellent body and texture; better flavour release.	Stabilizing	Milkshakes	Controls overrun and provides smooth, creamy body.
	Cooked puddings	Stabilizes pudding system, firms body, and reduces weeping.			
	Chiffons	Provides tender gel body that stabilizes instant (cold make-up) chiffons.			
	Pie and pastry fillings	Cold-water gel base for instant bakery jellies and instant lemon pie fillings. Develops soft gel body with broad temperature tolerance; improve flavour release.			
	Dessert gels	Produces clear, firm, quick-setting gels with hot or cold water.			
	Fabricated foods	Provides a unique binding system that gels rapidly under a wide range of conditions.			

(continued on the next page)

Table 6.8 Applications of Kelco algin products (from Kelco) (*continued*)

Industrial applications

Property	Product	Performance	Property	Product	Performance
Water-holding	Paper coating	Controls rheology of coatings; prevents dilatancy at high shear.	Emulsifying	Polishes	Emulsifies oils and suspends solids.
				Antifoams	Emulsifies and stabilizes various types.
	Paper sizing	Improves surface properties, ink acceptance, and smoothness.		Latices	Stabilizes latex emulsions, provides viscosity.
	Adhesives	Controls penetration to improve adhesion and application.	Stabilizing	Ceramics	Imparts plasticity and suspends solids.
	Textile printing	Produces very fine line prints with good definition and excellent washout.		Welding rods	Improves extrusion characteristics and green strength.
	Textile dyeing	Prevents migration of dyestuffs in pad dyeing operations. (Algin is also compatible with most fibre-reactive dyes).		Cleaners	Suspends and stabilizes insoluble solids.
Gelling	Air freshener gel	Firm, stable gels are produced from cold-water systems.			

(*continued on the next page*)

Table 6.8 Applications of Kelco algin products (from Kelco) (*continued*)

Industrial applications

Property	Product	Performance	Property	Product	Performance
Gelling	Explosives	Rubbery, elastic gels are formed by reaction with borates.	Gelling	Toys	Safe, nontoxic materials are made for impressions or putty-like compounds.
	Hydro-mulching	Holds mulch to inclined surfaces; promotes seed germination.		Boiler compounds	Produces soft, voluminous flocs easily separated from boiler water.

industry to fire-proof wood. Other building trade uses are in the manufacture of non-splinter glass, as a thickening for bitumen and in the production of a type of cement. It has been found that dried milk and cocoa can be rendered more soluble, and can also be made without any sediment if alginates are added to the milk before it is dried, or to the cocoa powder. Other uses for algin or alginates are the production of fire-retarding compounds consisting of chemicals dissolved in sodium ammonium alginate, in can-sealing compounds and in storage batteries where it is used for separating plates. In the production of insecticides, such as DDT and nicotine, it is used as a binding agent and it can be used for the same purpose in the fungicide, Bordeaux mixture, where it is said to be better than lime. It is also used as a binder for printer's ink and in cartridge primers.

It has extensive uses in medicine and dentisty. Alginates have largely replaced tragacanth and the natural gums in the manufacture of greaseless lubricating jellies, whilst according to Lillig (1928) alginoid arsenic and morphia alginate are other substances that can be employed. It is used as a suspending or stabilizing agent in a number of pharmaceutical preparations. Like agar, algin can be used as a binding material in the production of pills and pastilles and for emulsifying the petrolatum base of sulphanilamide ointments. Since alginates are not attacked by digestive juices of the stomach (Esdorn, 1934), if it is necessary to give a patient some substance that is to have an effect in the intestine, the material is enclosed in an alginate capsule which dissolves in the intestine. Gloess (1919) also recommends the use of alginic acid in the iodine form in place of cod-liver oil because it is said to be as efficacious but much less distasteful. Sodium alginate, when fed to rats, lowers the plasma cholesterol level. A recently reported additional use is algin's capacity to remove ^{90}Sr from the body and this could be of great significance in the future (Bhakuni and Silva, 1974; Skoryna *et al.*, 1964, 1965; Takagi, 1975). It is now largely replacing agar in the manufacture of dental moulds and models. The trade product 'Zelex' is used in Great Britain, and is said to be easier to work and it sets in 4·5 min. Another dental use is as a covering for dentures made of acrylic resin (Berger, 1961). In the USA a haemostatic agent, marketed as 'Hemo-pak', was produced in order to control bleeding in operations. It was made in a gauze-like form that could be left in the body where it gradually dissolved and was carried away by body fluids.

Algin is also probably the most useful seaweed product in the cosmetic industry (Berger, 1961). It is used as a base in creams, jellies, hair-sprays, hair-dyes etc. It is a valuable dispersal agent in shaving soaps and hair shampoos. The viscosity can be controlled by the amount of calcium ions added and it produces standard preparations which are transparent, white and almost odourless. Increasing use is being made of the organic derivatives, e.g. Triethanolamine alginate, propylene glycol alginate. The former compound has a wide use in dust protection creams for use in factories.

By itself it can be used for clarifying beer or as a water softener (Maass, 1962). When mixed with an inert siliceous material and concentrated sulphuric acid it makes an efficient colour-absorbent for decolourizing liquids. A potentially important use is as a means of sealing off porous formations in oil-well drilling, in dam foundations etc. Its successful use here involves a very low concentration of algin and its easy conversion to the gel form by the addition of formaldehyde. A very good mix for the purpose contains sodium alginate (algin) 59%, tricalcium phosphate 6%, di-ammonium phosphate 32% and Calgon 3% (McDowell, 1966). A low viscosity alginate is used in a 1% solution of the mixed product.

Some of the salts exert an oxidizing effect, e.g. sodium peralginate, and so they can be utilized for bleaching and washing. Soap to which the magnesium salt has been added can then be used in hard water or even in sea water. Crude algin can also be used as a water softener in boilers when it reacts with the scale-forming metallic ions to give an insoluble alginate, which forms flocculent masses that can be blown out of the boiler at intervals.

So far no mention has been made of any uses for alginic acid or alginates in agriculture. The wet seaweed contains much crude algin and there is the further possibility of using the crude salts. Very little work appears to have been carried out on this aspect (Waksman and Allen, 1934). It is known that the soil contains bacteria (*B. terrestralginicum, Alginobacter* and *Alginomonas*) which are capable of decomposing alginic acid (Steiner and McNeely, 1954). This decomposition is significant because in the process a primitive form of lignin is deposited in the soil in a form similar to natural humus. If the above researches are confirmed it means that algin, when put on the soil, will have a considerable effect upon both its chemical and physical properties. A summary of the uses of algin is provided in Table 6.8 and its uses as a gum are listed in Table 4.13.

7
Minor Uses of Algae and their Products

7.1 LAMINARAN

This is a reserve compound found in some quantity in fronds of the Laminariaceae and to a lesser extent in the rockweeds of the Fucaceae. It is a polysaccharide sugar with the general formula $(C_5 H_{10} O_5)_n$ and was discovered by Schmiedeberg in 1885. Barry (1939) and later Connell, Hirst and Percival (1950) and Percival and Ross (1951) showed that laminaran probably consisted of a branched chain of 20 β-D-glucopyranose units with 1, 3 glucosidic linkages, with the molecule bent in a spiral form. There are also mannitol end groups (2·4—3·7%). Subsequently it has been found that there are two forms, soluble and insoluble. The present situation is that laminaran exists in a reducing and non-reducing form, the latter possessing the mannitol end groups (Percival, 1969; Levring, Hoppe and Schmid, 1969).

$$G_1 \overset{\beta}{-} [_3G_1]_n \overset{\beta}{-}_6 G_1 \overset{\beta}{-} [_3G_1]_m \qquad G = \text{D-glucose}$$
$$\diagdown$$
$$1$$
$$\text{Mannitol}$$
$$2$$
$$\diagup$$
$$G_1 \overset{\beta}{-} [_3G_1]_n \overset{\beta}{-}_6 G_1 \overset{\beta}{-} [_3G_1]_m$$

non-reducing laminaran (= laminaritol, Dillon, 1964)

$$G_1 \overset{\beta}{-} [_3G_1]_n \overset{\beta}{-}_6 G_1 \overset{\beta}{-} [_3G_1]_m \overset{\beta}{-} 3G$$

Reducing laminaran (= laminarose, Dillon, 1964)

The soluble and insoluble forms differ in size of the colloidal particles and degree of branching, and it is thought that the soluble form contains a component that inhibits formation of large colloidal particles (Percival, 1964). The molecular weight of soluble laminaran, as prepared from *Laminaria digitata*, is about 5300, whilst that of the insoluble form obtained from *L. hyperborea* is about 3500 (Friedlaender,

Table 7.1 % Laminaran in brown algae (after Powell, 1964)

Species	% dry wt laminaran	
Agarum fimbriatum	3·16–4·16	
Alaria fistulosa	2·63–8·67	
leaf	2·95–8·67	
mid-rib	3·68	
sporophylls	3·39–22·4	
Costaria costata	2·63–4·67	
Cymathere triplicata	<2–5·37	
Desmarestia munda	2·5–5·3	
Desmarestia herbacea	3·15–4·66	
Desmarestia ligulata	6·30	
Fucus spp.	5·4–7·12	
Hedophyllum sessile	<2–3·36	
Sargassum muticum	4·62	
Pterygophora		
stipe	3·82	
frond	3·19–5·44	
Postelsia		
stipe	4·7	
frond	6·7	
Ecklonia radiata		
rib ‡	0–7·0	
stipe ‡	0–1·1	
lamina ‡	1·0–9·8	
Pleurophycus gardneri	<2–2·8	
Laminaria saccharina	1–24 (21·5 in loch)	Generally 10–30%
frond	<2·0–15·2	Generally 10–30%
tip	<2–29·6	Generally 10–30%
centre	2·97–33·9	Generally 10–30%
base	2·6–6·32	Generally 10–30%
L. cuneifolia	2·35–13·5	Generally 10–30%
L. hyperborea*	1·5–32·4	
Lessonia littoralis		
frond	4·6	
stipe	9·2	
Macrocystis		
frond	7·4	
stipe	5·8	
Nereocystis		
frond	<2–6·5	
stipe	4·5	
float	<2·0	

(*continued on the next page*)

Table 7.1 % Laminaran in brown algae (after Powell, 1964) (*continued*)

Species	% dry wt laminaran
Ascophyllum §	1·0−7·0
Fucus vesiculosus†	1·0−7·0
Fucus serratus‡	1·0−19·0
*Laminaria digitata**	0·75−16 (loch)
	0·75−27·5 (open)

*Black, 1950. † Kylin, 1915. ‡ Stewart *et al.*, 1961. § Jensen, 1966.

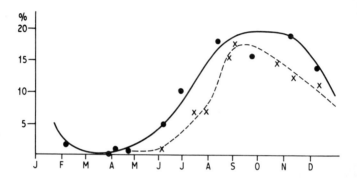

Fig. 7.1 Seasonal variation of laminaran in *Laminaria digitata*.

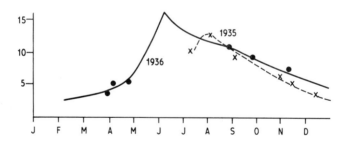

Fig. 7.2 Seasonal variation of mannitol in *Laminaria digitata*.

Cook and Martin, 1954). Both contain a small proportion of β−1, 6 glucosidic linkages (Fleming, Hirst and Manners, 1966). In possessing two components laminaran is comparable to alginic acid.

The amount of laminaran present in different algae varies considerably as Table 7.1 shows.

It appears that the richest sources of supply are the sporophills of *Alaria* and species of *Laminaria*. In the latter case plants from the open sea are rather richer

than those from lochs (Black, 1950), though in the case of open sea plants of
L. digitata, laminaran is absent for most of the year so that harvesting would have
to take place at the peak periods. In *L. hyperborea* the amount of laminaran present
in a plant increased with the age of the plant each year.

Investigations by Ricard (1931), Lunde (1937) and Black (1950) have shown that
the amount of laminaran varies according to the season of the year, e.g. in *Laminaria*
it reaches a maximum in the autumn and winter months and falls almost to zero in
the spring (Fig. 7.1). The exact time may depend upon habitat. Thus in *L. saccharina*
the maximum is reached in open sea plants in February but in loch plants in January.
According to Black (1949), laminaran in fucoids behaves in a somewhat similar
manner, reaching minimum values in *F. vesiculosus, F. serratus* and *F. spiralis* in
January and February, and again in September and October. This can apparently be
affected by region, because MacPherson and Young (1952), using *F. vesiculosus,
F. evanescens* and *Ascophyllum*, found little seasonal change. Jensen (1966) reports
minimum values for *Ascophyllum* in Norway as occurring in May. Depth at which
the sample is growing also appears to be significant. In the case of *Laminaria digitata*,
maximum contents were obtained from plants collected between 3–6m, and plants
down at 16 m were devoid of it. In the case of *L. longicruris* the content was 6–7·5%
at 6 m, 4–6·5 at 9 m and 1–4% at 18 m (Chapman and Craigie, 1978). The light
intensity at the latter depth is presumably inadequate for its manufacture. There are
a number of problems that remain unsolved so far as this compound is concerned.
Analyses must be conducted throughout the year in order to secure an adequate
picture and the effect of habitat needs to be more fully explored. Val and Montequi
(1951) have given an account of the isolation and extraction of laminaran as well
as the other workers quoted above.

Laminaran does not form any viscous solution nor does it gel. At present there
does not seem to be any major use, though Laminaran sulphate could gain therapeutical
significance as an anti-coagulant (Hoppe and Schmid, 1969), where it is said to be
one third as good as heparin (Hawkins and Leonard, 1958). It has also been stated
that grape sugar could readily be prepared from it on a commercial scale by acid
hydrolysis, but so far no attempts appear to have been made.

It is fairly readily decomposed by micro-organisms and it stimulates the micro-
flora of the bovine rumen to increased activity. Bacteria, actinomycetes and certain
fungi contain the enzyme laminaranase, which diffuses out in culture liquids. This is
in contrast to the degrading enzyme alginase which is not so widely distributed
(Chester, Apinis and Turner, 1956).

7.2 MANNITOL

Mannite or mannitol is a sugar alcohol which was first recorded in brown algae by
Stenhouse in 1844. It occurs as a cell sap food reserve in a number of brown algae,
and is quite abundant in the oarweeds. In earlier years it was only obtained

Table 7.2 Variations in mannitol content (after Mehta and Parekh, 1978)

Species	Locality	Month	% dry wt mannitol
Sargassum cinctum	Porbandar	December	11·53
Sargassum swartzii	Okha	November	5·61
Sargassum swartzii	Porbandar	December	11·11
Sargassum tenerrimum	Okha	November	3·56
Sargassum vulgare	Okha	November	3·79
Sargassum vulgare	Veraval	December	3·85
Sargassum vulgare	Porbandar	December	11·59
Stoechospermum marginatum	Okha	January	10·75
Stoechospermum marginatum	Okha	March	16·00
Stoechospermum marginatum	Okha	November	5·27

technically in small quantities (Lunde, 1937a and b). Hoffmann (1939) stated that mannitol is a 6 carbon sugar alcohol with the general formula ($C_6H_{14}O_6$). D-mannitol and its monoacetate have been isolated from *Sargassum natans* (Martinez *et al.*, 1962). From *F. vesiculosus*, Lindberg (1953) isolated 1-D-mannitol β-D-glucopyranoside and 1, 6 D-mannitol di β-D-glucopyranoside. It would seem therefore, like so many of these other algal products, there are at least two if not three chemical components in mannitol and a simple formula is not sufficient. Methods of extraction have been described in Sorensen and Kristensen (1950), Val and Montequi (1951) and Evitushenko (1965). The first named took out a USA patent in which the mannitol is extracted by water and then an aliphatic carbonyl compound, e.g. formaldehyde, added to give a triacetal compound. This is separated off and hydrolysed by dilute acid and the mannite separated. Because mannitol is soluble in water, dried algae wanted for its extraction should not be washed in fresh water (Garcia-Pineda, 1951).

Species of *Laminaria* contain maximum quantities in late summer and early autumn with minimal values in early spring (Fig. 7.2). *L. hyperborea* may have 9% in May and 18% in July, and in *L. digitata* it can range from 6% in March to 26% in August (Levring, Hoppe and Schmid, 1969). Cardinal and Breton-Provencher (1977) also found a maximum in August but the value (9%) was much lower. A comparable maximum was also recorded by them for *L. longicruris* in the same month (Fig. 6.2). Black reported the same seasonal trend in the British fucoids, but in Canada, Macpherson and Young (1952) found a maximum in spring and summer with minima in autumn, whilst Norwegian *Ascophyllum* exhibited little variation (Jensen, 1966). In *Ecklonia,* maximum values are found in the autumn and the minimum in late winter. In India the mannitol maximum in *Sargassum wightii* and *Turbinaria conoides* occurs between May and August (Rao, 1968). In *T. ornata* the mannitol content is high in the early stages of growth and is less when growth is maximal (Rao and Kalimuthu, 1972).

It is now quite evident that the mannitol content varies not only with the species but also there is a seasonal fluctuation which has a relationship with the reproductive

Table 7.3 Mannitol in brown algae (after Black, 1950)

Species	% dry wt mannitol
Laminaria digitata frond	
open sea	4–26.0 (4–19.5, Lunde, 1937; 9.35–19.4, Caraes, 1968)
loch	3.5–22.5
Laminaria saccharina	
open sea	6–21.0 (25, Lunde 1937; 5–23, Levring *et al.*, 1969)
loch	7–23
Laminaria hyperborea	6.1–25.7 (5–18, Levring *et al.*, 1969)
Fucus vesiculosus	8.3–16.1 (9, Levring *et al.*, 1969)
Fucus evanescens	
Nova Scotia	4.4–12.3
New Brunswick	5.6–10.9
Ascophyllum	
Nova Scotia	3.8–10.1
New Brunswick	5.8–9.4
Europe and Iceland	7–12 (10†; 3.17–10.2, Munda, 1964)
France	6.1–9.7 (Caraes, 1968)
Ecklonia radiata *	
stipe	6.7–13.6
rib	7.4–16.7
lamina	8.6–16.6
Durvillea antarctica	
disc	10.6–12.3 (Moss and Naylor, 1954)
stipe	11.9–13.3 (Moss and Naylor, 1954)
frond	6.7–9.9 (Moss and Naylor, 1954)
Durvillea willana	
stipe	8.5 (Moss and Naylor, 1954)
frond	6.3–11.4 (Moss and Naylor, 1954)
Sargassum spp.	
(Pillai and Varier, 1953)	7.3–9.4 (3.56–11.59, Mehta and Parekh, 1978)

* Stewart *et al.*, 1961.
† Jensen (1966) for Norwegian samples.

cycle, because the amount is low after reproduction (Mehta and Parekh, 1978). Table 7.2 gives mannitol variations related to place and season in Indian brown algae.

In the case of *Stoechospermum*, further work on seasonal changes would seem necessary because Kalimuthu *et al.* (1977) found that the mannitol content in plants from Padumadam reached a maximum value in October with secondary peaks in May and June.

Species of *Laminaria* can contain more mannitol than other algae though *Ecklonia radiata* may have a considerable amount (Table 7.3). Open sea plants of *L. digitata* were richer than loch plants, whereas the reverse was true for *L. saccharina*. The mannitol content increases with depth to 7–8 m after which it decreases, though this is not true for *L. longicruris* (Chapman and Craigie, 1978) where there is scarcely any fluctuation with depth. Analyses of *Durvillea* and *Ecklonia* indicate that there is a variation in the different organs. In the latter genus, seasonal changes in the frond content precede those of the rib and stipe. All these fluctuations can be accounted for on the basis that mannitol is a food reserve which is accumulated during the summer and used up during the winter.

Lunde (1937) considered that mannitol has many promising technical possibilities, but costs of extraction will need to be reduced still further. It can be used in pharmaceuticals (tablets, diabetic foods, etc.), paints, leather and in the preparation of lacquers, whilst plastic products obtained from it are said to be better than those obtained from glycerine. Mannitol can also be nitrated to form nitro-mannite, a powerful explosive similar to nitro-glycerine. At present it is manufactured in USA, UK, France (*Laminaria digitata* main source) and Japan. Production from kelp costs one third the amount of making it synthetically (Durrant, 1967).

7.3 FUCOIDAN (FUCOIDIN)

This compound, which is probably the calcium salt of a carbohydrate ethereal sulphate (see over), is found in the intercellular mucilage of rock weeds and also in *Ecklonia radiata*. Quantities that have been recorded are as follows: *Fucus spiralis*, 9–11%; *Ascophyllum*, 6–8%; *Pelvetia canaliculata*, 20%; *Laminaria*, 5–20% (lower to upper sublittoral respectively) (Levring, Hoppe and Schmid, 1969). It has been found that littoral rockweeds contain a higher quantity than do sublittoral plants.

After extraction fucoidan is extremely viscous, even in a 0·05% solution. The amount fluctuates during the year, figures for fucoids (Speakman and Chamberlain, 1944) showing a maximum of 9% dry weight in the autumn and dropping to about 4% in the spring. However, in *Laminaria* species the maximum occurs in the spring. In *Ecklonia* the quantity is much smaller and ranges from 0·4–1·4%, the maximum being recorded in mid-winter (Stewart *et al.* 1961). Percival and Ross (1950) obtained a highly purified sample and proposed the general formula $(C_6H_9O_3, SO_4, Ca_{1/2})_n$, but several structural variations have been considered (Mantner, 1954). Anno *et al.*, (1968, 1972) report the existence of two sulphated sugars in fucoidan made from *Pelvetia wrightii* as well as a third non-sulphated fraction. One of these

$$F = L\text{-fucose}$$
$$S = S\overset{\nearrow O}{\underset{\searrow O}{-}}OH$$

$$2F_1 \xrightarrow{\alpha} 2F_1 \text{---------} 4F_1 \text{---------} 2F_1 \xrightarrow{\alpha}$$
$$2 \text{ or } 3 \quad 4 \qquad\qquad 4 \qquad\qquad\qquad\qquad\qquad 4$$
$$\quad | \quad | \qquad\qquad\quad | \qquad\qquad\qquad\qquad\qquad\quad |$$
$$\quad S \quad S \qquad\qquad\quad S \qquad\qquad\qquad\qquad\qquad\quad S$$

is L-fucose₄ SO₄. Percival (1964) considered that the above units (see diagram, p. 232) were at least present.

Within the molecule there is said to be 38·3% of SO₄, 56·7% of L-fucose and 8.2% of metallic ion. Fucoidan from *Himanthalia* contains 57% fucose, 4% galactose, 15% xylose, 3% uronic acid; from *Pelvetia canaliculata* 35% fucose, 3% glucose, 22% galactose, 6% arabinose. In eleven algae the fucose content had a range of 31–72%, galactose 5–31% and xylose 3–29%, so that there may well be a family of fucoidans. Up to the present, however, no great effort has been made to extract it in quantity for commercial use. In the United Kingdom it is obtained on a small scale by a chloroform extraction process. It is believed that one of its potential uses would be as a blood anti-coagulant.

Finally, two other compounds that could be mentioned here are fucosan, a tanning substance obtained from *Ascophyllum* and *Sargassum,* and Ginnanso, a Japanese adhesive material derived from *Iridaea cornucopiae, Turnerella mertensiana* and *Rhodoglossum pulchrum.*

7.4 ALGAE IN MEDICINE

From historical times seaweeds have been employed for medicinal purposes, but it has been difficult to prove that any effect is, in fact, due to substances in the algae. Reference has already been made to the uses of carrageenan (p. 133), agar (p. 190) and alginates (p. 220) for medicinal purposes. First mention of such uses is probably that in the Chinese herbal 'Pen Tsae Kan Mu' (sixteenth century) where certain algae are listed as a cure for goitre.

The oarweeds and wracks are used in various remedies because of the iodine that they contain, e.g. *Fucus* is employed in the form of an infusion in cases of goitre. Another way, used over many years, of giving the iodine is in the form of kelp pills or to use the kelp ash or charcoal, which medicinally is known as 'Aethiops vegetabilis'. This was recommended by Meier (1935) as a remedy for Basedow's disease. The great consumption of seaweed by the Japanese is reflected by the low incidence of goitre in the country, which is related to the high iodine content of the algae, especially in kombu. The average consumption of dry seaweed per day per person in Japan is about 10 g (Kirby, 1953). Kombu contains about 240 ppm iodine and nori 18 ppm in contrast to other marine shell fish where it rarely exceeds 3 parts. Species of *Laminaria* have been imported into Malaya as a cure for goitre and it has been used in a New Zealand school (Moore, 1941).

Yarham (1944) says that the brown liquid which can be extracted from *Ascophyllum* is useful in the treatment of sprains, rheumatism and allied complaints. For those who want to lose weight there are special slimming teas which contain sea-weed iodine, usually extracted from one of the wracks. Hoffmann (1939) suggests that the effect of these might be much stronger if *Laminaria* were used because the iodine content is greater. *Laminaria* powder was given by Chaveaux (1927) in France in cases of consumption in order to strengthen the body and help to overcome the disease.

The use of an algal phycocolloid (Algasol T 331) has proved efficacious in the treatment of oncologic patients, producing a good recovery in 68% of 162 patients (Claudio and Stendardo, 1966). It also has a beneficial effect on patients with chronic bronchitis or emphysema (Cavi and Guiseppe, 1974).

Most of the algae are used in the form of drugs, but there is at least one surgical use. Short pieces of the stems of *Laminaria,* principally *L. hyperborea,*, known as 'Stipites Laminariae', are employed in surgery for widening fistulae and wound entrances. This use is, of course, based upon their large swelling capacity when moistened. In Germany most of the *Fucus* drugs and *Laminaria* stems were chiefly imported from France or Norway. Germany evidently intended to produce her own material because Esdorn (1934) reports that in 1931 Heligoland provided about 10 000 kg of *Laminaria* stems and in 1932 slightly more.

The Corsican worm moss, which is the red alga *Alsidium helminthochorton* of the Mediterranean, is regarded as an efficient vermifuge, and appears to have been first used in 1775. Before that, coralline algae (*Jania rubens* and *Corallina officinalis*) had been used. *Alsidium* is not very abundant, and in its preparation it became mixed, at first unintentionally and later intentionally, with other seaweeds. Some samples of the drug therefore contain very little *Alsidium* and it may even be absent. It is reported by Garcain (1906) that the drug is only effective so long as *Alsidium* is present, even though in small quantities. It is clear that our knowledge of the efficiency and action of this drug would be benefited by further study, because the active principle in the alga is a resinous material that apparently has not so far been carefully examined. An extract of the red seaweed *Digenea simplex* (Makuri in Japan) from Asia is also on the market under the name of 'Helminal' for use in the treatment of worms. Doctors say that this preparation is effective in some cases but not in others.

The active principle, first identified as α- (α-D-mannosido)-D-glyceric acid (Kawaguchi *et al.,* 1952) has now been shown to consist of α-kainic and α-allokainic acids. The formula for kainic acid is $C_{10}H_{15}NO_3$ (Fig. 7.3). *Digenea* is mainly collected from Pratas Island by divers at 5–20 ft (1·55–6·1 m). After collection the weeds are dried, but must not be washed, and are then shipped to Hong Kong or Taiwan (Kirby, 1953). Perrot and Gatin (1912) and Lillig (1928) quote the following additional algae that have been or are used as vermifuges: *Hypnea musciformis,* especially employed in Greece and Turkey, two species of *Chondria,* dulse (*Palmaria palmata*), *Sargassum vulgare, Ulva* sp. in Cuba and also a green fresh water alga (*Rhizoclonium rivulare*). In the rhodophyte *Chondria armata* the effective substance is domoic acid ($C_{15}H_2, O_6N-2H_2O$) (Takagi, 1975) (Fig. 7.3c). Chondriol, obtained from *C. oppositiclada,* with the formula $C_{15}H_{18}BrClO_2$, has an anti-viral activity (Baker and Murphy, 1976). This compound may also be the agent responsible for controlling the Herpes virus (Berg and Ehresmann, 1977). In fact an anti-herpes polysaccharide with the presence of α-D-glycosyl and sterically related residues has

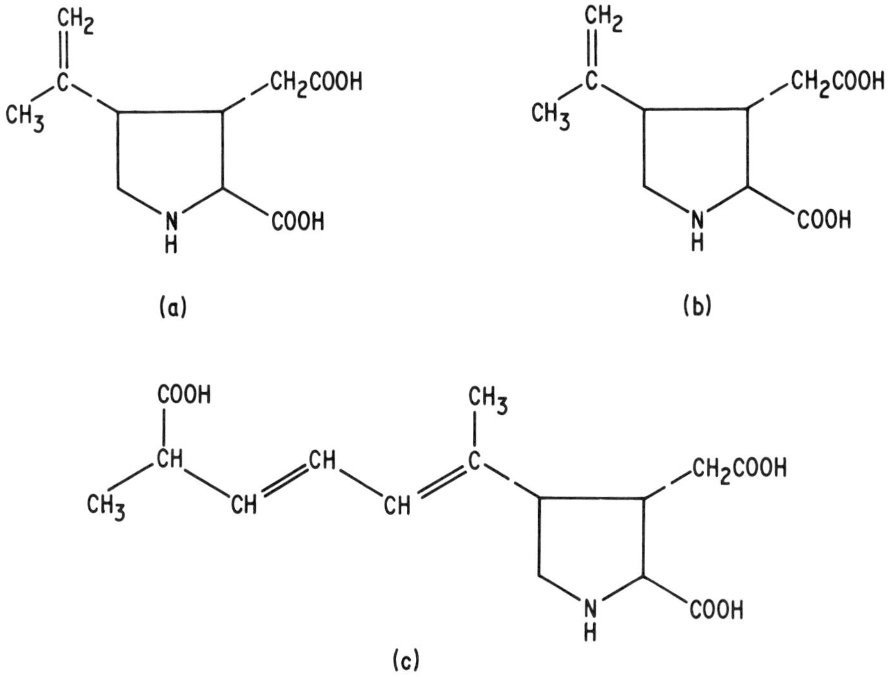

Fig. 7.3 Structural formulae for (a) L-α-kainic acid; (b) L-α-allokainic acid; (c) domoic acid.

been reported from a number of Rhodophyta (Hatch et al. 1979). Some 29 Pacific Rhodophycean species were examined and the extracts were found to be 99% effective in stopping viral multiplication and 50% effective in controlling spread. The polysaccharide apparently blocked the viral absorption point in the cell membrane.

Water extracts of certain algae, e.g. *Macrocystis, Gelidium cartilagineum, Pelvetia fastigiata* and *Egregia laevigata,* are toxic to mice (Habekost, Fraser and Halstead, 1955). There is no evidence, however, that consumption of alginate products at normal useage levels (see p. 216) presents any toxicological problems (McNeely and Kovacs, 1975).

A number of seaweeds have also been named by Lillig (loc. cit.) as useful in cases of lung disease and scrofula. In Japan *Gelidium cartilagineum* is used for this purpose, whilst in the Mediterranean, two brown seaweeds, *Dictyopteris polypodioides* and *Stilophora rhizoides,* have a similar use. A species of the brown genus *Sargassum* (*S. linifolium*) is said to be used in India in cases of bladder disorder. This is essentially a Mediterranean species and why it should be employed in India is a matter for speculation. The history of its introduction might perhaps provide a clue. Another species of the same genus (*S. bacciferum*) is reported to be employed in South America as a cure for goitre and kidney disease. In China *Laminaria bracteata*

(= *L. japonica* because it is apparently sometimes prepared like kombu, see p. 79) is used in the form of a viscous solution, known as 'Haitai' or 'Kwanpu', in menstrual troubles, whilst sugar wrack (*L. saccharina*) is used in India against goitre and in the Himalayas against syphilis. The mucilage of *Cutleria multifida* is said to be effective with stomach ulcers (Dizerbo, 1964).

A discovery by Elsner, Broser and Bunger (1937), which may yet prove to be important from the medical point of view, is that a water-soluble extract can be obtained from carrageenan which, even in very great dilution, acts as an anti-coagulating blood compound (p. 133). Agar may also have a use in this respect. Iridophycan, the carbohydrate-sulphate ester from *Iridophycus flaccidum* (p. 143) has a similar property. In solution the substance occurs principally as the sodium salt. The beautiful red seaweed *Delesseria sanguinea* also possesses a strong anti-coagulating action which is as good as or better than heparin. The effect of this extract can be stopped immediately by the injection of thionin. Burkholder *et al.* (1961) tested a number of West Indian algae for antibiotic activity and found about six that had a significant effect. These were *Chondria littoralis, Falkenbergia hillebrandii, Murrayella periclados, Wrangelia* spp. *Laurencia obtusa* and *Dictyopteris justii.* Rao *et al.* (1977) recently tested a number of Indian algae for antibacterial activity and they found that extracts from *Dictyota dichotoma, Gelidiella acerosa, Gracilaria corticata, Enteromorpha* spp. and *Padina gymnospora* were effective against *Bacillus megatherium* and *Staphylococcus aureus,* but were not effective against gram -ve bacteria. The extracts were active against a number of bacteria and the activity exceeded that of well-known antibiotics such as penicillin. A number of other algae have been tested for the presence of antibiotic compounds using hot or cold water, ether or acetone for the extraction. Some 11 algae, including *Sargassum thunbergii, Chrysymenia wrightii, Ulva pertusa, Undaria* and *Amansia* had active compounds (Levring, Hoppe and Schmid, 1969). *Laurencia obtusa* contains the compound oppositol, $C_{15}H_{25}BIO$, which exerts an antibiotic effect (Baker and Murphy, 1976). The blue-green alga, *Lyngbya maiuscula,* may also contain an anti-bacterial agent, but since the extract may also induce a dermatitis (Yamazoto, 1976), it is not likely to be useful.

A number of marine algae contain sterols, e.g. ergosterol, chondriosterol, poriferasterol etc., which all possess hypocholesterolemic activities. They can bring about depression of blood pressure in human atherosclerosis as well as lowering cholesterol levels in rabbits and rats. Species that heve been confirmed as possessing these steroids include *Enteromopha compressa, E. prolifera, Ulva pertusa, Monostroma nitidum, Laminaria* (6 spp.), *Costaria costata, Sargassum muticum, Fucus gardneri, Undaria* (2 spp.), *Arthrothamnus bifidus, Kjellmaniella* (2 spp.), *Heterochordaria abietina, Chondrus* and *Porphyra.*

In the species of *Laminaria* and *Undaria* (Takagi, 1975) the effective compound was laminine for which two chemical formulae have been given. Bhakuni and Silva (1974) give $C_{13}H_{24}N_2O_{10}$ whilst Baker and Murphy (1976) give $C_9H_{22}N_2O_3$. Further work is clearly needed to establish which is correct. Bhakuni and Silva (loc. cit.) suggest that laminine monocitrate could be a potentially useful pharmaceutical agent, whilst Baker and Murphy (loc. cit.) state that laminine has both antihelminthic and anti-hypertensive properties.

In the genus *Heterochordaria* the effective compounds appear to be phaeophytin and phaeophorbide. The most effective compound obtained from *Monostroma nitidium* is homobetaine though recently another, slightly less effective material, ulvaline, has also been extracted (Abe and Kaneda, 1975). It is evident that quite a number of organic compounds are involved and further study is clearly needed.

Species of the green algal genus *Caulerpa* contain caulerpin ($C_{24}H_{18}N_2O_4$) and caulerpicin ($C_{48}H_{87}O_2N$) which function as mild anaesthetics and which could therefore well have a clinical value. If this use is confirmed, *Caulerpa* species are widespread; beds are often extensive and they would probably lend themselves to cultivation (Chapman, 1978). Other antifungal and antibiotic compounds are pachydictiol A ($C_{20}H_{32}O$) from *Pachydictyon coriaceum*, isozonarol ($C_{21}H_{30}O_2$), zonarol ($C_{21}H_{30}O_2$), both from *Dictyopteris zonarioides*, and squalene ($C_{30}H_{50}$) from *Fucus vesiculosus*. The last named would certainly be available in adequate quantity from supplies of raw material. Trunova and Grintal (1977) found that of 15 species tested, *F. vesiculosus, F. serratus* and *Laminaria digitata* were most effective against *Salmonella* and the enteric bacteria. Squalene could again be the essential agent.

Among older remedies in *British Naturalist*, Turner recorded that in the middle of the 18th century dulse (p. 68) was used in Skye as a means of inducing sweating during an attack of fever. The weed was boiled in water and a little butter was added: prepared in this manner it also acted as purgative.

In Hawaii, the red seaweed *Hypnea nidifica* is used as a remedy for stomach toubles, whilst in the same place a cathartic is prepared in the form of an infusion from the small tufted red seaweed called *Centroceras clavulatum*, which is widely distributed in warm waters. A somewhat peculiar remedy, if remedy indeed it is, for headaches is reported from Alaska. Here the Indians of Sitka take the stipe of the bull kelp (*Nereocystis*) and place the thin end in one ear and put the bulb on a hot stone so that steam is generated and passes up the hollow stipe. Tseng (1944) has suggested that a seaweed diet could be a preventative for hay fever, but there appears to be no scientific evidence that marine algae would either prevent hay fever or give relief from it (Kirby, 1953).

Another primitive medicine is recorded in Maori lore. These people, according to Goldie (1904), used to employ *Durvillea* ('rimuroa') as an antidote to scabies: this was a common disease in the race partly because of close contact with their domestic animals. Another brown species was fermented by the Maoris with the juice of the poisonous Tutu shrub and then used as an aperient. According to Goldie the plant was a *Laminaria*, but this cannot be so as the genus does not exist in New Zealand. The reference gives no real clue to the nature of the alga, but it may have been another *Durvillea* species or *Ecklonia radiata*.

As long ago as the time of the Caesars, Roman ladies used a rouge extracted from the *Fucus* wracks, and a similar extract mixed with fish-oil is used today by the women of Kamchatka as a means of reddening the face. The word *Fucus* is in fact derived from the Latin word for rouge.

Because of the relative ease of mass cultivation, planktonic forms of algae ought to be investigated. At present only one species, *Gonyaulax catenella*, appears to have been studied in detail. This species produces the compound saxitoxin ($C_{10}H_{17}O_7O_4$) which acts as a block to nerve conduction. It is possible that other species of the genus also contain such toxins and this has recently been confirmed for *G. tamarensis* and *G. acatenella* from Chesapeake Bay (Winton, 1977). There is a related compound in *G. monilata*. Another planktonic form, *Prymnesium parvum*, is said to give rise to two toxins, one affecting the central nervous system and the other functioning as a haemolytic and anti-spasmodic agent. Recently Gauthier *et al.* (1978) have reported on the antibiotic function of two diatoms, *Asterionella japonica* and *Chaetoceras lauderi* and its modification by a dinoflagellate, *Procentrum micans*.

7.5 OTHER INDUSTRIES AND USES

Apart from medicine, algae or algal products are employed in small quantities in a number of other industries. In New Zealand, brewers have stated that *Gigartina decipiens* is much more efficient than Irish Moss as a clarifying agent, but in other industries it is not suitable as a substitute for *Chondrus*.

In the textile industry Irish Moss, funori, agar and algin are all valuable sizes and 'fixing' agents for mordants whilst *Chondrus* has been employed in wall plasters (Haas, 1921). *Fucus*, besides providing a rouge for the ladies, has also been utilized as a source of a red dye, whilst alginic acid after treatment with nitric acid is said to yield a brown dye which can be used in alkaline solutions.

One might also mention here the use of the bootlace weed, *Chorda*, on the French coast to make a kind of string. It was possibly used for a similar purpose in the Orkneys and Shetlands, where it is still known as sea cat-gut, or 'Lucky Minny's Lines'. These names are, at any rate, highly suggestive. On the opposite side of the world, the Indians in Alaska take the long stipes of the bull kelp, *Nereocystis*, and after a period of washing, drying and stretching, use them as fishing lines.

In the ceramic or pottery trade, soda from kelp was once important in glazing. Although the kelp industry no longer exists, seaweeds are employed for other purposes in ceramics. Thus, funori is used for decorating porcelains in Japan, whilst algin is valuable as a binder and plasticizer.

Paper manufacturers also find a use for seaweed products. The Chinese employ a mixture of agar and lacquer in order to strengthen paper, whilst in western countries agar is used alone as a coating for certain types of paper. At one time there was a plan to make a high-class paper from fibres of the *Yucca* plant, using kelp for binding and filling purposes. Irish Moss is also employed in the process known as the 'marbling' of paper, especially for the cut edges of books, because it does not cause the pages to stick together. The fresh-water alga (*Chaetomorpha* sp.) provides a product known as 'pelt' which is used in the West Indies as a packing material.

Yarham (1944) notes that the studios of Radio City in New York are insulated with specially packed seaweed, though he does not state whether it has undergone any manufacturing process. Some types of alga when washed and dried are employed as a cheap stuffing for furniture, though the marine 'grasses' are rather more useful for this purpose.

Algal products, mainly from red seaweeds, have been suggested as substitutes for gelatine in the preparation of photographic material, but they have never proved satisfactory though quite reasonable agar films can be made (see p. 192). It has also been suggested that they could provide a good source of galactose (Black and Cornhill, 1954).

Among other and more varied uses there is the Japanese custom of taking the brown seaweeds *Sargassum enerve* (Hondawarra), *Eckloniopsis radicosa* (Antokume) (in the literature on this subject it is usually referred to as *Laminaria radicosa*), and *Eisenia bicyclis* (Arame) for New Year decorations. The main reason for using these species appears to be that they turn green when dried. The stipes of the giant alga *Lessonia* in South America, which may be as thick as a man's thigh, form very useful knife handles because when dry they become as hard as bone. In southern California the stipes of *Pelagophycus* have a similar use in the manufacture of curios, but the trade is not considerable.

The literature only mentions one use of seaweeds in connection with alcoholic liquors. This refers to Kamchatka, where the natives regularly use dulse (*Rhodymenia*) for the production of a somewhat evil-tasting (at any rate to western palates) liquid. The alaskan Indians also produce an alcoholic drink called 'hoochenoo', but, though they do not use a seaweed to make it, they employ the hollow stipes of *Nereocystis* to form a worm condenser in which the distilled liquid is cooled.

The bull kelp of New Zealand, *Durvillea antarctica,* possesses an internal air tissue which is split by the Maoris and then used as a leathery bag in which 'mutton birds' can be preserved. The same species, when cleaned out, is used as water bottles by the natives of South America. The early New Zealand settlers used to burn the plant, pulverize it and mix with water when it served as a substitute for Chinese ink (Miller, 1938).

A potential use of unicellular algae is in the maintenance of life in space ships. For this purpose cultures of unicellular green algae, e.g. *Chlorella,* should be ideal. One man needs 600 l per day of oxygen and the amount of algae necessary to produce this will consume 720 l per day of CO_2. In the process 600 g per day of dried algae will be made and this can be consumed as a food. Freeze-dried *Chlorella* contains approximately 56% crude protein, 8% fat, 18% carbohydrate and 3% fibre. Making this food digestible is a problem that has yet to be fully solved. Rats can digest up to 72% of the carbohydrate and up to 86% of the protein (Myers, 1964). In order to manufacture the daily output of algae and oxygen, 800 W of light would be necessary and at present this provides the major technical problem. In both China and Russia *Chlorella* harvests have been used on a pilot scale as a hog food (Hua *et al.,* 1962;

Khokhlov, 1961). It has also been used in China as a food for sick persons; it is reported their recovery was hastened and their weight restored.

In concluding this chapter some reference must be made to the use of diatoms. These are minute brown algae living in fresh or marine waters which have a hard silica coat covered with very fine markings. These striations are so fine that they are used (two species in particular) by lens manufacturers in order to test the definition and angular aperture of microscope lenses. Another use for these algae is associated with fossil species. At some periods in the history of the earth diatoms were so abundant that they accumulated under certain conditions to form deposits which today are known as 'kieselguhr'. Kieselguhr is a fine earthy material composed of the silicified coats of these small organisms. Deposits have been found and are worked in Auvergne (France), Algeria, Bohemia, Virginia and Mount Lompoc in California (USA) and Australia. The powder is very hard and so it is used for the cleaning and polishing of metals, though the amount involved here is very small compared with its other applications. Its most important use is probably in the preparation of dynamite. The liquid nitro-glycerine, from which dynamite is made, is by itself very unstable and liable to explode on the least provocation. When, however, it is absorbed by kieselguhr it becomes quite safe to handle and it is then known as dynamite. Another extensive industrial use of kieselguhr is also found in the filtration of liquids, especially those of sugar refineries, whilst yet another important application is in the insulation of boilers and blast furnaces. For the latter purpose it is either used in the form of bricks or as a loose jacketing powder. Above a temperature of $1000°F$ ($537.5°C$) its efficiency is much greater than that of asbestos because it is more resistant to shrinkage. Considerable quantities of this material are excavated in the USA, the average annual production between 1933 and 1935 being 244 342 tons (248 100 tonnes). An unusual use for kieselguhr is found in eastern Europe and Asia where it is added to the flour because it gives a feeling of fullness. On account of this usage some of the races have been given the name of earth-eaters.

8
Mariculture of Seaweeds

The term 'mariculture' has different meanings to individual workers. This chapter will consider mariculture within a rather narrow definition, viz. the growing of marine algae in artificial environments often enriched by the addition of nutrients. Such systems may involve structural modifications, e.g. nets, tanks, substrates etc., but the utilization of such modifications in natural environments without the additional nutrients is not regarded as strict mariculture. The increase of substrate area, especially for attachment, and hence the increase in biomass production per land area is the principal behind this work. Mariculture is aimed at increasing the biomass yield per substrate area and achieving this under controlled or semicontrolled conditions. So far mariculture efforts have followed two distinct paths: the culture of macroscopic algae (e.g. *Eucheuma, Gigartina, Hypnea, Chondrus*) for direct commercial use, and the culture of planktonic algae for use as food for herbivorous animals (e.g. shrimps, oysters). This latter may be regarded as mariculture of a simple one-step food chain (Fig. 8.1).

8.1 NATURAL ENVIRONMENT CULTURE

This is the simplest form of 'mariculture' (Neish, 1976), and is aimed at increasing the yield per land area in natural bays and areas. Since available substrate is often a limiting factor, much of the work has concentrated on increasing substrate area with additional nets, rocks, concrete blocks and fences. This habitat manipulation has been the principal form for increasing algal yield in Asia, notably with *Laminaria* (Cheng, 1969; Hasegawa, 1976; Okazaki, 1971), *Cystoseira* (Gros and Knoepffler-Péguy, 1978), *Gracilaria* (Shang, 1976), *Porphyra* (Bardach *et al.,* 1972; Miura, 1975), *Eucheuma* (Doty, 1973; Doty and Alvarez, 1975; Parker, 1974), *Undaria* (Saito, 1975) and *Iridaea* (Mumford, 1977). Other approaches have involved the storage in seawater tanks of harvested algae. This method has been successfully employed by Atlantic Mariculture Ltd, New Brunswick, with *Palmaria* (*Rhodymenia*) (Neish, 1976). *In situ* management of existing beds to prevent overharvesting is of course an

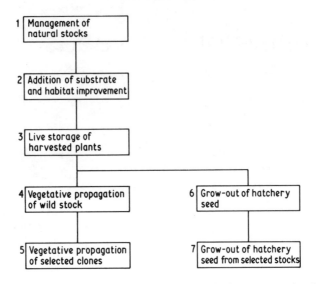

Fig. 8.1 Some broadly defined marine plant mariculture strategies. Increasing control over raw material supply and increased investment in capital and labour generally follow in the progression from top to bottom of the diagram. Any or all strategies may be employed for the production of any given species, and propagation of clones would be possible only with perennials (after Neish, 1976).

obvious component of any mariculture programme. However such an approach does not necessarily improve yield, so much as maintain it. The current trend therefore is towards artificial culture, or mariculture (see also Délépine, 1978).

8.2 ARTIFICIAL ENVIRONMENT CULTURE

As the name implies, this involves the culture of algae in ponds, impoundments or raceways using raw seawater, with or without added nutrients. It is these forms of culture conditions that we prefer to regard as mariculture. Mariculture received its main impetus in the early 1970s from three independent groups: the Atlantic Regional Laboratory in Halifax, Nova Scotia, headed by A.C. Neish; the Woods Hole group in Massachusetts headed by J.H. Ryther; and the Southern California group headed by W.J. North. Before considering the current status of the art it is advisable to consider some of the underlying principles, many of which have been diagrammatically illustrated by Neish (1976) (Fig. 8.2). One of the most important features is the ability of the algae in question to propagate vegetatively. Vegetative propagation enables workers to build up a standing crop of plants much more rapidly than dependence upon sexual or asexual reproduction, where growth from the single celled spore or zygote is time consuming, even in annuals, and where there would be a high

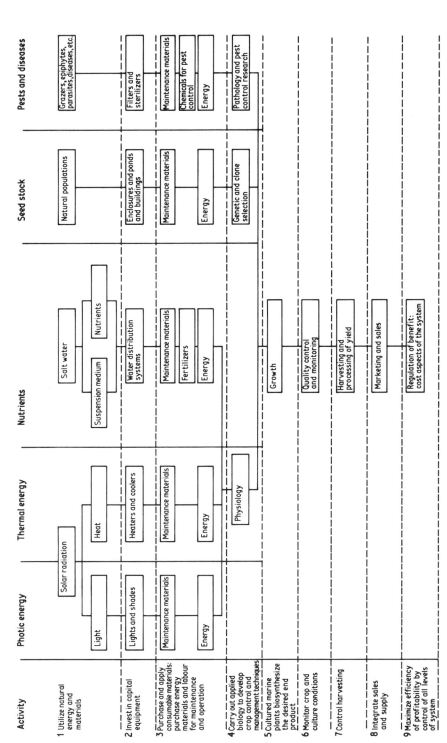

Fig. 8.2 Some key aspects of intensive marine plant mariculture (after Neish, 1976).

level of unpredictability in maintaining crop levels. In addition to those algal species that propagate principally by vegetative reproduction in the natural environment (e.g. marsh fucoids, pelagic *Sargassum,* marsh *Gracilaria*), *Chondrus* (Shacklock, 1975), *Eucheuma* and *Gigartina exasperata* (Waaland, 1977) all propagate well vegetatively. However, not all algae have been shown to propagate vegetatively. Two notable examples are *Iridaea* (Waaland, 1977) and *Palmaria* (*Rhodymenia*) (Neish, 1976). One of the most imaginative extensions of this work has been the application of cloning to vegetative reproduction. By this procedure one can select out clones with maximum yield or particular chemical characteristics. The potential for this becomes apparent when it is realised that the type of carrageenan in *Chondrus* (Chen et al., 1973; McCandless et al., 1973; McCandless and Craigie, 1974); *Gigartina* (Pickmere et al., 1973, 1975; Parsons et al., 1977), *Iridaea* and *Gigartina* (Waaland, 1975) varies between the tetrasporophytic or gametophytic generation. *Eucheuma,* however, (Dawes et al., 1977a and b) appears to lack this specificity though species may differ. Cloning has been practised principally with *Chondrus* by workers at the Canadian National Research Council Unit at Halifax (Allen et al., 1971; Shacklock et al., 1973, 1975). Their most notable isolate is the fast growing strain T-4 of *Chondrus.* More recently Waaland (1977) has isolated a strain M-11 of *Gigartina exasperata.* The advantages of cloning are obvious and it is hoped that this area of research will receive more attention.

The principal variables most likely to affect yields are incoming solar radiation, ambient temperature, carbon dioxide, nutrient supply, salinity, biological grazers and epiphytes, algal density and mechanical features such as mixing tank shape and water movement. It is generally accepted that a mariculture system must be a low-technology system, with low energy input. Usually no attempts are made to provide artificial lighting or heating, except possibly through the intermediacy of greenhouse conditions (e.g. Edelstein et al., 1976), or by warm seawater supplied from a power plant cooling system.

This means that there will be seasonal fluctuations which are illustrated in Table 8.1, taken from Waaland (1977). In this table it should be noted that seawater temperatures ranged from 7°C (late winter) to 14°C (mid-summer). In temperate latitudes, daily solar radiation shows a sevenfold fluctuation between winter and summer. Grazers and epiphytes could also present difficulties, though, up to the present, herbivore grazing has not been a problem. Under the usual conditions of seawater collection and retention times (normally a few days) very few herbivores have a chance to enter the system or attach themselves to the algae. Nicotri (1977) has indicated that moderate grazing by herbivores on *Gracilaria* and *Neoagardhiella* poses little problem, and would only be of concern in those uncommon instances when grazer density is high.

Similarly epiphytism, especially of diatoms, does not appear to be a major problem, although Mathieson and Prince (1973) have indicated that aeration can increase the degree of epiphytism on *Chondrus.* Prince (1974) has suggested however, that rapid turnover of the medium is all that is needed to eliminate *excessive* phytoplankton growth.

Table 8.1 Seasonal biomass per surface area from 1976–1977, tests at West Point, Seattle, Washington*

Month	% summer optimum biomass	Gigartina exasperata fresh wt (g m^{-2})	Iridaea cordata fresh wt (g m^{-3})
January	12	576	480
February	25	1200	1000
March	50	2400	2000
April	70	3360	2800
May	95	4560	3800
June	100	4800	4000
July	100	4800	4000
August	80	3840	3200
September	70	3360	2800
October	35	1680	1400
November	16	768	640
December	10	480	400

* This table shows the schedule used for seasonal biomass adjustment per lighted surface area. The summer optimum biomass density resulting in the maximum production per surface area was determined experimentally.

In tank culture of free floating forms, aeration is often used as the principal means of maintaining the algae in suspension. Aeration *per se* does not appear to be a critical variable, unless it is supplemented with carbon dioxide, which in itself may be a limiting factor.

Surprisingly, very little attention has been given to either the CO_2 supply or plant density in mariculture. Reductions in plant density should probably be brought about automatically in order to reduce mutual shading as a response to lowered light intensity (e.g. winter months) or reduction in nutrient levels. As a practical matter, density reduction will be achieved by periodic harvesting. Waaland (1977) has shown with *Gigartina exasperata* that major harvesting (7th and 35th day) with minor harvesting between, increased the yield 42% over an unharvested control tank.

Nutrient supply is probably the single most important variable that can be controlled with minimal cost or energy input. Most natural waters have nutrient levels well below the optimum, especially of nitrates or ammonia. Addition of industrially produced nutrients or 'fertilizer' is not feasible: very simply it is the cost factor. In recent years this problem has been neatly circumvented by the utilization of sewage effluent wastes as the nutrient source, or 'controlled eutrophication' as Ryther *et al.* (1972) have termed it. While it is inappropriate to talk of a typical sewage effluent, an examination of the data in Table 8.2 shows quite distinctly the availability of the key nutrients, nitrogen, phosphorus and silicon (if diatoms are to be grown), in sewage effluents compared to a sea water and a nutrient-rich upwelling water.

Natural seawaters can be considered nutrient limited, especially with regard to nitrogen. Sewage effluent (Table 8.2) is an excellent source of nitrogen and phosphorus and can serve as an enrichment source, even when diluted into seawaters. A treatment facility serving a large metropolitan area will have daily discharge volumes in the order of 60×10^6 litres, scaling down to discharge rates in the 2×10^6 litres per day range for small metropolitan areas. Considerable nitrogen and phosphorus enrichment is thus available. If a typical secondary effluent (Weinberger et al., 1966) contains 25 g m^{-3} of nitrogen, a small 2 million litre discharge system produces 50 kg of inorganic nitrogen per day. Even allowing for an effluent to seawater dilution of 1 to 8 (or less), there is still a considerable enrichment over undiluted seawater. Sewage effluents are typically nitrogen rich, whereas natural waters are often nitrogen limited. Relative amounts and ratios of N to P in effluent vary from place to place — and even from batch to batch, but these variations, at least in the long term analysis, do not appear to represent a serious problem. Toxicity of sewage effluents due to unexplained causes or toxic constituents can be a problem. Dunstan and Tenore (1972), for example, found that one batch of sewage effluent from Warwick, Rhode Island, was, for unexplained reasons, toxic, whereas other batches were not. It could also be expected that effluent from facilities serving highly industrialized areas would have a much higher toxicity potential than those from residential or mixed residential–industrial areas. As a practical matter it is to be expected that only effluent from the latter sources will be used in mixed system mariculture. Ryther et al. (1972) have indicated that freshly chlorinated effluent could be toxic, probably as a result of the residual chlorine. However, chlorinated effluent supported better growth of phytoplankton than non-chlorinated. Again, as a practical matter, Governmental regulations and standards will almost certainly require the use of chlorinated effluent in mixed mariculture systems.

8.3 MACROSCOPIC ALGAE

A number of macroscopic algae have been grown in mixed sewage effluent seawater systems. Most have been the typical agarophytes (see p. 149) and carragenophytes (see p. 112) and have included *Hypnea musciformis* (Haines, 1975; La Pointe et al., 1976), *Chondrus crispus* (Prince, 1974) *Ulva* (*Enteromorpha*) *linza* and *Fucus vesiculosus* (Prince, 1974), *Gracilaria* spp. (La Pointe et al., 1976; De Boer and Ryther, 1977), *Neoagardhiella baileyii* (De Boer et al., 1976a and b; De Boer and Ryther, 1977; Nicotri, 1977), *Gracilaria foliifera, Chaetomorpha linum* and *Enteromorpha clathrata* (Ryther et al., 1976). Interestingly enough *Eucheuma* mariculture has not been attempted in a mixed system. Part of the answer may lie in the fact that *Eucheuma* mariculture in the Philippines is labour-intensive (Parker, 1974; Doty, 1973), whereas the systems in North America are capital-intensive. One further macroscopic alga that has been cultured in artificial sea-water tanks, some miles from the sea, is the source of Japanese 'nori' *Porphyra amansii* (see p. 104).

Table 8.2 (After Ryther et al. 1972, Atlas et al. 1977, and Dunstan and Tenore, 1972).

Concentration ($\mu mol\,l^{-1}$)

	National USA average	Auckland* New Zealand	Cranston,* Rhode Island	Plymouth†	N.E. Pacific sea water	Upwelling ocean water
NO_3^-	242	0·30	9–16	1042	5	35
NH_4	1180	1000	1577–2156	133	—	—
PO_4^{3-}	264	4·8	169–213	243	0·7	2·4
SiO_3^{2-}	656	—	110–171	70	10	45

* Residential industrial.
† Residential.

8.4 PHYTOPLANKTON

Similar sewage effluent seawater systems have been used for phytoplankton studies. Indeed, such studies were often the forerunners of macroscopic algae mariculture. Recovery costs and value per yield preclude mariculture of phytoplankton *per se*. The emphasis has been in using maricultured phytoplankton as food for invertebrates, notably *Crassostrea* (oyster), *Mytilus* (mussel), *Mercenaria* (clam) and *Argopecten* (scallop) (cf. Fig. 8.3).

Most of this food chain research originated with Ryther at Woods Hole Oceanographic Institute. Diatoms were the principal phytoplankton cultured. The research can be considered in two phases, the culturing of the algae in the mixed mariculture system and the utilization of these cultures directly as the first trophic level in the food chain. Much of this work has been summarized in a series of publications (Dunstan and Menzel, 1971; Ryther *et al.*, 1972; Goldman *et al.*, 1973; Dunstan and Tenore, 1972; Goldman and Stanley, 1974; Goldman *et al.*, 1974) and hence need not be repeated here.

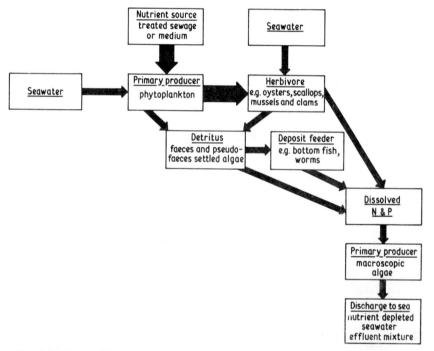

Fig. 8.3 Model of a multi-species aquaculture food web. There are an infinite number of combinations possible with such systems. Only a very few have been explored in any depth. Research is leading not to a single well-defined food web but rather to the development of a spectrum of possibilities.

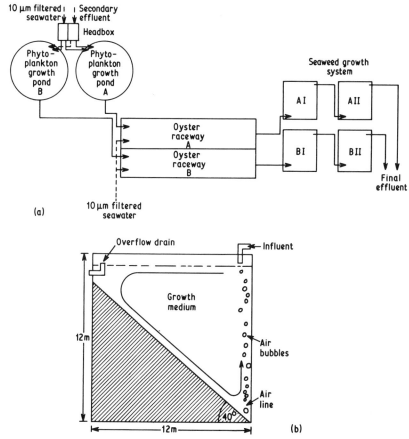

Fig. 8.4 (a) Schematic diagram of a marine aquaculture–advanced wastewater treatment system in Fort Pierce, Florida; (b) Cross section of seaweed growth tanks.

8.5 MARICULTURE IN THE OPEN OCEAN

The potential use of large brown algae as energy biomass as well as providing valuable farm fertilizer and food has been described on p. 55 and need not be repeated here. The value of producing upwelling currents in order to augment nutrient supply cannot be over-emphasized.

8.6 TANK AND SYSTEM DESIGN

Initially systems were designed to enable both plants and animals to be cultured. Basically seawater was mixed with nutrients, either artificially prepared, or else

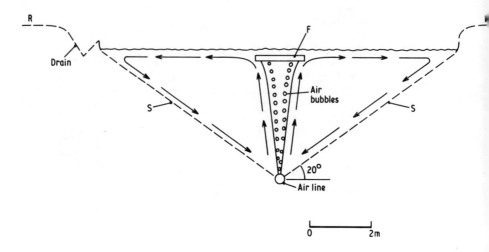

Fig. 8.5 Cross-section of seaweed culture race way. R = Road, top pitched to drain; F = Float with built-in nutrient line; S = Race Way surface of expoxy-coated asphalt.

mixed with rich sewage effluent. The model of this kind of system is shown in Fig. 8.3. The type of system utilized at Woods Hole and Florida is shown in Fig. 8.4 together with a cross section of the seaweed tank, designed so as to secure adequate water circulation. Subsequently, it was found desirable to eliminate the animal culture and to concentrate upon commercially valuable seaweeds. This eliminated the initial phytoplankton tanks and animal tanks. The idea is to use unattached seaweeds that will multiply by vegetative reproduction, e.g. *Chondrus, Gracilaria.* They can be grown in circular outdoor tanks, as is being tried at Auckland, but most workers prefer to use long raceways with either a V-shaped cross section (Fig. 8.5) or a farm like that in Fig. 8.4. Such raceways continuously provided with nutrients show great potential, with growth rates of 15–20% a day (Huegenin, 1976). Much more work remains to be done in order to establish the best kind of raceway, the best kind of construction material and the optimal means of air circulation. In view of the fact that unattached algae do best in such outdoor culture tanks or raceways, it would seem that attention should be given to naturally occurring free-living forms that could have a commercial application. Among the brown algae there are free-living forms of *Ascophyllum* (*A. mackii*), *Hormosira* (*H. banksii* ecad *libera*), *Ecklonia* (*E. brevipes*) and the Sargasso weed (*Sargassum natans*). *Palmaria* and *Pachymenia* are two Rhodophyte genera that could also be studied.

8.7 DEEP WATER ENRICHMENT

It has been known for some time that deep water from the oceans is richer in nutrients than the surface waters, and it was therefore natural to consider its potential use in culture work. Using *Hypnea musciformis* Haines (1975) has shown that growth in deep water is about three times faster than in surface water and this rate can be further increased by adding ammonia, nitrogen and EDTA (Table 8.3).

Table 8.3

Growth and carrageenan content of *Hypnea*

Growth conditions	Growth rate (g wet wt day^{-1})	Carrageenan yield (% dry wt)
Deep water	12·1	29
Deep water and low NH_4 and EDTA/vitamins	21·7	27
Deep water and high NH_4 and EDTA/vitamins	35·7	25
Deep water and low NH_4 and EDTA/vitamins	23·6	19
Deep water and 4% primary treated sewage	7·6	19
Shellfish tank effluent	64·5	16

Table 8.4

St. Croix water properties

	Nutrient			
Source	$NO_3 + NO_2$ N	NH_4 –N	PO_4 –P	SiO_4 –Si
Surface water	0·2	0·9	0·2	4·9
Reef water	0·2	0·8	0·3	2·1
870 m deep water	31·3 ± 0·31	0·7 ± 0·13	2·1 ± 0·1	20·6 ± 0·37

Table 8.4 from Roebs, Haines and Sunderlin (1975) gives an indication of the richness of the deep water used at St. Croix.

8.8 CARRAGEENAN ALGAL CULTURING (see also pp. 138–141)

Physiological studies on *Eucheuma uncinatum* (Davies, Stanley and Moon, 1977) indicated that cool temperatures, high salinities and shallow tanks were likely to provide optimal culture conditions. In the case of *E. nudum*, optimum conditions for mariculture were found to be at temperatures of 20–24°C, at a light intensity of 100–300 ft. candles and in seawater containing at least 50 ppm nitrate and 3·00 ppm phosphate. Thus within a single genus it becomes necessary to determine the optimum conditions for each species. The importance of trace elements has become apparent from the work of Prince (1974) and whether sewage effluent or deep water is used increased growth can be secured by adding trace metals.

A comparison of growth rates of *Chondrus crispus* and *Gigartina stellata* (Burns and Mathieson, 1972) showed that the former species exhibited a more rapid growth rate up to light intensities of 440 ft candles after which it levelled off, whilst the *Gigartina* continued increased growth rate up to 770 ft candles. The most promising algae are probably to be found in the genus *Gracilaria,* though here again latitude and hence climate can be important. Thus, *Gracilaria foliifera* grown at Woods Hole had an annual mean growth rate of 9 g dry wt m^{-2} day^{-2} as compared with the same species in Florida with a growth rate of 20 g dry wt m^{-2} day^{-1} (= 7^3 tonnes ha^{-1} $year^{-1}$. The yield per day in Florida is about double that of *Hypnea musciformis* (Ryther *et al.*, 1976). Pilot plant experiments in Aukland using *G. secundata* f *pseudoflagellifera* have given yields of 10–20 g dry wt m^{-2} day (Johnson, personal communication). All results exceed commercial yields obtained from ponds in Taiwan where the average output is 10 tonnes ha^{-1} $year^{-1}$ (Shang, 1976), though this may be partly explained by the fact that only seawater is used in Taiwan and there is no sewage effluent addition.

8.9 CONCLUSION

It should be evident that mariculture of marine algae possesses an important future for a variety of reasons, and it is likely to be extended in future years. It will probably prove most successful in warm temperature and tropical areas and hence should be of considerable value to developing countries.

9

Looking for Seaweeds – the World's Supplies

In the preceding chapters it has been shown that seaweeds play a considerable part in the economic life of some nations, and that in certain circumstances, e.g. stress of war or famine, their use increases and may extend even to countries that normally do not employ them. It is perhaps surprising that no attempt was made to survey the available tonnage until the beginning of the present century, possibly because there appeared to be such an abundance of seaweed that it did not seem worthwhile attempting to estimate the quantities available.

An initial indication of important areas of actual or potential seaweed usage was provided at the First International Seaweed Symposium (1953) and maps prepared then are reproduced in Figs. 1.13 and 1.14. Increasing pressure on the world's supplies has recently resulted in a publication by FAO (Michanek, 1975) putting together all the presently known information. As the author points out, one has to distinguish between the standing stocks, or biomass, and the potential harvest. Some of the richest kelp and seaweed growths are to be found on extremely exposed coasts, e.g. *Durvillea antarctica* on New Zealand shores, where collection would be very difficult.

Up to the present, seaweed resources have not been fully exploited although this situation is being approached in respect of agarophytes. However, due to the continued growth of the world's population, resulting in increasing pressure for food and energy, seaweeds, which form an annually renewable resource, are likely to become increasingly important (cf. Chapters 2 and 3). In the following pages the areas considered will be those used by FAO in compiling fishery statistics.

9.1 AREA 18: ARCTIC SEA

Because of the unfavourable environment, especially the grinding action of ice in shallow waters, the greatest quantity of algae is to be found in the sublittoral at depths of 20–25 m. Although substantial sublittoral beds of *Laminaria* spp. have been reported from the Kara Sea, the north Alaskan coast and the northern

Canadian coast, where *Alaria* and *Agarum* are associated with the *Laminaria,* as well as some areas of rockweed in sheltered Canadian localities, conditions make harvesting almost impossible.

9.2 AREA 21: NORTH WEST ATLANTIC

In west Greenland there are no records on standing crops or on any seaweed harvested. Michanek (1975) estimates that the algal biomass of the littoral between 70° and 76° 30′ N ranges from 1·2—2·0 kg m^{-2} and in the sublittoral from 4·2—7·5 kg m^{-2}. As may be expected rather higher figures (1·1—4·4 and 3·3—9·2 kg m^{-2} respectively) are quoted for Area 18 at lower latitudes (60° 25′ to 70°N).

In neither Labrador nor Newfoundland have any quantitative estimates been made, though harvesting occurs in the latter region. In Labrador, moderately exposed coasts are richest in rockweed species but there are also sublittoral beds of *Laminaria nigripes* and *Alaria grandifolia.* Harvesting of *Chondrus* commenced in 1941—43 and it has been re-exploited over the last decade. The biomass is sufficient for commercial utilization but is not sufficient to keep a factory fully employed (Humm, 1948). Of the brown algae, *Ascophyllum* is present in sufficient quantity for local residents to collect it and use it as a manure. Dulse is present but not in a quantity that would enable it to compete with regions to the south (New Brunswick, Nova Scotia).

The Nova Scotia—St. Lawrence area is extremely rich in seaweeds and the annual harvest is estimated at more than $1 000 000 US. There are four drying plants in the Province as well as an alginate factory using *Ascophyllum* and *Laminaria* (see p. 204). The Irish Moss (*Chondrus crispus*) is the most important industry in the area (see pp. 4, 117), and on the Canadian Atlantic it comprises around 86% of the weight of all seaweeds harvested and 96 per cent of the total value (Michanek, 1975). In the South Western Sub-area — the Fundy approaches — the density varies from 5—12 kg m^{-2}. In the Bay of Fundy itself, *Chondrus* is largely replaced by *Gigartina stellata* with a density of 5—12 kg m^{-2}. In the Northumberland Strait area there are quite substantial beds but they do not compare with those on Prince Edward Island. An estimated 11 500 to 13 500 tons (11 684—13 716 tonnes) wet weight of weed is harvested annually (Morrison, 1973) as compared to 450 tons (45 720 tonnes) in New Brunswick. *Furcellaria* also occurs in considerable quantity on the eastern shores of Prince Edward Island and up to 9000 tons (9141 tonnes) year^{-1} are collected (Idler, 1971; Christensen, 1971). More recent figures (Moss, 1979) indicate that some 60 000 tons (60 960 tonnes) wet weight of carrageenan algae are harvested annually throughout the Canadian maritimes. Details of the *Chondrus* beds are set out in Fig. 9.1 (after Ffrench, 1970).

The Nova Scotia area is also extremely rich in rockweeds and oarweeds. The Fundy approaches have an average density of 14 kg m^{-2} of *Ascophyllum* and there is a total of about 200 000 tons (203 200 tonnes). *Laminaria* beds are estimated as

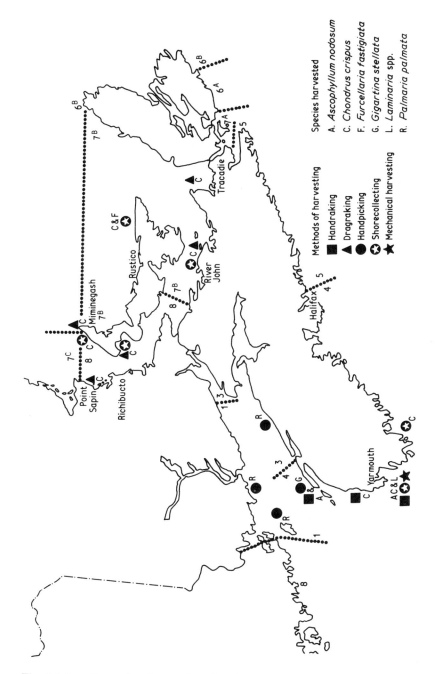

Fig. 9.1 Locations of major marine plant harvesting sites in the Maritime Region, species harvested at each site and method employed in harvesting. The dotted lines and their respective numbers denote the Marine Plant Harvesting District (after Pringle, 1976).

containing 900 000 tons (914 400 tonnes) with a density of 18·5 kg m^{-2} (Macfarlane, 1952). The adjacent coastline indicated the existence of rockweed (*Ascophyllum* and *Fucus*) beds of 80 000 tons (81 280 tonnes) and oarweed beds of 8000–10 000 tons (8128–10 160 tonnes) (Macfarlane, 1956). Along the Atlantic coast the *Laminaria* beds are narrow and because of the exposure harvesting would probably not be economical. In St. Margaret's Bay, *Laminaria* and *Agarum* have a maximum density of 16 kg m^{-2} and a metre of shoreline could yield 1481 kg fresh weight (Mann, 1972a and b).

Colinvaux (1966) has reported substantial rockweed and oarweed beds on the New Brunswick side of the Bay of Fundy, but no tonnage or densities are given. In the Grand Manan Archepelago, Neish (1972, 1973) estimated that the standing crop of *Laminaria* was about 13 000 tons (13 208 tonnes).

Apart from the algae mentioned above, Dulse (*Palmaria palmata*) also occurs in this region, most of the harvesting being in New Brunswick where 200 tons (203·2 tonnes) were collected in 1960. This weed is also collected at Grand Manan (Fig. 9.1). Craigie (1972) believes that all these algae are at present under-utilized and that there are others, e.g. *Ahnfeltia, Gracilaria, Phyllophora, Alaria*, that have a potential for development.

Chondrus occurs also in quantity along the New England Coast, and Ffrench (1970) has estimated the annual sustainable yield as between 5 500 to 9000 tons (5588–9144 tonnes) wet weight. In New Hampshire there is sufficient *Gigartina stellata* (7–8 kg m^{-2}) to justify careful summer harvesting (Burns and Mathieson, 1972). In the Chesapeake Bay area, *Gracilaria verrucosa* has a biomass of about 60 g m^{-2} and as it occupies about 44 km^2 of the bay it could be a valuable economic crop.

9.3 AREA 27: NORTH EAST ATLANTIC (INCLUDING THE NORTH SEA)

The main regions of seaweed utilization in this area are Norway and Iceland (*Ascophyllum*), Denmark (*Furcellaria*), Scotland, Eire and Northern France (*Laminaria, Ascophyllum, Chondrus*), Spain and Portugal (*Gelidium*).

Although rockweed and oarweed occur on the Greenland shores, up to the present they have not been utilized. In Iceland rockweed is harvested and distribution maps have been prepared for some of the species (Munda, 1972). The density of *Fucus vesiculosus* depends upon exposure, but it can be up to 7 kg m^{-2}, *Fucus distichus* up to 5·6, and *Ascophyllum* up to 7·7 kg m^{-2}. Of the red algae *Gigartina stellata* could have a density of 5·6 and *Palmaria palmata* 3·6 kg m^{-2} (Munda, 1970).

In the USSR, attention has been given to estimating the quantities in the Arctic Sea of *Ascophyllum* and *Laminaria* and there is little doubt that the resources are very rich. *Ascophyllum* occurs in beds of up to 28 kg m^{-2} density and the Murman littoral could yield 500 000 tons (508 000 tonnes) of wet weed. On the same coast the sublittoral is estimated to contain 500 000–600 000 tons (508 000–609 000 tonn wet weight of *Laminaria digitata* and *Laminaria saccharina*. In the White Sea there is an estimated 250 000 tons (254 000 tonnes) of *Fucus*, 800 000 tons (812 800 tonnes) of *Laminaria* and 14 000 tons (14 164 tonnes) of *Ahnfeldtia* (Michanek, 1975).

On the Norway coasts the Norwegian Institute of Seaweed Research has over many years assessed the quantities of *Laminaria* spp. and *Ascophyllum*. In southern Norway the weight ratio between *Ascophyllum* and *Fucus vesiculosus* was 100 to 30; farther north (64–67° N) the ratios were 100 to 39. *Ascophyllum* density could be as much as 26 kg m^{-2} with a mean of 5·2 kg m^2, and based on this value Baardseth (1970) has estimated there could be available 1·8 million tons (1·83 x 10^6 tonnes) of wet weed. Recent figures show that about 50 000 tons (50 800 tonnes) of *Ascophyllum* are harvested annually from which 7000 tons (7112 tonnes) of dried meal are produced, the remainder of the harvest being used for alginate production (Jensen, personal communication). Increasing demand for seaweed meal for manure, feed or liquid extract, should prove no problem as Norway could provide up to 50 000 tons (50 800 tonnes) of dry meal annually (Baardseth, 1970). The main problem is labour in harvesting, and this has decreased in recent years with the transfer of labour to offshore oil recovery. Most of the *Ascophyllum* harvesting is carried on around Trondheim.

The other Norwegian harvest is of *Laminaria hyperborea*, which has been going on since 1964, and at present about 50 000 tons (50 800 tonnes) are collected annually, primarily for alginate production. In 1977 some 2000–3000 tons (2082–3048 tonnes) of alginate were produced (Jensen, personal communication).

The west coast of Sweden is also rich in rockweeds and oarweeds but no industry has yet developed. Michanek (1975) suggests that this is due to little tidal fluctuation which makes harvesting more difficult, the existence of undesirable epiphytes and finally the high cost of labour. Densities of the major weeds compare favourably with those from other regions. Average wet weight densities reported by Gislen (1930) were *Ascophyllum* 15·3 kg m^{-2}, *Fucus serratus* 8·4, *Fucus vesiculosus* 7·6, *Laminaria hyperborea* 5·7, *Laminaria digitata* 8, *Laminaria saccharina* 3·6 and *Furcellaria* 3·2 kg m^{-2}.

The principal industrial alga in Denmark is *Furcellaria fastigiata*. Harvesting commenced in the 1940s and rose from an initial few thousand tons annually to 30 000 tons (30 480 tonnes) in the 1960s when stocks started to be depleted. Harvesting still continues but over a wider area than that originally used. In order to meet demand, the industry also imports *Chondrus* and *Eucheuma* (Michanek, 1975).

In the Baltic Sea, the Russians have estimated supplies of *Furcellaria* between Klaipeda and Ventspils as up to 90 000 tons (91 410 tonnes) and in the Irban Strait and Gulf of Riga as up to 10 000 tons (10 160 tonnes). Loose-lying resources of the same alga around the islands of Hiumaa and Saaremaa were estimated at 150 000 tons (152 400 tonnes). This abundant supply of a red alga is only exceeded by *Phyllophora* from the Black Sea (see p. 141). Kireeva (1965) reports that up to 40 000 tons (40 640 tonnes) of *Furcellaria* have been collected. The same author

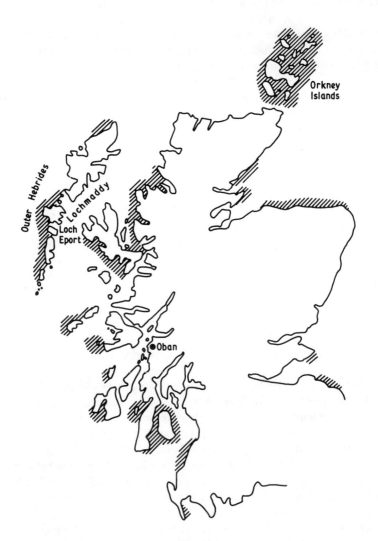

Fig. 9.2 The distribution of the Laminariaceae around the coast of Scotland. (For details on quantities see Walker, 1954).

estimates the standing crop of *Fucus* at 70 000—80 000 tons (71 120—81 280 tonnes).

No harvesting is carried out in Finland and on the Polish coast up to 20 tons (20·3 tonnes) dry weight of *Furcellaria* can be cast ashore (Czapke, 1964).

On the Baltic coast of the Federal German Republic, the total *Fucus* beds were estimated at 40 000 tons (40 640 tonnes) over a length of 184 km. Because the *Fucus* beds extend down to 6 m, densities are low (1—2·5 kg m^{-2}) and profitable harvesting may not be possible (Michanek, 1975). On the North Sea coast the amounts of *Fucus* and *Laminaria* are not sufficient to justify harvesting, though on Heligoland, Hoppe and Levring (1969, 1970) recorded densities of *Laminaria digitata* of 3·7—5·6 kg m^{-2}, *Laminaria saccharina* of 7·1 kg m^{-2} and *Laminaria hyperborea* of 11·1 kg m^{-2}.

The coasts of the Netherlands and Belgium do not lend themselves to the growth of large algae and no weights have been recorded.

In the UK, extensive assessments of seaweed quantities have been carried out. The first official survey in 1945 indicated that there are at least 180 000 tons (182 880 tonnes) of rockweed, mostly *Ascophyllum*, distributed over 870 km of coastline with an average density of 4 kg m^{-2}. Using stereoscopic aerial photography, with grab sampling in order to check weights, it was estimated that there were some 10 million tons (10·16 x 10^6 tonnes) of *Laminaria* species (mostly *Laminaria hyperborea*) between 0—18 m in depth with 4 million tons (4·06 x 10^6 tonnes) located along 2700 km of shoreline (Walker, 1947—58). It was further estimated that at least one million tons (1·016 x 10^6 tonnes) could be harvested annually without damaging the beds (Fig. 9.2). The average density in the *Laminaria* beds was 3·7 kg m^{-2}, but there was a variation with depth, it being greatest near the surface and minimal at 10 m down. Between 1963 and 1973 the quantities harvested varied between 22 500 and 29 700 tons (22 860—29 700 tonnes) so that there is considerable scope for further development.

Eire no longer harvests the amount of weed that it used to. The annual harvest of *Laminaria hyperborea* stipes fell from 3000 tons (3048 tonnes) in the 1950s to not more than 1700 tons (1727 tonnes) in the early 1960s (Michanek, 1975). The standing *Ascophyllum* crop was estimated at 150 000 tons (152 400 tonnes) in 1974 and has been harvested increasingly since 1963 so that at present the total *Laminaria—Ascophyllum* harvest is about 40 000 wet tons (40 640 tonnes) for algin production. *Chondrus crispus* (Irish moss) is abundant all along the west coast and could yield, if required, up to 500 tons (508 tonnes) annually, but at present this figure is not always approached. *Chondrus* is also collected along the northern French coast and in 1965 1883 tons (1913 tonnes) and in 1970 3192 tons (3423 tonnes) were landed, though both these figures are less than the 4400 tons (4470 tonnes) landed in the 1950—51 era.

A study by Audoin and Perez (1970) of the Calvados shoreline showed that there was a standing crop of at least 100 000 tons (101 600 tonnes). There is much more elsewhere along the French coast because it was estimated that 100 000 tons

(101 600 tonnes) were landed in 1969. It is stated that France produces 1200 tons (1219 tonnes) of alginate per year (see p. 207) and this would involve at least 30 000 tons (30 480 tonnes) of the 100 000. Some of the balance must refer to the large amounts used by the Brittany peasants for land manure purposes (see p. 39). There are no figures available for the quantity of *Lithothamnion* used for maerl production (p. 40).

In Spain, where there is a major agar industry, 8000–9000 tons (8128–9144 tonnes) of raw *Gelidium* are harvested annually, and at least another 1000 tons (1016 tonnes) of the carrageenan algae *Chondrus* and *Gigartina stellata*. Portugal is also a producer of agar and it has been estimated that 6000 tons (6096 tonnes) dry weight of agarophytes (*Gelidium* spp.) could be collected from the mainland and a further 2000 tons (2032 tonnes) (*Pterocladia capillacea*) from the Azores. Moss (1979) reports an export of 500 tons (508 tonnes) dry weight annually to France. The carragheen algae (*Gigartina* spp. and *Chondrus*) together could yield at least 1200 tons (1219 tonnes). In support of this, Palminha (1971) reports that in 1970, 23 500 tons (23 876 tonnes) of wet weeds were forwarded to six factories.

9.4 AREA 31: WESTERN CENTRAL ATLANTIC

The seaweed resources in this area are not great and have scarcely been utilized. Most of the commercially important algae are agarophytes. *Gracilaria verrucosa* and *Gracilaria foliifera* both occur on the North Carolina coast, the latter species ascending further up the estuaries than the former. It also has a much better growth rate at high light intensities. The value of *Gracilaria foliifera* as an agarophyte is currently recognized by its use in mariculture experiments at Harbor Branch in Florida (see p. 252). On the Florida Atlantic coast, agarophytes are present but not in abundance, as the coast morphology does not favour seaweed growth. *Gracilaria, Hypnea* and *Agardhiella* occur in some abundance from Cape Sable northward to Tampa Bay. The north coast of the Gulf of Mexico, whilst possessing more algae than was previously thought (Earle, 1969; Edwards, 1970), may have in the Laguna Madre commercially useful weights of the two *Gracilaria* species mentioned above. The coast from Mexico through to Panama does not appear, from the very little information available, to possess algae in any great quantity. However, Moss (1979) states that Mexico exports some 600 tons (609 tonnes) dry weight ($\equiv$ 4000 tons (4064 tonnes) wet weight) of *Gigartina* species to the USA. Southward, *Gracilaria cylindrica, G. mamillaris* and *G. domingensis* are reported as common on Colombian shores (Schnetter, 1967). On the Venezuelan coasts there are numerous agarophytes, e.g. *Gelidium, Gracilaria, Hypnea, Digenea,* of which *Hypnea* species are probably the most important. *Hypnea musciformis* and *Gracilaria verrucosa* are very abundant around Cubagua Island. The former species is also abundant on the shores of the Netherlands Antilles (Van den Hoek, 1969; Van den Hoek *et al.,* 1972).

In Puerto Rico, Diaz-Piferrer and Perez (1964) have reported highest yields from agarophytes of 1·4 kg m^{-2} dry weight for *Gracilaria debilis* and 1·0 kg m^{-2} dry weight for *G. verrucosa, G. domingensis,* and *Hypnea musciformis* (see p. 142). These algae also occur on the other larger islands of the Caribbean, but there is no data on density or total quantity. A study in Cuba by Baardseth (1968) showed that although agarophytes were present, it was likely that the maximum yield would be around 10 tons dry weight, which is about one fifth of Cuba's needs.

There are, of course, the deep waters of the Gulf of Mexico with considerable amounts of the free-floating *Sargassum natans.* Parr (1939) estimated the total biomass of this species as around 90 000 tons (91 440 tonnes), which could produce almost 200 tons (203 tonnes) of alginate (Davis, 1950). This same alga eventually finds its way into the Sargasso Sea, where Parr (1939) estimated the quantities as ranging from 0·8–2·15 tons km^{-2} and this, over the entire sea, could mean the presence of several million tons. There would, however, be some problems in harvesting.

9.5 AREA 34: EASTERN CENTRAL ATLANTIC

In Morocco the main harvest of *Gelidium sesquipedale,* an important agarophyte, is between Casablanca and Qualidia where some 3000 tons (3048 tonnes) wet weight ($\equiv$ 300 tons (304·8 tonnes) dry weight) a year are collected and exported. It is also known that there are *Gelidium* fields around Taifaya in the south, but exploitation is difficult because of access problems. Despite earlier statements (Rodriguez, 1953) that *Gelidium* was rich in the Canary Islands, later work (Johnston, 1969a and b) has discounted this, though the alga does occur in some quantity at Ifni and Cape Juby.

In Senegal, species of *Hypnea* occur in considerable quantities and much is thrown ashore as castweed. The extraction of and the nature of the carrageenan was studied by Mollion in 1973 and a factory constructed in 1972 had a capacity of 12 000 tons (1219 tonnes) wet weight of weed. Species of *Gelidium* continue to occur from here southwards to the Cameroons but not in any great quantity (Michanek, 1975).

9.6 AREA 37: MEDITERRANEAN AND BLACK SEA

The seaweed resources of the Mediterranean are not rich, except possibly for species of the brown seaweed *Cystoseira* from which, however, it is not at present possible to manufacture algin economically. South-east of Marseilles, *Cystoseira stricta* has a summer density of 13·2 kg m^{-2} and a winter one of 6·9 kg m^{-2}. This is in clean disturbed water. In clean calm water *Cystoseira crinita* has a summer density of 5·16 kg m^{-2} and a winter one of 5·88 kg m^{-2} (Bellan-Santini, 1969). Likewise a population of *Cystoseira mediterranea* at Banyuls-sur-mer ranged from 10·1 kg m^{-2}

in summer to 1·8 kg m^{-2} in winter, so that if these species ever become economically useful summer harvesting would seem essential.

In Italy *Gracilaria verrucosa* appears to be the only alga currently used commercially The main source for the agar industry in Italy is in the lagoon at Venice (Michanek, 1975). Extensive beds of the same alga have also been reported from Goro Bay near Ferrara in the Po delta. This area receives extensive nutrients (phosphate and nitrate) from agricultural irrigation channels and with a very turbid water *Gracilaria* is almost the only occupant. Algal densities of 2—4 kg m^{-2} have been reported. The total mean productivity of the bay has been estimated at about 20 000 tons (20 320 tonnes) per year giving an agar yield of around 1000 tons (1016 tonnes) (Simonetti, Giaccone and Pignatti, 1970). *Gracilaria dura* and *Hypnea musciformis* also occur in Sicily (Cavaliere, 1969), the latter reaching a density of 1·5 kg m^{-2} (Giaccone, 1970), but at present there is no evidence of any collection. If *Cystoseira* should become a source of raw material for alginate production there are considerable beds in the Adriatic where densities have been determined (Table 9.1). Another potentially useful brown alga for algin production from the Inner Adriatic is *Fucus virsoides* which has a density of 2·5 kg m^{-2}.

Table 9.1

Species	Locality	Biomass (kg m^{-2})
Cystoseira barbata	Inner Adriatic	1·1—2·2
Cystoseira fimbriata	Inner Adriatic	1·1—2·2
Cystoseira spicata	Tremiti Islands	6·9
Cystoseira corniculata	Tremiti Islands	5·2
Cystoseira spinosa	Lower Tyrrhenian Sea	74·2
Cystoseira fimbriata	Lower Tyrrhenian Sea	6·7

The green alga *Cladophora prolifera* occurs in great quantities in the Gulf of Taranto at depths of 29—34 m between Gallipoli and Porto Cesareo. Parenzan (1970) has estimated that there may be as much as 1·4 million tons (1·42 x 10^6 tonnes). Analyses show that it has a high protein value (25% of dry weight). It is reported that industry has become interested because of the micro-elements and a production of up to 50 tons (50·8 tonnes) per month is envisaged for fertilizer and animal fodder.

In Yugoslavia the principal alga of commercial importance is the genus *Cystoseira*. However, the amounts are not great, as Span (1969) estimated that there were only 9850 tons (10 000 tonnes) fresh weight over 1300 km of shoreline. The biomass was 2·5 kg m^{-2} on mainland shores and 2·0 kg m^{-2} on island shores. Munda (1972), in a study of the shores of Western Istria, recorded somewhat lower values (1·2—2·08 kg m^{-2} Span (loc. cit.) also reported an annual growth rate for *Cystoseira barbata* of 3—8 cm which would not give a very good recovery rate after harvesting. *Cystoseira* also occurs on the shores of Greece but there is no indication that the genus is present in any real quantity. The situation is apparently different in Bulgaria where the total standing crop of algae is put at 100 000 tons (101 600 tonnes), half of which grows in beds that could be harvested commercially (Michanek, 1975).

The Black Sea represents what is probably the world's greatest accumulation of red algae. The red algal genus *Phyllophora* occurs on all the Black Sea shores, but the greatest quantity is in the north west in a region appropriately named Zernov's Phyllophora Sea. Here it occurs at 30–60 m below the surface and occupies an area of about 15 000 km^2. There are three species involved: *Phyllophora nervosa* in the northern part, *Phyllophora brodiaei*, occupying about 59% of the area, and *Phyllophora membranifolia*. The average density is 1·7 kg m^{-2} but densities as high as 13 kg m^{-2} have been reported. The total quantity of *Phyllophora* available for the local Russian industry is estimated at 5–6 million tons (5·08–6·11 x 10^6 tonnes), with a possible annual production of 90 000 tons (91 440 tonnes). On the Turkish coast of the Black Sea, *Phyllophora* has been used for the production of iodine. The Turkish shores also carry a well-developed flora of *Cystoseira* species but there is no record of their utilization (Michanek, 1975).

Egypt has a range of algae along its shores, and some 4000–6000 tons (4064–6096 tonnes) are cast up annually around Alexandria. Tidal harvesting has not been carried out, but cast weed of red algae (*Pterocladia pinnata*, *Gracilaria* spp.) is collected and the agar extracted. In Tunisia, there are considerable areas occupied by the green alga *Caulerpa prolifera* and Michanek (1974) reported a Belgian firm as being interested in its exploitation. Analyses of the alga showed 29·3% protein, 25% cellulose, 17·5% starch and 1·65% fats on a dry weight basis. Stein (1968) has suggested that the large quantities of *Ulva* that develop in autumn in Tunis lake could be used as an agricultural fertilizer and this would seem sensible in such an arid area.

9.7 AREA 41: SOUTH WEST ATLANTIC

In northern Brazil, three factories have been built in the states of Rio Grande da Norte, Ceará and Paraiba. There are nearly 200 collection centres over 2000 km of shoreline, the principal algae harvested being *Hypnea* and *Gracilaria*, though *Eucheuma*, *Digenea*, *Gelidiella*, *Gelidium*, *Pterocladia* and *Sargassum* are also collected. Production figures reported by Michanek (1975) for 1973 are 144 tons (146·3 tonnes) of agar, 960 tons (975 tonnes) alginates and 11 520 tons (11 700 tonnes) seaweed meal, which is equivalent to harvests of 14 400 tons (14 630 tonnes) agarophytes, 96 000 tons (97 506 tonnes) of alginophytes and 55 000 tons (55 980 tonnes) of mixed seaweeds for meal. There are probably further supplies of algae that could be used for commercial purposes on other parts of the Brazilian coast.

In Argentina, seaweed harvesting, particularly of red seaweeds, has been quite significant. In 1958 it amounted to 2000 tons (2032 tonnes) but this had risen to

24 800 tons (25 196 tonnes) in 1968. In 1973, 1700 tons (1727 tonnes) of kelp were harvested and 21 400 tons (21 740 tonnes) of red algae, all in wet weight. By 1976 the amount of kelp harvested may have increased because 1000 dry tons (1016 tonnes) were exported. On the other hand, the quantity of red algae harvested would seem to have decreased as only 2000 tons (2032 tonnes) dry weight of *Gracilaria* were exported (Moss, 1979). It is known that there are very extensive *Macrocystis* beds in Patagonia, southern Argentina and Chile, and although no figures appear to be available this region must represent one of the largest seaweed resources.

9.8 AREA 47: SOUTH EAST ATLANTIC

Very little information is available here, other than from South Africa. It is known that there are some agarophytes such as *Gelidium* and *Hypnea* but there are no details of whether the amounts would justify an industry.

In South Africa, *Gracilaria verrucosa* and *Gelidium pristoides* are the two most important agarophytes commercially (see p. 175). Michanek (1974) reports *Gracilaria* casts of 1000 tons (1016 tonnes) dry weight along 26 km of coast north-west of Cape Town (Saldanha–Langebaan lagoon), and states that there may be as much as 100 tons (101·6 tonnes) dry weight of *Gelidium*. *Hypnea spicifera* is also abundant as well as other potential red agarophytes (*Gelidium cartilagineum, Gelidium amansii, Suhria vittata, Iridophycus capensis*). In addition there are extensive kelp beds, dominated mainly by *Ecklonia maxima* and *Laminaria pallida,* the former occurring widely between Cape Agulhas and the Tropic of Capricorn on the west coast. Isaac and Molteno (1953) estimated that from some five localities up to 6000 tons (6096 tonnes) dry weight, mostly *Ecklonia,* could be harvested annually. Harvesting is proceeding, but whether on this scale it is not possible to say. Moss (1979) suggests that as much as 2000 tons (2032 tonnes) dry weight kelp is being exported annually for alginate manufacture, but he gives no figures for any internal consumption.

9.9 AREA 51: WESTERN INDIAN OCEAN

Information from much of this area is scanty. However it is stated that 300 tons (304·8 tonnes) dry weight of *Gelidium* are harvested from the south coast of Madagascar and that 1000 tons (1016 tonnes) are collected annually from the coast of the Sudan. Tanzania collects up to 3000 tons (3048 tonnes) wet weight of *Eucheuma striatum* and exports the dried weed (700 tons, 7112 tonnes). Michanek (1975) points out that the presence of some 2500 tons (2540 tonnes) of *Hypnea* and 28 000 tons (28 448 tonnes) of *Sargassum* on the coasts of Oman could be a basis for a fertilizer and animal feed industry.

The Indian coastline has a range of seaweeds but their potential is nowhere near fully exploited. At present the industrial demand exceeds the annual harvest but India should be able to meet internal demands and export also. At present about 60 tons (61·1 tonnes) of alginates are produced from 700 tons (711·2 tonnes) of dry *Sargassum* and an equal quantity is imported. 100 tons (101·6 tonnes) of agar are required annually with present production of about half the amount (see p. 179). The main resources are between Mandapam and Cape Comorin at the southern tip and also around the Kathiawar peninsula (Rao, 1970). Estimates of quantities vary but it is likely that at least 19 000 tons (19 304 tonnes) of *Sargassum* and other kelps are present in the Gulf of Kutch with a density of up to 4 kg m^{-2}. The principal algal resources are agarophytes with several species of *Gracilaria* on all coasts where there are lagoons and protected areas. Another agarophyte is *Gelidiella acerosa* which occurs in the Mandapam area at a density of around 15 kg m^{-2}. In the Gulf of Mannar area, Varma and Rao (1964) estimated harvestable *Gracilaria* at 335 tons (340·5 tonnes) and in Palk Bay, Rao (1968) calculated a potential harvest of 140 tons (142·2 tonnes). Krishnamurthy *et al.* (1967), as a result of their studies at Pambam and to the south, concluded that the total standing crop of agarophytes could be as much as 3000 tons (3048 tonnes) wet weight. Very much greater quantities are estimated for the coast between Mandapam and Tuticorin, where 20 000 tons (20 320 tonnes) of wet *Gracilaria* and 2000 tons (2032 tonnes) of wet *Gelidium* could be available annually. Further north, around Madras, a harvest of 500 tons (508 tonnes) wet weight of *Gelidiella* has been collected together with 155 tons (157·5 tonnes) *Gracilaria* (Rao, 1969).

In Sri Lanka, 'Ceylon Moss' (*Gracilaria lichenoides*) has been collected in a small way since 1913 but it has been estimated that in the Kalpitiya district some 126 tons (128 tonnes) wet weed could be collected and in Trincomalee a further 250 tons (254 tonnes) (Durairatnam, 1961). The south-west coast also has some areas of extensive *Sargassum* with a harvestable amount of about 775 tons (787·4 tonnes) wet weight (Durairatnam, 1966).

In the southern Indian Ocean, Kerguelen possesses a very long coastline with fjords and islands and rich kelp beds of *Macrocystis* and *Durvillea antarctica*. The total biomass of *Macrocystis* in the Baie de Morbihan was calculated by Grua (1964) as being about 6·3 million tons (6·4 x 10^6 tonnes) with weights ranging from 95 kg m^{-2} in clear waters to 137–606 kg m^{-2} in least transparent waters. There is little doubt that Kerguelen has one of the most extensive, at present untapped, seaweed resources.

9.10 AREA 61: NORTH WEST PACIFIC

In China recent years have seen a great increase in the quantities of seaweed harvested and used. Michanek (1975) reports a dry weight harvest of 54 tons (54·8 tonnes) kelp in 1952 and 24 000 tons (24 384 tonnes) in 1959 (= 328 tons

Table 9.2 *Laminaria japonica* production in China (after Cheng, 1968)

Year	Production in Liaoning Province (tons)*		Production in Shantung Province (tons)		Total	% by Raft culture
	Raft	Natural	Raft	Natural		
1952	0	72	62	0	134	46·4
1953	0	451	169	68	688	24·5
1954	15	813	457	244	1 529	31·0
1955	147	1 411	1 089	520	3 167	39·0
1956	243	701	2 107	305	3 356	70·0
1957	3 763	2 604	4 873	881	12 981	65·4
1958	18 397	4 539	13 207	1 378	37 521	84·2

* 1 ton = 1·016 tonnes.

(333·2 tonnes) and 145 000 tons (148 000 tonnes) wet weight respectively). More recently it has been thought that to meet demand an annual production of 1 million tons ($1·016 \times 10^6$ tonnes) dry weight is necessary. The principal seaweed is *Laminaria japonica* which is collected wild and also cultivated on floating rafts, the greatest bulk coming from the latter (Table 9.2). Raft culture can yield up to 8 tons (8·1 tonnes) dry wt ha^{-1} $year^{-1}$, whilst wild populations give 5–10 tons (5·1–10·2 tonnes) ha^{-1} $year^{-1}$ depending upon fertility of the sea water. In many places artificial fertilizers are added to the seawater, and it has also been suggested that nitrogen-fixing blue-green algae and bacteria could be used (Bardach, Ryther and McLarney, 1972). Raft cultivation has now been extended into southern provinces of China and nurseries in Lüta produce some 3800 million young plants for use on rafts (Cheng, 1968). Green algae (*Ulva pertusa, Ulva lactuca*), used for medical purposes grow abundantly around Hsiamen (Amoy) and some 500 tons (508 tonnes) of *Gloiopeltis* (see p. 145) are collected annually.

Whilst seaweeds are used in Hong Kong and Macao there does not appear to be any quantitative data.

In the Republic of Korea, kelp, laver (*Porphyra*) and agarophytes (*Gelidium, Gigartina*) are all collected wild and cultivated. Much of the crop is exported to Japan. Laver culture has increased from some 10 000 tons (10 160 tonnes) in 1965 to 36 000 tons (36 648 tonnes) in 1970 (FAO, 1971). Michanek (1975) states that 1800 tons (1829 tonnes) wet weight of *Gelidium* were collected annually around 1970, but this amount has now increased substantially since Moss (1979) records 600 tons (609 tonnes) dry weight ($\equiv$ 4000 tons (4064 tonnes) wet weight) exported annually. In addition some 900 tons (914 tonnes) of other agarophytes (*Gigartina*) are also exported (Moss, loc. cit.).

The brown algal harvest (*Sargassum, Undaria*) is around 47 000 tons (47 752 tonnes) fresh weight, only 6600 tons (6706 tonnes) coming from aquaculture. The great bulk of the harvest is represented by *Undaria* and a very small poroportion (500 tons (508 tonnes)) by the alga *Hijikia fusiforme*.

The only statistics from Taiwan (Moss, 1979) refer solely to *Gracilaria* collected and exported for agar production and the annual amount is about 500 tons (508 tonnes) dry weight.

Table 9.3 Average yield of principal algae grown in Japan (after Okazaki, 1971 and Oohusa, 1971)

		Tons* dry wt (x 10^3)	Tons wet wt (x 10^3)
Green:	Ulva	4 018	Ca 31
	Enteromorpha	3 973	
Brown:	Laminaria	36 600	
	Undaria	12 586	
	Eisenia, Ecklonia	5 427	
	Hijikia	2 369	
	Alaria	1 300	246
	Eckloniopsis	428	
	Sargassum	354	
	Endarachne	152	
	Mesogloia	115	
	Heterochordaria	74	
Red:	Porphyra	6 660	134
	Iridophycus	776	
	Gloiopeltis	608	
	Chondrus	565	Ca 16
	Pachymeniopsis	492	
	Digenea	281	
	Others	147	
Agaroids:	Gelidium, Beckerella	4 050	
	Gracilaria	1 654	
	Acanthopeltis	179	
	Campylaeophora	132	17
	Pterocladia	131	
	Ceramium	70	
	Others	233	
		73 374	444 x 10^3

* 1 ton = 1·016 tonnes.

Japan is the leading nation in the use of seaweed, both wild and cultivated, and the demand is so great that the country imports dried seaweed from Korea, Chile, the

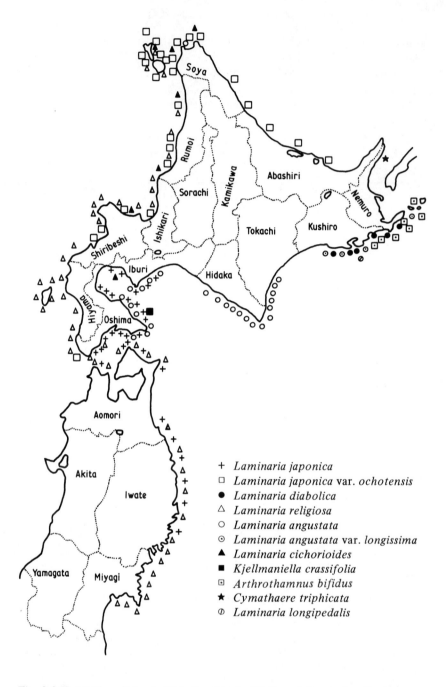

Fig. 9.3 Haversting sites of edible *Laminaria, Kjellmaniella, Arthrothamnus* and *Cymathaere* (from Okazaki, 1971).

Argentine and South Africa. Table 9.3 gives the average yield of the principal algae.

The principal harvesting sites for the major brown algae are given in Fig. 9.3 and the distribution of the agarophytes is depicted in Fig. 5.1. Okazaki (1971) points out that the entire harvest of *Laminaria, Heterochordaria* and *Alaria* is restricted to the subarctic shores (Fig. 9.3), whereas 80% of *Digenea* and 5·5 per cent of *Gelidium* is collected from the far south (Fig. 5.1). According to Okazaki (loc. cit.) the average havest and the seaweeds involved in the various Prefectures is set out in Table 9.4.

Table 9.4

Region		Average harvest (tons* dry wt)	Species
Subarctic	Hokkaido	40 667	*Laminaria, Undaria, Alaria*
	Miyagi	3 993	*Laminaria, Undaria, Porphyra*
	Iwate	3 847	*Laminaria, Undaria*
	Aomori	1 831	*Undaria*
Temperate	Mie	7 424	*Ulva, Eisenia, Ecklonia, Enteromorpha*
	Chiba	5 976	*Porphyra, Enteromorpha, Gracilaria*
	Shizuoka	1 833	*Ulva, Gelidium, Eisenia, Ecklonia*
	Archi	1 784	*Porphyra*
	Tokyo	1 659	*Gelidium, Porphyra*
Subtropics	Kagoshima	919	*Undaria, Digenea, Gelidium*

* 1 ton = 1·016 tonnes.

In the case of *Porphyra* (nori) Michanek (1975) states that by 1970 there were 8·7 million nets occupying a total sea area of 191 million m^2 and 171 000 blinds (see p. 100) occupying a sea area of 18 000 000 m^2. In addition to this there is artificial factory production (see p. 104).

At present there are brown algal resources in excess of those actually harvested. Thus the standing crop around the island of Hokkaido has been estimated at 1·5 million tons ($1·52 \times 10^6$ tonnes). Whilst *Undaria* is collected mainly on the west side of Hokkaido, where it forms 11·15% of the total seaweed harvest, it can also be grown artificially on bamboo rafts where an area of 66 m^2 will produce about 1 ton (1·016 tonnes) wet weight of weed. According to Bardach *et al.* (1972) there are over 1000 rafts on the Ojika Peninsula which give rise to 145 tons (147·3 tonnes) dry weight of Wakame.

Mention should be made of the green algal genus *Monostroma* which, like *Porphyra,* is grown on nets suspended in river estuaries. In 1973 the output amounted to some 700 tons (711·2 tonnes) which was made into sheets ('aonori') like nori.

Whilst Japan has huge crops of seaweed, there are also substantial crops on the

Siberian coast of the USSR where the cold waters promote the growth of *Laminaria japonica* and other species. The extent of these beds is shown below:

N.E. Vladivostock	429 000	tons (435 800 tonnes) wet weight
Strait of Tartary	552 000	tons (560 800 tonnes) wet weight
Sea of Okhotsk	117 000	tons (118 900 tonnes) wet weight
Kamchatka	784 000	tons (796 400 tonnes) wet weight
Lesser Kurile Island	70 000	tons (71 120 tonnes) wet weight
Total	1 952 000	tons (1 983 000 tonnes) wet weight

In the Sea of Okhotsk *Laminaria* densities range from $2 \cdot 5 - 2 \cdot 8$ kg m^{-2} on exposed shores to $4 \cdot 0 - 6 \cdot 5$ kg m^{-2} in the sublittoral (Vozzhinskaya, 1966). The other important algal genus collected on these coasts is the red seaweed *Ahnfeltia plicata*, the source of 'Russian' or 'Sakhalin' agar. These supplies total some 170 000 tons (172 700 tonnes) of which 104 000 tons (105 664 tonnes) are in Peter the Great Bay (Vladivostock), Izmen Bay (48 000 tons, 48 768 tonnes) and Busse lagoon (24 000 tons, 24 304 tonnes). In the last named locality densities range from $1-3$ kg m^{-2} and the algal bed can be up to 20 cm deep.

9.11 AREA 67: NORTH EAST PACIFIC

There are no recent estimates of the beds in the extreme north but it is unlikely that there has been very great change from the quantities estimated by Rigg (1912) and Cameron (1915). At that time western Alaska was estimated to have 3 500 000 tons (3 556 000 tonnes) wet weight, south-east Alaska 15 666 000 tons (15 920 000 tonnes) wet weight and Puget Sound 520 000 tons (528 320 tonnes). Later, around Vancouver Island it was estimated that 392 000 tons (398 200 tonnes) of *Macrocystis* and *Nereocystis* were available on an annual basis, but these figures were later increased (Scagel, 1948) to between 750 000 to 1 000 000 tons (762 000–1 016 000 tonnes). Michanek (1975) states that a private firm studied the north west and north east coasts of Vancouver Island and reached the conclusion that over 500 000 tons (508 000 tonnes) of kelp were available commercially. The most recent data come from an official survey commenced in 1972 which has resulted so far in producing a total of 173 160 tons (175 900 tonnes) biomass around the Queen Charlotte Islands. So far as Canada is concerned it would seem that the kelp beds could yield around one million tons annually. The red alga, *Iridaea,* which yields a phycocolloid, had a biomass of 1650 tons (1676 tonnes) in the Central Georgia Strait.

9.12 AREA 71: WESTERN CENTRAL PACIFIC

The amount of seaweed (mainly for human consumption or agar manufacture) has risen from 1300 tons (1320·8 tonnes) raw seaweed in 1940 to 2300 tons (2336·8 tonnes) in 1971. At the present time *Eucheuma* is the genus that is most important as an export source for carrageenan. Soerjodinote (1969) has suggested that the species could well be cultivated as they are in other parts of the Pacific. There is no data available at the present time for the other countries in S.E. Asia although it is known that coastal dwellers eat seaweed and that algae are sold in the markets.

In the Philippines *Eucheuma* is probably the most important commercial alga, being eaten locally as well as being exported for carrageenan manufacture (see p. 138). It is sufficiently important that, because of over-picking of the wild crop (Caces-Borja, 1973), artificial cultivation has now been introduced. Since the introduction of this new technique, the amount of weed exported has increased from 570 tons (579 tonnes) in 1973 (Caces-Borja) to the present rate of some 7000 tons (7112 tonnes) dry weight of *Eucheuma spinosum* and *Eucheuma cottonii* (Moss, 1979). Depending upon species and locality the annual yield from artificial farms is likely to range from 13–30 tons ha^{-1} (13·2–30·4 ha^{-1}) (Doty, 1973). There is no estimate available of seaweeds gathered for human consumption, but it is known that several tons of *Gracilaria* are harvested daily for use as supplemental food for fish maintained in fish ponds.

9.13 AREA 77: EASTERN CENTRAL PACIFIC

So far as kelps are concerned this is one of the richest areas in the world and off the coast of California there is regular harvesting. The importance of these beds is recognized by the efforts made to restore them in recent years after they had been decimated by sea-urchins and bacterial attack arising from an inflow of warm water from southern California.

The area is famous for the giant kelp beds of *Macrocystis pyrifera*, *Macrocystis integrifolia* and *Nereocystis luetkeana*. The first-named occurs from the Monterey Peninsula to the centre of Baja California, and the other two north of Monterey. The annual harvest for the Californian and Mexican coast probably ranges from 125 000 to 150 000 tons (127 000–152 400 tonnes) in a year, the great bulk being used for algin manufacture, the balance being processed for seaweed meal. The 1911–13 survey produced figures for kelp from Cape Flattery to Cedros Island based on a biennial cut. Michanek (1974) calculates that the corresponding figure at that time for the Eastern Central Pacific would have been 36·7 million tons (37·29 x 10^6 tonnes).

The beds off the Californian coast, despite every care, have varied greatly over the years (North, 1971), and this has been reflected in the amounts reported as harvested. Thus, in 1917 400 000 tons (406 400 tonnes) were harvested between

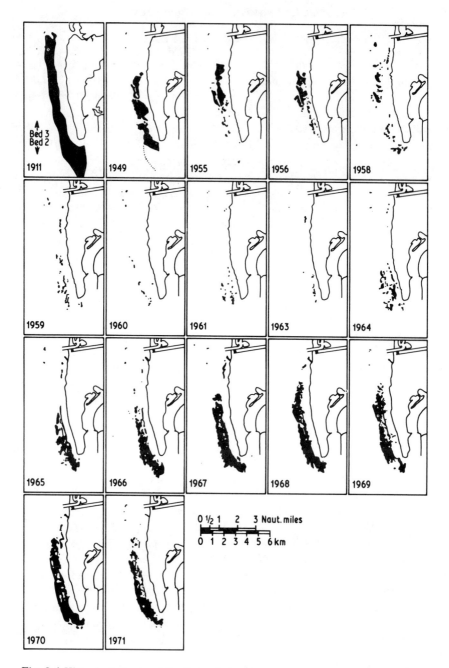

Fig. 9.4 Historical charts of the Point Loma kelp bed (shown as black). 1963–66 from oblique photos, courtesy Kelco Company. Years 1963–71 represent fall conditions.

Point Conception and San Diego whilst between 1940–45 the mean Californian harvest was 56 000 tons (56 896 tonnes). In the 1940s and 1950s there was a serious depletion of the kelp beds, principally from overgrazing by sea urchins, though there was also some evidence of a bacterial or fungal infection, probably associated with increases in water temperature in 1957–59. The urchins have been controlled by application of quicklime (North, 1965–74) or by divers using hammers. The beds have been restored by cultivation, transplantation and seeding with sporelings raised in an aquarium (North, loc. cit.). Because of these fluctuations the available amount of kelp has varied greatly from year to year and this is reflected in the maps of the Point Loma kelp beds (Fig. 9.4).

As an example of the fluctuations the annual kelp programme reports give the following information:

1970–71 Area and harvest of Point Loma beds declined.
Area and harvest of La Jolla beds increased.

1971–72 Harvest at Point Loma increased.
Harvest at La Jolla highest since 1958.

1972–73 Harvest at Point Loma highest since 1918.
Harvest at La Jolla highest since 1953.

1973–74 Harvest at Point Loma fell by 75% (winter storm effect).
Harvest at La Jolla fell by 60%.

Aleem (1973) estimated the standing crop of *Macrocystis* beds at La Jolla at 6–10 kg m^{-2} or an average annual yield of 10–15 tons ha^{-1} (10·16–15·2 tonnes ha^{-1}). This compares with an estimated yield of 60 tons ha^{-1} (60·9 tonnes ha^{-1}) per year off the Monterey Peninsula (Towle and Pearce, 1973).

The other important algae in this region are the agarophytes, of which *Gelidium cartilagineum* is the most important. Tseng (1947) estimated that up to 1500 tons (1520 tonnes) of fresh *Gelidium* were available in Southern California, but very little of this is actually harvested. Most of the weed processed (pp. 160–162) comes from the Mexican coast where 800 tons (812·8 tonnes) dry weight were harvested in 1965 as compared with 59 tons (59·9 tonnes) in 1959 (Guzman del Próo, 1969). This harvest has further increased to 2450 dry tons (2489 tonnes) in 1974 (Guzman del Próo, 1974). If necessary *Eucheuma uncinatum* could also be used, though Guzman del Próo (1969) estimated that the amount was only 1650 tons (1676 tonnes) wet weight ($\equiv$ 165 tons (167·8 tonnes) dry weight).

Macrocystis pyrifera is also harvested along the Mexican coast. Starting with 10 000 tons (10 160 tonnes) wet weight in 1956 the annual harvest in 1974 was 37 000 tons (37 580 tonnes) and it is probably still increasing. Guzman del Próo *et al.* (1971) have given maps of the beds and biomass data. Season is important as biomass can vary from 14–35 kg m^{-2} in spring and summer to 5–10 kg m^{-2} in winter. The maximum harvestable tonnage is estimated at 147 000 tons (149 300 tonnes) but it is likely that the readily exploitable weight is about 65 000 tons (66 040 tonnes).

There do not appear to be any data for the Pacific coastline of countries south of Mexico nor for Samoa or Hawaii.

9.14 AREA 81: SOUTH WEST PACIFIC

Macrocystis grows in some quantity around the shores of South Australia and Tasmania. An alginate industry was established in the area but has since closed down (see p. 205). The beds available were mapped for both Australia and New Zealand. On the east coast of Tasmania the beds were calculated to yield 355 000 tons (360 600 tonnes) fresh weight on the basis of three cuts per annum. This was probably an over-estimate and may have contributed to the failure of the venture. Since 1975 a new collection industry has been established exporting dried weed to Great Britian (see p. 205). In South Australia *Macrocystis angustifolia* has been estimated to have a biomass of about 1 400 000 tons (1 422 400 tonnes) but since the species is smaller it presents harvesting difficulties. In New Zealand, *Macrocystis* beds in Cook and Foveaux Straits were estimated to yield about 5200 tons (5283 tonnes) dry weed per annum (Rapson, Moore and Elliot, 1943) whilst a more recent survey indicates about 8500 tons (8636 tonnes) wet weight available of which, under recent New Zealand regulations, only a third could be harvested in any one season.

The other important resource in Australia is agarophytes, in particular *Gracilaria verrucosa.* In the Botany Bay area a yield of 270–360 kg ha^{-1} has been reported and there is probably sufficient weed to manufacture 100 tons (101·6 tonnes) agar annually. In New Zealand *Pterocladia capillacea* and *Pterocladia lucida* are harvested for agar manufacture, an average amount of 126 tons (128 tonnes) dry weight per year being harvested during 1962–71 (Watkinson and Smith, 1972) (see also p. 176). At the present time experiments are under way to ascertain whether *Gracilaria secundata* var. can be cultivated using oxidation pond effluent as a source of nutrients. The red alga *Pachymenia* spp. provides a good carrageenan and is a potentially valuable crop.

The bull kelp, *Durvillea antarctica,* occurs in considerable quantity on exposed coasts where individual plants can weight up to 50 kg. Recently a trial harvest of 10 000 tons (10 160 tonnes) was made for export to Japan, but the labour costs, since the weed has to be harvested manually, proved uneconomical.

9.15 AREA 87: SOUTH EAST PACIFIC

In Chile, species of *Gracilaria* provide the most important economic seaweed resource. Kim (1970) states that there are 11 principal beds covering 7·7 km^2 and estimated to contain 130 000 tons (132 080 tonnes) wet weight. In 1968, 1600 tons (1625·6 tonnes) dry weight were exported and in 1976, 2000 tons (2032 tonnes) (Moss, 1979). Other red algae exported from Chile are *Gelidium, Iridaea* and *Gigartina.* In 1968 some 374 tons (379·9 tonnes) dry weight were exported but by 1976 this had risen

Table 9.5 Actual and potential harvests of seaweeds

	Red algae		Brown algae	
Area	Recent harvests (1971–1973) (tons* wet wt × 10³)	Potential (tons wet wt × 10³)	Recent harvests (1971–1973) (tons wet wt × 10³)	Potential (tons wet wt × 10³)
Arctic Sea	—	—	—	—
N.W. Atlantic	35	100	6	500
N.E. Atlantic	72	150	208	2 000
W.E. Atlantic	—	(10)	1	1 000
E.C. Atlantic	10	50	1	150
Mediterranean/Black Sea	50	1000	1	50
S.W. Atlantic	23	100	75	2 000
S.E. Atlantic	7	100	13	100
W. Indian Ocean	4	120	5	150
E. Indian Ocea	3	100	10	+1 000 kerguelen 500
N.W. Pacific	545	650	825	1 500
N.E. Pacific	—	10	—	1 500
W.E. Pacific	20	50	1	50
E.C. Pacific	7	50	153	3 500
S.W. Pacific	1	20	1	100
S.E. Pacific	30	100	1	1 500
	802	2600	1293	15 600

* 1 ton = 1·016 tonnes.

Table 9.6 World production by countries and product type

Country	Estimated no. of commercial plants	Algin (tons* dry wt)	Carrageenan (tons dry wt)	Agar (tons dry wt)	Furcelleran (tons dry wt)
Canada	1	1 200			
Chile	2	200		200	
Denmark	3			100	1500
England	2	6 500	2 300	?	
France	3	1 200	100		
India	1	200	2 000	1000	
Japan	10–20	?	300	2000	
Korea	1/2			200	
Mexico	1			100	
Morocco	1			250	
Norway		3 000			
New Zealand	1			100	
Philippines	1		3 530†	100	
Portugal (and Azores)	2			400	
Spain	8–10		1 000	800	
USA	5	5 500	4 800	100	
	43	17 800	14 030	4400	1500

* 1 ton = 1·016 tonnes.
† Ricohermoso and Deveau, 1977.

to 3600 tons (3658 tonnes) (Moss, loc. cit.). Southern Chile is also rich in large kelps (*Macrocystis, Lessonia, Durvillea*), but problems are presented in collection, communications and great distance from markets. Even so, kelps are collected and exported as Moss (1979) reports an export of 9000 dry tons (9141 tonnes) (= 90 000 tons (91 410 tonnes) wet weight) for 1976. These kelps also occur in Peru but no attempt to exploit them or assess their quantities seems to have been made. There is a modest export of some 400 tons (406·4 tonnes) dry weight of *Gigartina* species (Moss, 1979).

9.16 ANTARCTIC

Macrocystis, Lessonia, Durvillea, Phyllogigas all occur in this region and rich beds are to be found down to 42 m. Many of the beds have been investigated but there are no estimates of quantity, and it is unlikely that any attempt to exploit them will be made until other, more accessible, beds have been fully exploited.

9.17 ACTUAL AND POTENTIAL HARVESTS

Table 9.5 sets out the actual and potential harvests of seaweeds in the regions just discussed (Michanek, 1975). From this table, it is evident that in the case of the red algae the Mediterranean/Black Sea region, the Indian Ocean, the South East Atlantic and East Central Pacific have an enormous potential for future exploitation. In the case of the kelps all the shores of the Atlantic Ocean could yield far greater quantities than are at present harvested. The same applies to the Pacific Ocean with the exception of the North-West region.

Table 9.7 (after Moss, 1979)

Product	Present output (tons* dry wt)	Potential harvest (tons wet wt)
Seaweed meal	30 000	100 000
Algin	18 000	400 000
Agar	4 500	150 000
Carrageenan	14 000	130 000
Furcelleran	1 500	10 000
Ncri	18 000	220 000
Wakame	7 000	60 000
Kombu	100 000	700 000

* 1 ton = 1·016 tonnes.

In a recent contribution, Moss (1979) summarized the world production of seaweed products for 1976 (Table 9.6).

In order to gauge the importance of the industry, Moss (1979) summarized the present production of various products and compared it with the potential (Table 9.7).

As Moss (1979) points out, the future of the seaweed industry depends on four factors: (a) quality of product, (b) chemical composition and consistency, (c) markets, (d) economy. Looking to the future and potential needs of a growing world population one could expect a demand for up to 50 000 tons (50 800 tonnes) of algin per year which could means a harvest of over 1 000 000 tons ($1 \cdot 016 \times 10^6$ tonnes) wet weight. Demand for carregeenan could rise to 30 000 tons (30 480 tonnes) dry weight and for agar to 20 000 tons (20 320 tonnes). Pressures to provide fertilizer and animal feed could require several million tons of brown rockweed and oarweed to be collected annually. There is also the potential demand to use large brown algae for energy production. The world's natural harvests will not be sufficient to meet all these demands and hence there will be an increasing need to cultivate the most important species, *Macrocystis, Undaria, Laminaria, Gracilaria, Eucheuma,* etc., artificially.

Bibliography*

Abbott, I.A. and Williamson, E.H. (1974), *Pacific Tropical Botanical Gardens.* Hawaii Publ.
Abdel-Fattah, A.F., Abed, N.M. and Edrees, M. (1972), *Aust. J. Mar. F.W. Res.,* **24**, 177–182.
Abe, K. Sakamoto, T., Sasaki, S.F. and Nisizawa, K. (1973), *Bot. Mar.,* **16**, 229–234.
Abe, K. and Kaneda, T. (1975), *Bull. Jap. Soc. Sci. Fish.,* **41**, 567–571.
Aberdeen, C. (1967), *Am. Perf. Cosm.,* **82**, 45.
Adams, R.W. and Austin, A. (1979), *Proc. 9th Int. Seaw. Symp.,* Santa Barbara, Science Press, Princeton, 499–508.
Adolph, W.H. and Whang, P.C. (1933), *Chinese J. Phys.,* **6**, 345.
Akiyama, K. (1965), *Bull. Tohoku Fish. Res. Lab.,* **25**, 143.
Albert, R. and Krause, M. (1919), *Chem, Zg.,* **43**, 97.
Aleem, A.A. (1973), *Bot. Mar.,* **16**, 83–95.
Alfimov, N.N. (1963), *Bot. Zh.,* **48**, 132.
Allen, J.H., Neish, A.C., Robson, D.R. and Shacklock, P.F. (1971), Atlantic Regional Lab. Halifax. Rep. 15, 15 pp. NRC. 12254, Canada.
Allsobrook, A.J.R., Nunn, S.R. and Parolis, H. (1968), *Proc. 6th Int. Seaw. Symp.,* La Marina Merchante, Madrid, Spain, 417–420.
Altmann, P.E. (1924), *Chem. Zg.,* **48**, 777.
American Gas Assoc. (1977), *Energy from Marine Biomass,* Arlington.
Anderegg, F. (1940), *Photo. Tech.,* **2**, 78.
Anderson, N., Frennette, E. and Webster, G. (1977), *Global village? Global pillage,* Social Action Commission, Charlottetown, P.E. 1., Canada.
Anderson, N.S. (1968), *Carbohydrates Res.,* **7**, 468.
Anderson, N.S., Dolan, T.C.S. and Rees, D.A. (1966), *Nature, Lond,* **205**, 1060.
Anderson, N.S., Dolan, T.C.S. and Rees, D.A. (1968a) *J. Chem. Soc.,* 596.
Anderson, N.S., Dolan, T.C.S. and Rees, D.A. (1968b) *J. Chem. Soc.,* 602.
Anderson, N.S. and Rees, D.A. (1966), *Proc. 5th Int. Seaw. Symp.,* p. 243, Pergamon Press, Oxford.
Anderson, W. (1969), *Proc. 6th Int. Seaw. Symp.* , La Marina Merchante, Madrid, pp. 417–420.
Andrew, T.R. and Macleod, W.C. (1970), *Food Product Development,* Aug.-Sept.

* The literature is now so extensive that it is not possible to provide a complete bibliography.

Anglo, P.G. (1977), *J. Phycol.,* **13**, Suppl. Abst. 10, p. 4.
Anglo, P.G., Baens-Arcega, L., Arguelles, A. and Sarabia, N. (1973), *Phil. J. Sci.,* **102**, 55–65.
Angst, E.C. (1929), *Publ. Puget Sd. Biol. Sta.,* **7**, 49.
Annett, H.E., Darbishire, F.V. and Russell, E.J. (1907), *J.S.E. Agric. Coll. Wye,* **16**.
Anno, K. Seno, N. and Ota, M. (1968), *Proc. 6th Int. Seaw. Symp.,* La Marina Merchante, Madrid, Spain, 421–426.
Anno, K. and Uemuria, K. (1972), *Proc. 7th Int. Seaw. Symp.,* Univ. Tokyo Press, Japan, pp. 439–442.
Anon (1907), *J. Am. Med. Assoc.,* **48**, 142.
Anon (1918a), *Chem. Met. Eng.,* **18**, 566.
Anon, (1918b), *Chem. Met. Eng.,* **19**, 450.
Anon (1921a), *Farben. Zg.,* **26**, 1849.
Anon (1921b), *J. Ind. Engng. Chem.,* **13**, 413.
Anon (1926), *Wochb. Papierfarb.,* **57**, 1186.
Anon (1927), *Chem. Met. Engng.,* **34**, 294.
Anon (1928a), *Chem. Zg.,* **52**, 686.
Anon (1928b), *Chem. Drugg.,* **108**, 668.
Anon (1930), *U.S. Bur. Fish. Spec. Memo.,* 2315.
Anon (1934), *Oil, Paint Drug Rep.,* **125**, 30.
Anon (1937a), *Deutsche, Fis. Rundschau.,* **20**.
Anon (1937b),*Deutsche Fishereiwert,* **21**.
Anon (1937c), *Gardeners' Chron.,* **102**, 3rd ser., 264.
Anon (1938a), *Keeplgs. Textil. Ztsch.,* **41**, 690.
Anon (1938b), *Centrb. Bakt. Parasitenk. Abt. 1.,* **143**, 142.
Anon (1939), *Centrb. Bakt. Parasitenk. Abt. I.,***144**, 67.
Anon (1939–40), Report, *Dept. Ind. Comm., Ind. Res. Counc.,* Dublin.
Anon (1940), *Sci News Letter,* Feb. 3, 68.
Anon (1941a), *Sci. News Letter,* **39** and **40**, 201.
Anon (1941b), *Foreign Comm. Weekly,* **4**, 71.
Anon (1941c), *J. Counc. Sci. Indu. Res. Australia,* **14**, 221.
Anon (1941d), *Fisheries News Letter, J. Counc. Sci. Indu. Res. Australia,* **I**, 11.
Anon (1942a), *Pop. Sci.,* **141**, 52.
Anon (1942b), *Foreign Comm. Weekly,* **8**, 29.
Anon (1942c), *Foreign Comm. Weekly,* **8**, 21.
Anon (1942d), *Fishery Market News,* **4**, 23.
Anon (1942–3), *Report, Dept. Ind. Comm., Ind. Res. Counc.,* Dublin.
Anon (1943a), *Library Bibliog. Ser. London, Sci. Mus.*
Anon (1943b), *Bull. Imp. Inst. Gt. Brit.,* **41**, 163.
Anon (1944a), *Textile World,* **94**, 130.
Anon (1944b), *Times Trade and Eng. Suppl.,* May, p. 18.
Anon (1944c), *Chem. Age,* **50**, 84.
Anon (1944d), *Nature, Lond.,* **154**, 247.
Anon (1944e), *Country Life,* Oct.
Anon (1944), *Mon. Sci. News.,* Dec.
Anon (1946), *Ann. Rep. Fis. Res. Bd.,* Canada.
Anon (1967), A new type of food algae. *Inst. Franc. Petrole.* 14237A.
Anon (1968), *Nutrit. Rev.,* **26**, 182.

Araki, C. (1937a), *J. Chem. Soc. Japan,* **58**, 1085.
Araki, C. (1937b), *J. Chem. Soc. Japan,* **58**, 1214.
Araki, C. (1937c), *J. Chem. Soc. Japan,* **58**, 1338.
Araki, C. (1937d), *J. Chem. Soc. Japan,* **58**, 1351.
Araki, C. (1937e), *J. Chem. Soc. Japan,* **58**, 1362.
Araki, C. (1938a), *J. Chem. Soc. Japan,* **59**, 304.
Araki, C. (1938b), *J. Chem. Soc. Japan,* **59**, 434.
Araki, C. (1944), *J. Chem. Soc. Japan,* **65**, 725.
Araki, C. (1966), *Proc. 5th Int. Seaw. Symp.,* p. 3, Pergamon Press, Oxford.
Araki, C. and Arai, K. (1956), *Bull. Chem. Soc. Japan,* **29**, 339.
Araki, C. and Hirase, S. (1953), *Bull. Chem. Soc. Japan,* **26**, 463.
Arasaki, S. (1949), *Bull. Jap. Soc. Sci. Fish.,* **15**, 137.
Arasaki, S., Fujiyama, T. and Saito, Y. (1956), *Bull. Jap. Soc. Sci. Fish.,* **22**, 167.
Aravio-Torre, J. (1953), *Proc. 1st Int. Seaw. Symp.,* Edinburgh, p. 65.
Armstrong, E.F. and Miall, L.M. (1946), *Raw materials from the Sea,* Leicester.
Astbury, W.T. (1945), *Nature, Lond.,* **155**, 667.
Aston, B.C. (1916), *N.Z.J. Agric.,* **13**, 446.
Atlas, E.F., Gordon, L.I. and Tomlinson, R.D. (1977), In: *Marine Plant Biomass of the Pacific Northwest,* (Krauss, R.W., ed.), pp. 57–59, Oregon State University Press.
Atsuki, K. and Tomoda, (1926), *J. Soc. Chem. Ind. Japan,* **29**, 509.
Audoin, J. and Perez, R. (1970), *Sci. Pêche,* **199**, 1–11.
Augier, H. (1976a), *Bot. Mar.,* **19**, 127–143.
Augier, H. (1976b), *Bot. Mar.,* **19**, 351–377.
Augier, H. (1976c), *Bot. Mar.,* **19**, 245–254.
Augier, H. and Santimone, M. (1978), *Bot. Mar.,* **21**, 337–341.
Ausman, L.H. (1942), *Commercial Intelligence J.,* **61**, 129, 409.
Austin, A.P. (1960), *Hydrobiol.,* **14**, 255.
Ayers, S.H., Madge, C.S. and Rupp, P. (1920), *J. Bacteriol.,* **5**, 589.

Baardseth, E. (1955), *Inst. Indust. Res. Stand. Rep., Eire.*
Baardseth, E. (1966), *Proc. 5th Int. Seaweed Symposium,* p. 19, Pergamon Press, Oxford.
Baardseth, E. (1968), *Algologia marina,* UNESCO BMS-RD/AVS. (553), 13 pp.
Baardseth, E. (1970), *F.A.O. Fish Synopsis* (38), Rev. 1.
Baardseth, E. and Grenager, B. (1962), *Norsk. Inst. Tang-og tareforsk. Rep. 25.*
Baier, W.R. and Manchester, T.C. (1943), *Food, Ind.,* **15**, 94.
Bailey, E.M. (1929), *Conn. Agr. Exp. Sta. Bull.,* **307**, 807.
Baily, N.A. and Kelly, S. (1955), *Biol. Bull.,* **109**, 13.
Baker, J.J. and Murphy, V. (1976), *Handbook of Marine Science Compounds from Marine Organisms,* Vol. 1, CRC Press.
Bal, S.N., et al., (1946), *Pharm. J.,* **157**, (4th Ser. **103**), 152.
Banerji, S.N. and Ghosh, S. (1939a), *Proc. Natl. Acad. Sci. India,* **9**, 144;
Banerji, S.N. and Ghosh, S. (1939b), *Proc. Natl. Acad. Sci. India,* **9**, 148.
Bardach, J.E., Ryther, J.H. and McLarney, W.O. (1972), *Culture of Seaweeds,* In: *Aquaculture – The Farming and Husbandry of Freshwater Marine Organisms,* pp. 790–819, J.W. Wiley–Interscience, New York.

Barendamm, W. (1931), *Ber. Deut. Botan. Ges.*, **49**, 288.
Barilotti, G.D. and Silverthorne, W. (1972), *Proc. 7th Int. Seaw. Symp., Japan*, pp. 255–261.
Barry, V.C. (1938), *Sci. Proc. Roy. Dublin. Soc.*, **21**, 615.
Barry, V.C. (1939), *Sci. Proc. Roy. Dublin. Soc.*, **22**, 59.
Barry, V.C. and Dillon, T. (1936), *Sci. Proc. Roy. Dublin Soc.*, **21**, 285.
Barry, V.C. and Dillon, T. (1944), *Chem. Ind.* 167.
Barry, V.C., Dillon, T. and O'Muineachain, P. (1936), *Sci. Proc. Roy. Dublin Soc.*, **21**, 289.
Barry, V.C., Dillon, T. and McGettrick, W. (1942), *J. Chem. Soc.*, 183.
Barry, V.C., Halsall, T.G., Hirst, E.L. and Jones, J.K.N. (1949), *J. Chem. Soc.*, 1468.
Bartlett, H.H. (1940), *Bull. Tor. Bot. Club*, **67**, 347.
Bauche, J.D. (1906), *Agricola Club J.*
Bauder, C.S. (1920), *Calif. Fish Game*, **6**, 31.
Bauer, R.W. (1884), *J. Prakt. Chem.*, **30**, 367.
Bayley, S.T. (1955), *Biochim. Biophys. Acta.*, **17**, 194.
Beckmann, E. (1915), *Sitzber. Akad. Wiss.* Berlin 645.
Beckmann, E. and Bark, E. (1916), *Sitzber. Akad. Wiss.* Berlin 1009.
Bellan-Santini, O. (1969), *Rec. Trav. Stat. Mar. Endoume*, **47**, 5–249.
Benitez, L.V. and Macaranas, J.M. (1979), *Proc. 9th Int. Seaw. Symp., Santa Barbara*, Science Press, Princeton, 353–360.
Benoit, R.J. (1964), *Algae and Man*, Pergamon Press, Oxford.
Berard-Therriault, L. and Cardinal, A. (1973), *Bot. Mar.*, **16**, 96–102.
Berger, F. (1961), *Acta Phytother.*, **8**, 182.
Bergeron, M. (1949), *Bull. Tech. Inf. Ingen. Serv. Agric.*, **44**, 613.
Bhakuni, D.S. and Silva, M. (1974), *Bot. Mar.*, **17**, 40–51.
Biggar, (1917), *Trans. Am. Inst. Chem. Engng.*, **10**, 85.
Bird, G.M. and Haas, P. (1931), *Biochem. J.*, **25**, 403.
Bisserie, M. (1907), *Ann. Inst. Pasteur*, **21**, 235.
Bjorndal, H., Eriksson, K.E., Garegg, P.J., Lindberg, B. and Swan, B. (1965), *Acta. Chem. Scand.*, **19**, 2309.
Black, M. and Edelman, J. (1970), *Plant Growth*, Heineman, 193 pp.
Black, W.A.P. (1948), *J. Soc. Chem. Ind.*, **67**, 165, 169, 355.
Black, W.A.P. (1949), *J. Soc. Chem. Ind.*, **68**, 183.
Black, W.A.P. (1950), *J. Mar. Biol. Assoc., U.K.*, **29**, 45, 379.
Black, W.A.P. (1953), *Agriculture*, **60**, 126.
Black, W.A.P. (1955a), *Agriculture*, **62**, 12, 57.
Black, W.A.P. (1955b), *Chem. Ind.*, 1640.
Black, W.A.P., Blakemore, W.R. Colqwhoun, J.A. and Dewar, E.T. (1966), *Proc. 5th Int. Seaweed Symposium*, p. 251, Pergamon Press, Oxford.
Black, W.A.P. and Cornhill, W.J. (1954), *Chem. Ind.*, **18**, 514.
Black, W.A.P. and Dewar, E.T. (1949), *J. Mar. Biol. Assoc., U.K.*, **28**, 673.
Black, W.A.P., Dewar, E.T. and Woodward, F.N. (1952), *Food Agric.*, **3**, 122.
Black, W.A.P., and Mitchell, R.L. (1952), *J. Mar. Biol. Assoc. U.K.* **30**, 575.
Black, W.A.P., Richardson, W.D. and Walker, C.T. (1959), *Econ. Proc. Roy. Dubl. Soc.*, **4**, 8.

Black, W.A.P. and Woodward, F.N. (1957), *Empire J. Expt. Agric.*, **25**, 51.
Blaine, G. (1951), *Postgrad. Med. J.* **27**, 613.
Blasdale, W.C. (1899), *U.S. Dept. Agric. Bull.*, **68**, 46.
Bleich, M. (1895), *Centrb. Bakt. Parasitenk. Abt. I.*, **17**, 360.
Blunden, G. (1972), *Proc. 7th Int. Seaw. Symp., Japan*, pp. 584–589.
Blunden, G. (1977), *Cytokinin Activity of Seaweed Extracts*, In: *Marine Natural Products Chemistry*, (Faulner, D.J. and Fenical, W.H., eds.), pp. 337–344. Plenum Press.
Blunden, G. and Woods, D.L. (1969), *Proc. 6th Int. Seaw. Symp.*, La Marina Merchante, Madrid, Spain, p. 647.
Board of Agriculture, Scotland (1914), *The Kelp Industry*.
Bodard, M. and Christiaen, D. (1978), *Bull. Phycol. de France*, **23**, 19–28.
Bogan, E.J. and Moyer, H.V. (1942), *Ind. Engng. Chem. Anal. Edit.*, **14**, 849.
Boney, A.D. (1966), *Biology of Marine Algae*, Hutchinson, London.
Booth, E. (1964), *Proc. 4th Int. Seaw. Symp.*, p. 385, Pergamon Press, Oxford.
Booth, E. (1966), *Proc. 5th Int. Seaw. Symp.*, p. 349, Pergamon Press, Oxford.
Booth, E. (1969), *Proc. 6th Int. Seaw. Symp.*, La Marina Merchante, Madrid, Spain, p. 655.
Bose, J.L., Siddiqui, K. and Siddiqui, S. (1943), *J. Sci. Indust. Res., India*, **1**, 98.
Bouthillier, L.P. and Cosselin, G. (1937), *Naturaliste Canadien*, **64**, 65.
Brandt, R.P. and Turrentine, J.W. (1923), *U.S. Dept. Agric. Bull.*, 1191.
Braud, J.P. and Perez, R. (1979), *Proc. 9th Int. Seaw. Symp., Santa Barbara*, Science Press, Princeton, 533–540.
Brin, F. (1926), *J. d'Agric. Pract.*, **90**, 234.
Britten, G.F., *S. African J. Sci.* (*Ref. in Scient. Am.*, **118**, 475, 1918).
Brodie, J. and Stiven, D. (1942), *J. Hyg.*, **42**, 498.
Bronfenbrenner, J. and Hetler, D. (1928), *Proc. exp. Biol. Med.*, **25**, 480.
Brown, O.H. and Sweek, W.O. (1917), *J. Am. Med. Soc.*, **69**, 467.
Buchanan, J., Percival, E.E. and Percival, E.G.V. (1943), *J. Chem. Soc.*, 51.
Buchner, E.H. and Kleijn, D. (1927), *Proc. Akad. Wettenschapen Amsterdam*, **30**, 740.
Burgvic, G. (1934), *Bull. Acad. Sci. U.R.C.S.* 7 ser. **6**, 837.
Burns, R. (1971), *An Autecological Study of the Marine Red Alga Gigartina stellata*, PhD. Thesis, University of New Hampshire, pp. 1–95.
Burns, R.L. and Mathieson, A.C. (1972a), *J. Exp. Mar. Biol. Ecol.*, **8**, 1–6.
Burns, R.L. and Mathieson, A.C. (1972b), *J. Expt. Mar. Biol. Ecol.*, **9**, 77–96.
Burkill, H.M., Greenwood-Barton, L.H. and Crowther, P.C. (1968), *Gard. Bull. Singapore*, **22**, 429.
Burkholder, P.R., Burkholder, L.M. and Almodovar, L.R. (1961), *Bot. Mar.*, **2**, 150.
Burlew, J.G. (1953), *Carneg. Inst. Wash. Pub.* 600.
Butler, M.R. (1931), *Plant Physiol.*, **6**, 295.
Butler, M.R. (1934), *Biochem. J.* **28**, 759.
Butler, M.R. (1935), *Biochem. J.* **29**, 1025.
Butler, M.R. (1936), *Biochem. J.* **30**, 1338.
Butler, M.R. (1937), *Biol. Bull.*, **73**, 143.
Buyankina, S.K. (1977), *Proc. all-Union Res. Inst. Mar. Fish Oceanog.*, **124**, 52–56.

Caces-Borja, P. (1973), *The Seaweed Industry of the Philippines*, S.E. Asia Fish. Devel. Center Tech. Sem. on S. China Sea Fish Resources, Bangkok 7 pp.

Cady, W.H. (1948), *Am. Dyes. Rep.*, **37**, 283.
Calabek, J. (1927), *Protoplasma*, **3**, 17.
Calabek, J. (1929), *Protoplasma*, **7**, 541.
Cameron, F.K. (1913), *J. Franklin Inst.*, **176**, 347.
Cameron, F.K., Crandal, W.C., Rigg, G.B. and Frye, T.C. (1915), *U.S. Dept. Agric. Rep.* 100, Washington.
Cameron, M.C., Ross, A.G. and Percival, E.G.V. (1948), *J. Soc. Chem. Ind.*, **67**, 161.
Campbell, W.L. (1946), *N.Z. Weekly News*, Sept. 18.
Caraes, A. (1968), *Proc. 6th Int. Seaw. Symp.*, La Marina Merchante, Madrid, Spain, 663–669.
Cardinal, A. and Breton-Provencher, M. (1977), *Bot. Mar.*, **20**, 243–257.
Carey, C.L. (1921), *Bull. Torrey. Bot. Club*, **48**, 173.
Cauer, H. (1938), *Biochem. Z.*, **299**, 69.
Causey, N.B., Pryterch, J.P., McCaskill, J., Human, H.J. and Wolf, F.G. (1945), *Duke Univ. Mar. Sta. Bull.*, **3**, 19.
Cavaliere, A. (1969), *Boll. Pesca. Piscio. Idiobiol.*, **22**, 167–174.
Cavi, G. de and Guiseppe, S. (1974), *8th Inst. Seaw. Symp. Bangor, Abst.*, B. p. 54.
Cerezo, A.S. (1974), *Carb. Res.*, **36**, 201–204.
Challen, S.B. and Hemingway, J.C. (1966), *Proc. 5th Int. Seaw. Symp.*, p. 359, Pergamon Press, Oxford.
Chamberlain, N.H., Johnson, A. and Speakman, J.B. (1944), *J. Soc. Dyers Colourists.*
Channubhotla, V.S.R., Kalimuthu, S., Najmuddin, M., Slavarag, M. and Panigraphy, R. (1977), Abst. 455, *9th Seaw. Symp.*, Santa Barbara.
Chapman, A.R.O. and Craigie, J.S. (1978), *Mar. Biol.*, **46**, 209–213.
Chapman, A.R.O. and Doyle, R.H. (1979), *Proc. 9th Inter. Seaw. Symp.*, Santa Barbara, Science Press, Princeton, 125–132.
Chapman, V.J. (1944), *J. Mar. Biol. Ass.*, **26**, 37.
Chapman, V.J. (1945), *Nature, Lond.*, **155**, 673.
Chapman, V.J. (1948), *Econ. Bot.*, **2**, 363.
Chapman, V.J. (1962), *The Algae*, Macmillan, London.
Chapman, V.J. (1969), *Explorer II*, **3**, 4.
Chapman, V.J. (1980), *Organic Chemicals from the Sea, Proc. CHEMSEC Conf.*, Brisbane (in press).
Charles, A.C. and Neushal, M. (1979), *Aquatic Bot.*, **6**, 67–78.
Chase, F.M. (1942), *Ann. Rept. Smith. Inst. Wash. 1941*, 401.
Chaveaux, H. (1927), These Fac. Med., Paris 9.
Cheeseman, T.L. (1888), *Am. Nat.*, **22**, 472.
Chemin, E. (1929), *C.R. Acad. Sci., Paris*, **188**, 1624.
Chen, L.C.M. and McLachlan, J. (1972), *Canad. J. Bot.*, **50**, 1055–1060.
Chen, L.C.M., McLachlan, J., Neish, A.C. and Shacklock, P.F. (1973), *J. Mar. Biol. Ass. U.K.*, **53**, 11–16.
Cheney, D.P. and Mathieson, A.C. (1977), *J. Phycol.*, **13**, Suppl. Abst. 61, p. 13.
Cheng, T.H. (1968), *Econ. Bot.*, **20**, 215–236.
Chesmeau, R. (1931), *Tiba*, **9**, 1265.
Chester, C.G.C., Apinis, A. and Turner, M. (1956), *Proc. Linn. Soc. Lond.*, **166**, 87.
Chidambaram, K. and Unny, M.H. (1958), *Proc. 1st Int. Seaweed Symposium*, Edinburgh, p. 67.

Christensen, J. (1971), *Proc. Canad. Atl. Seaw. Indust.*, 1971. Industrial Development Branch Environment, Canada pp. 23–27.
Cioglia, L. (1940), *Ric. Sci. Prog. Tech. Econ. Naz.*, **11**, 179.
Claudio, F. and Stenardo, B. (1965), *Minerva Medica*, **56**, 3617–3622.
Claudio, F. and Stenardo, B. (1966), *Proc. 5th Int. Symp. Halifax*, p. 369, Pergamon Press, Oxford.
Clark, A.H. (1887), *Fish and Fish Ind. U.S.*, **2**, 219.
Clarke, B.L. (1935), *Am. Chem. Soc. J.*, **47**, 1954.
Clement, G. (1978), *Soc. Phycol. de France*, **23**, 21–57.
Clingman, A.L., Nunn, J.R. and Stephen, A.M. (1957), *J. Chem. Soc.*, 197.
Cmelik, C. (1948), *Acta Adriat.*, **3**, 127.
Coassini-Lokar, L. and Bruni, G. (1970), *Falconara Marittima, Techografica*, pp. 26–37, CNR Lab. technol. delta pesca, Ancona.
Cobb, J.B. (1918), *Cal. War. Papers*, State Council of Defence.
Cogswell, I.F. (1942), *Am. Mag.*, **5**, 42.
Coker, R.E. (1908), *Bull. Bur. U.S. Fish*, **28**, 335.
Colin, H. and Gueguen, E. (1930), *C.R. Acad. Sci. Paris*, **190**, 884.
Colin, H. and Ricard, P. (1930), *Bull. Soc. Chim. Biol.*, **12**, 1392.
Colinvaux, L.H. (1966), *Proc. 5th Int. Seaw. Symp. Halifax*, Pergamon Press, Oxford, pp. 91–98.
Congested Districts Board for Ireland (1898), *The Kelp industry*.
Conn, H.J. and Dotterer, W.D. (1916), *N.Y. Agr. Exp. Sta. Bull.*, **53**, 12.
Connell, J.J., Hirst, E.L. and Percival, E.G.V. (1950), *J. Chem. Soc.*, 3494.
Cook, H. and Smith, D.B. (1954), *Can. J. Biochem. Physiol.*, **32**, 227.
Cooper, N.C. and Johnstone, G.R. (1944), *Am. J. Botany*, **31**, 638.
Cooper, W.F. and Nuttall, W.H., *Br. J. Photog.*, **55**, 62, 109.
Cooper, W.F., Nuttal, W.H. and Freak, G.A. (1909), *7th Int. Cong. Appl. Chem., Sect.*, **9**, 62.
Cotton, A.D. (1910), *Kew Bull. Misc. Inf.*, p. 15.
Cotton, A.D. (1912), *Proc. R. Irish Acad.* **31**, 1–178.
Cotton, A.D. (1914), *Kew Bull.*, 219.
Cotton, A.D. (1915), *J. Linn. Soc. Botany*, **43**, 168.
Cottrell, I.W. and Kovacs, P. (1977), 'Algin' in *Food Colloids*, (Graham, H.D. ed.), Avi. Publ. Co., Connect., pp. 438–463.
Craigie, J.S. (1971), *Proc. Can. Atl. Seaw. Indust. Meeting, P.E.I. 1971.* Indust. Devlop. Branch. Environ., Canada, pp. 17–22.
Craigie, J.S. and Wong, K. (1979), *Proc. 9th Int. Seaw. Symp., Santa Barbara*, Science Press, Princeton, 369–378.
Cranwell, L.M. and Moore, L.B. (1933), *Trans. R. Soc. N.Z.*, **67**, 375.
Crossman, E.G. (1918), *Scient. Am.*, **118**, 475.
Cullen, I.A. (1914), *J. Ind. Engng. Chem.*, **6**, 581.
Cunningham, J. (1919), *India J. Med. Res.*, **6**, 560.
Czapke, K. (1964), *Proc. 4th Intern. Seaw. Symp., Brarritz*, p. 393–437, Pergamon Press, Oxford.
Czapke, K. (1966), *Proc. 5th Int. Seaweed Symposium*, p. 293, Pergamon Press, Oxford.

Dahlberg, A.C. (1926), *N.Y. Agr. Exp. Sta. Bull.* No. 536.
Dahlberg, A.C. (1927), *J. Dairy Sci.*, **10**, 106.
Davidson, C.J. (1906), *Bull. Imp. Inst.*, **4**, 125.
Davis, F.W. (1950), *Science, N.Y.*, **111**, 150.
Davis, F.W. (1950), *Proc. Gulf. Carib. Fish. Inst.*, **2**, 102–103.
Dawes, C.J. (1977), *Proc. 9th Int. Seaw. Symp.*, Santa Barbara, Science Press, Princeton, 199–208.
Dawes, C.J., Lawrence, J.M., Cheney, D.P. and Mathieson, A.C. (1974), *Bull. Mar. Sci.*, **24**, 286–299.
Dawes, C.J., Mathieson, A.C. and Cheney, D.P. (1974a), *Bull. Mar. Sci.*, **24**, 235–273.
Dawes, C.J., Mathieson, A.C. and Cheney, D.P. (1974b), *Bull. Mar. Sci.*, **24**, 273–285.
Dawes, C.J., Stanley, N.F. and Moon, R.E. (1977), *Bot. Mar.*, **20**, 437–442.
Dawes, C.J., Stanley, N.F. and Stancioff, D.J. (1977), *Bot. Mar.*, **20**, 137–147.
Dawson, E.Y. (1966), *Marine Botany. An Introduction,* Holt Rhinehart, New York.
Day, D. (1942), *Bull. Torrey Bot. Club,* **69**, 11.
De, P.K. (1939), *Proc. R. Soc. Ser. B.*, **127**, 121.
De Boer, J.A. (1979), *Proc. 9th Int. Seaw. Symp.*, Santa Barbara, Science Press, Princeton, 263–272.
De Boer, J.A., Lapointe, B.E. and D'Elia, C.F. (1976), In: *Marine polyculture based on natural food chains and recycled wastes,* (Ryther, J.H. ed.), Woods Hole Oceanog. Inst. Tech. Rept. WHO 1–76–92.
De Boer, J.A. and Ryther, J.H. (1977), In: *Marine Plant Biomass of the Pacific northwest Coast,* (Krauss, R.W., ed.), pp. 231–249, Oregon State University Press.
De Boer, J.A., Ryther, J.H. and Lapointe, B.E. (1976), In: *Marine Polyculture based on Natural Food Chains and Recycled Wastes,* (Ryther, J.H., ed.), Woods Hole Oceanog. Tech. Rept. WHO 1–76–92.
Delf, E.M. (1940), *South E. Nat.*, **45**, 1.
Delf, E.M. (1943), *Nature, Lond.*, **152**, 149.
Delf, E.M. (1940), *Proc. R. Soc. Arts.*, **91**, 505.
Delf, E.M. (1944), *Nature, Lond.*, **153**, 223.
De Loach, W.S., Wilton, O.C., Humm, H.J. and Wolf, F.A. (1946), *Duke Univ. Mar. Sta. Bull.*, **3**, 25–39.
Deraniyagala, P.E.P. (1933), *Ceylon J. Sci.*, **5**, 49.
Deschiens, M. (1926), *Chim. et Ind., Paris,* **15**, 675.
Deschiens, M. (1928), *Bull. Soc. Oceanog. Fr.*, **43**.
Deschiens, M. (1930), *Rev. Gener. Mat. Plast.*, **6**, 261.
De Toni, G.B. (1907), *Bot. Ulf. Min. d'Agric.*, **6**, 249.
Diaz-Piferrer, R.M. (1961), *Inst. Cubano Invest. Tecnol.*, No. 17.
Diaz-Piferrer, R.M. (1967), Algas de importanica economica, In: *El Farol,* No. 222, p. 18.
Diaz-Piferrer, R.M., De la Campa, J.M.N. and Losa, C.S. (1961), *Instit. Cubano Invest. Tecnol.*, No. 16.
Diaz-Piferrer, R.M. and Perez, C.C. de (1964), *Taxonomía, Ecología y Valor Nutrimental de Algas Marinas de Puerto Rico: Algas Productoras de Agar,* Hato Rey, P.R., Administracíon de Fomento Ecómico de Puerto Rico y Universidad de Puerto Rico, 145 pp.

Dickover, E.F. (1927), *U.S. Dept. Comm. Rept.*, **5**, 67.
Dijachkorskii, I.S. and Dudevov, V. (1940), *Colloid J., U.S.S.R.*, **6**, 333.
Dillon, T. (1929), *Nature, Lond.*, **123**, 161.
Dillon, T. (1938), *Chem. Ind.*, **57**, 616.
Dillon, T. and Cavelle, E.F. (1935), *Econ. Proc. R. Dublin Soc.*, **2**, 407.
Dillon, T. and McGuiness, A. (1931), *Sci. Proc. Dublin Soc.*, **20**, 129.
Dillon, T. and O'Colla, P. (1940), *Nature, Lond.*, **145**, 749.
Dillon, T. and O'Tuama, T. (1935), *Sci. Proc. R. Dublin, Soc.*, **21**, 147.
Dizerbo, A.-H. (1964), *Penn ar Bed.*, **4**, 173.
Dokau, S. (1924a), *Kolloid Z.*, **34**, 155.
Dokau, S. (1924b), *Kolloid Z.*, **35**, 11.
Dolan, T.C.S. and Rees, D.A. (1965), *J. Chem. Soc.*, 3534.
Doty, M.S. (1973), *Micronesica*, **9**, 59–73.
Doty, M.S. (1979), *Proc. 9th Int. Seaw. Symp., Santa Barbara*, Science Press, Princeton, 35–58.
Doty, M.S. and Alvarez, V.B. (1973), *Mar. Techn. Soc. Proc. Ann. Conf.*, **9**, 701–798.
Doty, M.S. and Alvarez, V.B. (1974), *Marine, Tech. Soc. J.*, **9**, 30–35.
Doty, M.S. and Santos, G.A. (1978), *Aquatic Bot.*, **4**, 143–150.
Drummond, D.W., Hirst, E.L. and Percival, E. (1958), *Chem. Ind.*, 1088.
Duff, R.B. and Percival, E.G.V. (1941), *J. Chem. Soc.*, 830.
Dumon-Tondo, F.U. (1930), *Bull. Sta. Biol. Arcachon.*, **27**, 175.
Dunstan, W.M. and Menzel, D.W. (1971), *Limnol. and Oceanogr.*, **16**, 623–632.
Dunstan, W.M. and Tenore, K.R. (1972), *Aquaculture*, **1**, 181–192.
Durairatnam, M. (1961), *Economic Marine Algae of Ceylon*, Occas. Pap. IPFC. (61), 1–5.
Durairatnam, M. (1966), *Bull. Fish. Res. Sta. Ceylon*, **19**, 1–2.
Durrant, N.W. (1967), *Comm. Fish. Rev.*, **29**, 65.

Earle, S.A. (1969), *Phycologia*, **7**, 71–254.
Eddy, B.P. (1956), *J. exp. Botany*, **7**, 372.
Edelstein, T., Bird, C.J. and McLachlan, J. (1976), *Canad. J. Bot.*, **54**, 2275–2290.
Edward, D.W., Paul, T.M. and Skoryna, S.C. (1964), *Canad. Med. Ass. J.*, **91**, 1006–1070.
Edwards, R. (1970), *Contrib. Mar. Sci.*, **15**, Suppl. 128 pp.
Effront, J. (1935), *C.R. Acad. Sci. Paris*, **180**, 29.
Ehresmann, D.W., Deig, E.F., Hatch, M.T., Disalvo, L.H. and Vedros, N.A. (1977), *J. Phycol.*, **13**, 37–40.
Ehresmann, D.W., Deig, E.F., Hatch, M.T., Riedlinger, D.L. (1975), *Abst. Ann. Meet. Am. Soc. Microbiol.*, A7.
Elenkin, A.A. (1934), *Sovietskaia. Bot.*, **4**, 89.
Elin, V. (1932), *Vrach. Delo.*, **15**, 186.
Elin, V. et al. (1931), *Byull. Nauch. Issfledovatel. Khim-Farm. Inst.*, 140.
Elsner, H. et al. (1937), *Z. physiol. Chem.*, **246**, 244.
Elsner, H. et al. (1938), *Arch. exp. Path. U. Pharm.*, **190**, 510.
Endo, T. and Matsudaira, Y. (1960), *Bull. Jap. Soc. Sci. Fish.*, **26**, 871.

Englar, J.R., Whyte, J.N.C. and Kung, M. (1979), *Proc. 9th Int. Seaw. Symp., Santa Barbara,* Science Press, Princeton, 319–328.
Esdorn, I. (1935), *Heil Gewisspflanzen,* **86**, 16.
Esikov, L.A. (1938), *Lib. Prakt., U.S.S.R.,* **13**, 9.
Etcheverry, H. (1953), *Proc. 1st Int. Seaweed Symposium,* Edinburgh, p. 108.
Evenden, W. and Schuster, C.E. (1938), *Stain Technol.,* **13**, 145.
Evitushenko, V.A. and Pankratova, O.I. (1963), *Ref. Zh. Biol.,* No. 21V9.
Fairbrother, F. and Mastin, H. (1923), *J. Chem. Soc.,* **123**, 1412.
Feldmann, J. (1953), *Proc. 1st Int. Seaw. Symp.,* Edinburgh, p. 88.
Fellers, C.R. (1961a), *Soil Sci.,* **2**, 255.
Fellers, C.R. (1916b), *J. Ind. Engng. Chem.,* **8**, 1128.
Fenical, W., McConnell, O.J. and Stone, A. (1979), *Proc. 9th Int. Seaw. Symp., Santa Barbara,* Science Press, Princeton, 387–400.
Ferguson Wood, E.J. (1941a), *J. Council Sci. Ind. Res., Australia,* **14**, 221.
Ferguson Wood, E.J. (1941b), *J. Council Sci. Ind. Res., Australia,* **14**, 315.
Ferguson Wood, E.J. (1942), *J. Council Sci. Ind. Res., Australia,* **15**, 295.
Ferguson Wood, E.J. (1945), *J. Council Sci. Ind. Res., Australia,* **18**, 263.
Ffrench, R.A. (1970), *The Irish Moss Industry,* Canadian Atlantic Dept. of Fisheries and Forestry, Ottawa, 230 pp.
Field, I.A. (1921a), *Dept. Comm. Bur. Fish. Econ. Inc.,* **51**, 1.
Field, I.A. (1921b), *Chem. Age, N.Y.,* **29**, 485.
Field, I.A. (1922), *U.S. Bur. Fish. Doc.,* **929**.
Filho, E.C. de and De Paula, E.J. (1979), *Proc. 9th Int. Seaw. Symp., Santa Barbara,* Science Press, Princeton, 479–486.
Filho, E.C. de O. and Quege, N. (1978), Pesquisa e Desenvolvimento, Instit. Pesquisas Tecnol. do Est-, Saõ Paulo, 19 pp.
Fischer, F.G. and Dorfel, H. (1955), *Z. physiol. Chem. Hoppe-Seyler,* **302**, 186.
Fischer, H. (1904), *Centrb. Bakt. Parasitenk. Abt.* I., **35**, 527.
F.A.O., (1971), Yearbook of Fishery Statistics: Catches and Landings, 1970, 30 pp.
Fitzgerald, G.P. and Skoog, F. (1952). *Seaweed Ind. Wastes,* **24**.
Fitzgerald, G.P. and Skoog, F. (1954), *Seaweed Ind. Wastes,* **26**.
Fleming, M., Hirst, E. and Manners, D.J. (1966), *Proc. 5th Int. Seaw. Symp.,* p. 255, Pergamon Press, Oxford.
Forbes, L.A. and Percival, E.G.V. (1938), *Nature, Lond.,* **142**, 214, 797, 1076.
Forbes, L.A. and Percival, E.G.V. (1939), *J. Chem. Soc.,* 1844.
Forsdike, J.L. (1950), *J. Pharm. Pharmacol.,* p. 796.
Fourment, R. and J. (1941), *Bull. Soc. Hist. Nat. Afr. du Nord,* **32**, 176.
Fox, F.W. and Stephens, E., (1943), *S. African J. Sci.,* **39**, 147.
Fraser, M.J. (1942), *Fishery Market News,* **4**, 24.
Frei, E. and Preston, R.D. (1962), *Nature, Lond.,* **196**, 130.
Freundler, P. (1922), *Bull. Soc. Chim. Fr.,* **4** ser. **31**, 1341.
Freundler, P. (1925), *Bull. Soc. Chim. Fr.* **4** ser. **37**, 1466.
Freundler, P. (1928), *Bull. Soc. Chim. Biol.,* **10**, 1123.
Freundler, P., Menager, Y. and Laurent, Y. (1921), *C.R. Acad. Sci. Paris,* **173**, 931, 1116.
Friedlaender, M.H.G., Cook, W.H. and Martin, W.G. (1954), *Biochem. Biophys. Acta,* **14**, 136.

Friedman, L. (1930), *Am. Chem. Soc. J.*, **52**, 1311.
Fries, L. (1973), *Bot. Mar.*, **16**, 19–31.
Fries, L. (1977), *J. Phycol.* **13**, Suppl. Abst. 121, p. 23.
Fritsch, F.E. (1935), *The Structure and Reproduction of the Algae*, Vol. I. Cambridge.
Fritsch, F.E. (1943), *Endeavour*, **2**, 142.
Fritsch, F.E. (1945), *The Structure and Reproduction of the Algae*, Vol. II, Cambridge.
Frye, T.G., Rigg, G.B. and Crandall, W.C. (1915), *Botan. Gaz.*, **60**, 473.
Fujikawa, T., Yahiso, M., Higuchi, T. and Wada, M. (1971), *Bull. Jap. Soc. Sci. Fish.*, **37**, 654.
Fujita, Y. (1978), *Bull. Jap. Soc. Sci. Fish.*, **44**, 15–20.
Fujita, Y. and Zenitani, B. (1977), *Bull. Jap. Soc. Sci. Fish.*, **43**, 921–928.
Fulton, C.O. and Metcalfe, B. (1945), *Can. J. Res.*, **23**, 273.
Funaki, K. and Kojima, Y. (1951), *Bull. Jap. Soc. Sci. Fish.*, **16**, 401, 405, 411, 419.
Funck, E. (1937), *Klin. Wochenschr.*, **16**. 1546.

Gail, H. (1930), *Bull. Pacif. Scient. Fish. Inst. Vladivost.*, **4**, 1.
Galetti, A.C. (1931), *L'Agricultura Colonial*, **25**, 243.
Gall, J. le. (1936), In: *Mem. de l'Off. des Peches marit. fasc.*, **4**, 12, 65.
Gall, J. le. (1937), *Handb. Seef. Nordeuropas*, **7**, 153.
Garcain, J.E. (1906), Theses de Montpellier.
Garcia-Pineda, D. (1951), *Biol. Inst. Espanol. Oceanog.*, **39**, 1.
Gardiner, A.C. (1939), *Proc. Ass. Appl. Biol.*, **26**, 165.
Gardiner, A.C. (1942), *Nature, Lond.*, **148**, 115.
Gardner, N.L. (1927), *Univ. Cal. Pub. Bot.*, **13**, 273.
Gauthier, M.J., Bernard, P. and Aubert, B. (1978), *J. Expt. Biol. Ecol.*, **33**, 37–50.
Ghosh, S. (1931), *Bull. Acad. Sci. United Provinces Agra., Oudh Allahabad*, **I**, 12.
Giaccone, G. (1970), *Stato Attuale Della Ricerca Algologica in Italia e le Sue Possibilità di Applicazione Industeriale* in *Possibilità de Utilizzazione Industriale delle Alge in Italia*, Vol. 5. Dagli Incontri Technici, CNR Lab. di Techol, Pesca ed Ente Fierà Internaz. Della pesca, Ancona, p. 5–13.
Gislen, T. (1930), *K-Svensk. Vetens. Akad. Handl.*, (Skr. ser.), **4**, 1–380.
Glaze, H.L. (1916), *Met. Chem. Engng.*, **14**, 355.
Glicksman, M. (1962), *Advances in Food Research*, **11**, 124.
Gloess, P. (1919), *Bull. Inst. Ocean Monaco*, 350.
Gloess, P. (1920), *Monit Sci.*, **10**, 30.
Gloess, P. (1932), *Rev. Chem. Ind.*, **162**, 190, 218.
Goh, K.M. (1971), *Sci.*, **14**, 734–748.
Goldie, W.H. (1904), *Trans. N.Z. Inst.*, **37**, 1.
Goldman, J.C., Ryther, J.H. and Williams, L.D. (1974), *Nature*, **254**, 594–595.
Goldman, J.C. and Stanley, H.I. (1974), *Mar. Biol.*, **28**, 17–25.
Goldman, J.C., Tenore, K.R., Ryther, J.H. and Corwin, N. (1974), *Water Research*, **8**, 45–54.
Goldman, J.C., Tenore, K.R. and Stanley, H.I. (1973), *Science*, **180**, 955–956.
Goldman, J.C., Tenore, K.R. and Stanley, H.I. (1974), *Water Research*, **8**, 55–59.
Goresline, H.E. (1933), *J. Bact.*, **106**, 485.

Gran, H.H. (1902), *Bergens Mus. Aarbog*, **2**, 1.
Grasdalen, H., Larsen and Smidsrød, O. (1979), *Proc. 9th Int. Seaw. Symp.*, *Santa Barbara*, Science Press, Princeton, 309–318.
Gray, P.H.H. and Chambers, C.H. (1924), *Ann. Appl. Biol.*, **11**, 324.
Greenish, H. (1881), *Ber. deutsch. Chem. Ges.*, **14**, 2253.
Grenager, B. and Baardseth, E. (1966), *Proc. 5th Int. Seaw. Symp.*, p. 129, Pergamon Press, Oxford.
Griffiths, A.B. (1903), *A Treatise on Manure*, London.
Grimmett, R.E.P. and Elliot, A.G. (1940), *N.Z.J. Agric.* **61**, 167.
Gruzewska, Z. (1921), *C.R. Acad. Sci. Paris*, **173**, 52.
Gryuner, V.S. (1931), *Issledo. Inst. Psich. Ukers, Prom. I.*
Gryuner, V.S. (1939), *Issledo. Inst. Kondit. Prom. I.*, 153.
Gryuner, V.S. and Tanson, N. (1936), *Colloid J., U.S.S.R.*, **2**, 783.
Gryuner, V.S. and Veronyan, L. (1939), *Colloid J., U.S.S.R.*, **5**, 851.
Guéguen, F. (1904), *Bull. Sci. Pharm.*, **10**, 225.
Guerrero, P.G. (1954), *An Inst. Bot. A.J. Cavanilles*, **10**, 511.
Guffroy, M.Ch. (1915), *L'emploides Engrais en Bretagne*, Saint-Brienc.
Gunther, R. (1928), *Spinner and Weber*, **9**, 40.
Guyen, K.C., Bora, A., Aktin, E., Oker, C., Hatemi, H. and Cetin, E.T. (1972), *Bot. Mar.*, **15**, 46–48.
Guzman del Próo, S.A. (1967), *Programa Nacional Sobre Algas Marinas Mexicanas*, Trabajo de Divulgacion No. 130, Vol. 13.
Guzman del Próo, S.A. (1969), *Proc. 6th Int. Seaw. Symp.*, La Marina Merchante, Madrid, Spain, pp. 685–690.
Guzman del Próo, S.A., Guzman, de la Campa de and Barrera, J.P. (1974), *Ser. Divulg. Inst. Nac. Pesca, Mex.* 6.
Guzman del Próo, S.A., Guzman, G. de la C. de and Gallegos, J.L.G. (1971), *Rev. Soc. Mex. Hist. Nat.*, **32**, 15–49.

Haas, P. (1921a), *Farm. J.*, **106**, 485.
Haas, P. (1921b), *Biochem. J.*, **15**, 409.
Haas, P. and Hill, T.G. (1921), *Ann. Appl. Biol.*, **7**, 352.
Haas, P. and Hill, T.G. (1933), *Ann. Botany*, **47**, 55.
Haas, P. and Russell-Wells, B. (1923), *Biochem. J.*, **17**, 696.
Haas, P. and Russell-Wells, B. (1929), *Biochem. J.*, **23**, 425.
Habekost, R.C., Fraser, I.M. and Halstead, B.W. (1955), *J. Wash. Acad. Sci.*, **45**, 101.
Haegler, S. (1895), *Centralb. Bakt. parasitenk. Abt. I.*, **17**, 558.
Haines, K.C. (1976), In: *10th European Symposium on Marine Biology*, (Persoone, G. and Jaspers, E. eds.), pp. 17–23, Universa Press, Wettesen, Belgium.
Hallsson, S.V. (1964), *Proc. 4th Int. Seaw. Symp.*, p. 398, Pergamon Press, Oxford.
Hand, C.J.E. (1953), *Proc. 1st Int. Seaw. Symp.*, Edinburgh, p. 68.
Hand, C.J.E. and Tyler, C. (1953), *J. Sci. Fd. Agric.*, **6**, 743–754.
Hands, S. and Peat, S. (1938a), *Soc. Chem. Ind. J.*, **57**, 937.
Hands, S. and Peat, S. (1938b), *Nature, Lond.*, **142**, 797.
Hanic, L.A. and Pringle, J.D. (1978a), *Brit. Phycol. J.*, **13**, 25–33.
Hanic, L.A. and Pringle, J.D. (1978b), *J. Fish. Res. Board. Canada*, **35**, 336–338.

Hardy, A.C. (1941), *Nature, Lond.,* **147**, 695.
Hart, F.L. (1937), *J. Assoc. Off. Agr. Chem.,* **20**, 527.
Hart, M.R., De Fremery, D., Lyon, C.K. and Kohler, G.O. (1979), *Proc. 9th Int. Seaw. Symp., Santa Barbara,* Science Press, Princeton.
Harvey, M.J. and McLachlan, J. (Ed.) (1973), *Chondrus crispus,* Nova Scotia Inst. Sci. 153 pp.
Harvey, W.H. (1852), *Ner. Bor. Am.,* pt. I. p. 35.
Harwood, F.C. (1923), *J. Chem. Soc.,* **123**, 2254.
Hasegawa, Y. (1962), *Bull. Hokkaido Reg. Fish. Res. Lab.,* **24**.
Hasegawa, Y. (1971a), *Bull. Hokk. Reg. Fish. Res. Lab.,* **37**, 46–48.
Hasegawa, Y. (1971b), *Bull. Hokk. Reg. Fish. Res. Lab.,* **37**, 49–52.
Hasegawa, Y. (1971c), *Proc. 7th Int. Seaw. Symp.,* Univ. Tokyo Press, Japan, p. 391–393.
Hasegawa, Y. (1976), *J. Fish. Res. Bd. Can.,* **33**, 1002–1006.
Hasegawa, Y. (1979), *Proc. 9th Int. Seaw. Symp., Santa Barbara,* Science Press, Princeton.
Hashimoto, Y. (1954), *J. Vitaminol,* **1**, 49.
Hassid, W.Z. (1933), *J. Am. Chem. Soc.,* **55**, 4163.
Hassid, W.Z. (1935), *J. Am. Chem.,* **57**, 2046.
Hassid, W.Z. (1936), *Plant Physiol,* **11**, 461.
Hatch, M.T., Ehresmann, D.W., Deig, E.F. and Veros, N.A. (1979), *Proc. 9th Int. Seaw. Symp., Santa Barbara,* Science Press, Princeton.
Hatchek, E. and Humphrey, R.H. (1924–5), *Trans. Faraday Soc.,* **20**, 18.
Haug, A. (1964), *Norsk. Instit. Tang-og Tareforsk, Rept. 30.*
Haug, A. and larsen, B. (1962), *Acta Chem. Scand.,* **16**, 1908.
Haug, A. and Larsen, B. (1966), *Proc. 5th Int. Seaw. Symp.,* p. 271, Pergamon Press, Oxford.
Haug, A., Larsen, B. and Baardseth, E. (1968), *Proc. 6th Int. Seaw. Symp.,* La Marina Merchante, Madrid, Spain, pp. 443–452.
Haug, A., Larsen, B. and Smidsrød, O, (1966), *Acta. Chem. Scand.,* **20**, 183.
Haug, A., Larsen, B. and Smidsrød, O. (1967), *Acta. Chem. Scand.,* **21**, 691.
Haug, A., Myklestad, S., Larsen, B. and Smidsrød, O. (1967), *Acta. Chem. Scand.,* **21**, 768.
Haug, A. and Smidsrød, O. (1962), *Acta Chem. Scand.,* **16**, 1569.
Hawkins, W.W. and Leonard, U.G. (1958), *Can. J. Biochem, Physiol.,* **36**, 161.
Hay, J.M. (1947), *The Engineer,* Oct. 10, 17, 24.
Healey, D.J. (1926), *J. Bact.,* **12**, 179.
Heen, E. (1938), *Kolloid. Z.,* **83**, 204.
Heilbron, I.M. (1942), *J. Chem. Soc.,* **72**, p. 79.
Heincke, F. (1889), *Mitt. Seefisher. Ver. Yahr.,* 136.
Hendrick, J. (1916a), *J.Soc. Chem. Ind.,* **35**, 365.
Hendrick, J. (1916b), *J. Board Agric.,* **22**.
Hendrick, J. (1919), *Nature, Lond.,* **102**, 494.
Hennig, T. (1932), *Papeterie,* **54**, 1122.
Hercules Powder Co. (1918), *Met. Chem. Eng.,* **18**, 576.
Hercus, C.E. and Aitken, H.H.A. (1933), *J. Hyg.,* **33**, 55.
Hesse, F. (1897), *Centrb. Bakt. Parasitenk. Abt.I.,* **21**, 932.
Hill, A.F. (1937), *Economic Botany,* New York.

Hill, J. (1941), *Wild Foods of Britain*. London.
Hirase, S. and Watanabe, K. (1977), *Proc. 7th Int. Seaw. Symp.*, Univ. Tokyo Press, Japan, pp. 451–454.
Hirst, E.L. (1939), *Nature, Lond.*, **143**, 856.
Hirst, E.L. (1958), *Proc. Chem. Soc.*, 177.
Hirst, E.L., Jones, J.N.K. and Jones, W.O. (1939), *J. Chem. Soc.*, 1880.
Hirst, E.L., Percival, E.G.V. and Wold, J.K. (1964), *J. Chem. Soc.*, 1493.
Hirst, M.S. (1965), *Okeanologiya*, **5**, 14.
Hitchens, A.P. and Leikind, M.C. (1939), *J. Bacteriol.*, **37**, 485.
Hjerten, S. (1961), *Biochim. biophys. Acta*, **53**, 514.
Hjort, J. (1922), *Proc. R. Soc. Ser. B.*, **93**, 440.
Hoagland, P.R. (1915), *J. Ind. Engng. Chem.*, **7**.
Hoagland, P.R. (1916), *J. Agric. Res.*, **4**, 39.
Hoek, C. van den, A.M. (1969), *Proc. K. Ned. Akad. Wet. (c)*, **72**, 537–577.
Hoek, C. van den, A.M. (1972), *Verh. K. Ned. Akad. Wet. (Afd. Nat. Ser. 2)*, **61**, 1–72.
Hoffman, C. (1939), *Kiel, Meeresfors.*, **3**, 165.
Hoffman, W.F. and Gortner, R.A. (1925), *J. Biol. Chem.*, **65**, 371.
Hog, D.R. and Lieb, L.L. (1915), *J. Biol. Chem.*, **23**, 287.
Høie, J. and Sandvik, O. (1955), *Sci. Rept. Agric. Coll. Norway. Rept. 8*, Inst. Poultry and Farm Animals, p. 122.
Høie, J. and Sannay, F. (1960), *Sci. Rept. Agric. Coll. Norway. Rept. 12*. Inst. Poultry and Farm Animals, p. 1.
Holman, W.L. (1912), *J. Infect. Dis.*, **10**, 129.
Holmes, E.M. (1906), *Pharm. J.*, **3**, 735.
Homai, L. and Hotta, H. (1949), *Bull. Jap. Soc. Sci. Fish.*, **15**, 434.
Hong, K.C. and Cruess, R.L. (1977), *J. Phycol.*, **13**, Suppl. Abst. 170, p. 31.
Hong, K.C. and Yaphe, W. (1968), *Proc. 6th Int. Seaw. Symp.*, La Marina Merchante, Madrid, Spain, 473–482.
Hoppe, H.A. (1962), *Bot. Marina*, **3**, Suppl., 12.
Hoppe, H.A. (1966), *Bot. Marina*, **9**, Suppl., 19.
Hoppe, H.A. and Schmid, O.J. (1962), *Bot. Marina*, **3**, Suppl., 16.
Hoppe, H.A. and Schmid, O.J. (1969), *Bot. Mar. Handb.*, **1**, 288–368.
Hosada, K. (1970), *Bull. Jap. Soc. Sci. Fish.*, **36**, 698.
Hosada, K. (1975), *Bull. Jap. Soc. Sci. Fish.*, **41**, 739–742.
Hosada, K. (1979), *Bull. Jap. Soc. Sci. Fish.*, **45**, 163–166.
Howe, M.A. (1917), *J.N.Y. Bot. Gard.*, **18**, 1.
Hoygaard, A. and Rasmussen, H.W. (1939), *Nature, Lond.*, **143**, 943.
Hoyle, M.D. (1978), *Bot. Mar.*, **21**, 343–345.
Hua, Ju-Ch'eng and Li, Ping-Tao. (1962), *Engl. Abs. Soviet bloc and Mainl. China Tech. Jour.* Ser. VI: Biol. Sci., 145 (22) 61–11.
Huang, S.M., Shen, T.H., and Tang, F.F. (1941), *Chin. Med. J.*, **59**, 176.
Huerta, M.-L. (1960), *Biol. Soc. Bot. Mex.*, **25**, 62.
Huguenin, J.E. (1976), *Aquaculture*, **9**, 313–342.
Humm, H.J. (1942), *Science, N.U.*, **96**, 230.
Humm. H.J. (1944), *Science, N.U.*, **100**, 209.
Humm, H.J. (1946), *Duke Univ. Mar. Sta. Bull.*, 3.

Humm, H.J. (1947), *Econ. Botan.*, **1**, 317.
Humm, H.J. (1948), *Seaweed Resources of Newfoundland.* Dept. to Newfoundland Indust. Dev. Board.
Humm, H.J. (1962), Marine Algae of Virginia as a Source of Agar and Agaroids. *Spec. Sci. Rept. Va. Inst. Mar. Sci.* (37) 13 pp.
Humm, H.J. and Shepard, R.S. (1946), *Duke Marine Univ. Bull.*, **3**, 76–80.
Humm, H.J. and Wolf, F.A. (1946), *Duke Marine Univ. Bull.*, **3**, 2–18.
Hussain, A. and Bona, A.D. (1973), *New Phytol.*, **72**, 403–410.
Hutchinson, A.H. (1949), *Proc. 7th Pac. Sci. Cong.*, 62.

Idler, D.R. (1971), *Proc. Canad. Atl. Seaw. Indust.*, Oct. Industrial Development Branch, Environment, Canada, pp. 3–12.
Idler, D.R. and Wiseman, P. (1970), *Comp. Biochem. and Phys.*, **35**, 1679–1688.
Imada, O., An, S., Saito, Y., Maeki, S. and Seki, T. (1970), *Bull. Jap. Soc. Sci. Fish.*, **36**, 889.
Imada, O., Saito, Y. and Maeki, S. (1970), *Bull. Jap. Soc. Sci. Fish.*, **36**, 369–376.
Irmascher, (1892–3), *Mitt. Inst. Alg. Bot. Hamb.*, **5**, 8.
Isaac, W.E. (1942), *J.S. Afric. Bot. Soc.*, **8**, 225.
Isaac, W.E., Finlayson, M.H. and Simon, M.G. (1943), *Nature, Lond.*, **151**, 532.
Isaac, W.E. and Molteno, C.J. (1953a), *Proc. 1st Int. Seaw. Symp.*, Edinburgh, p. 103.
Isaac, W.E. and Molteno, C.J. (1953b), *J.S. Afric. Bot.*, **19**, 85–92.
Ishibashi, M. and Yamamoto, T. (1960), *Rec. Oceanog. Works Jap.*, **5**, 55.
Ishio, S., Yano, T. and Nakagawa, H. (1972), *Proc. 7th Seaw. Symp.*, Univ. Tokyo Press, Japan.
Isimatu, K. (1940), *Sci. Ind.*, **15**, 598.
Itallie, L. van. (1889), *Archs. Pharm.* **27**.
Itano, A. (1932), *Proc. Imp. Acad. Tokyo*, **9**, 398.
Itano, A. (1933), *Ber. Ohara. Inst. Landw. Forsch., Japan*, **6**, 59.
Itano, A. and Tusji, Y. (1934a), *Nogaku Kenkyu*, **22**, 168.
Itano, A. and Tusji, Y. (1934b), *Bull. Agr. Chem. Soc., Japan*, **10**, 111.
Itano, A. and Tusji, Y. (1935), *Ber. Ohara, Inst. Landw. Forsch., Japan*, **6**, 575.
Itano, A. and Tusji, Y. (1936), *Ber. Ohara, Inst. Landw. Forsch., Japan*, **7**, 529.
Itano, A. and Tusji, Y. (1938), *Ber. Ohara, Inst. Landw. Forsch., Japan*, **8**, 249.
Iwasaki, H. (1961), *Biol. Bull.*, **121**, 173.
Iwasaki, H. and Matsudaira, C. (1954), *Bull. Jap. Soc. Sci. Fish.*, **20**, 380.
Iwase, K. (1920), *J. Chem. Soc. Tokyo*, **41**, 468.
Iwase, K. (1927), *Bull. Chem. Soc., Japan*, **2**, 61.

Jackson, P. (1948), *Scott. Geog. Mag.*, **64**, 136.
Jensen, A. (1966), *Norsk. Inst. tang-og Tareforsk, Rept.* 31.
Jensen, A. (1972), *Proc. 7th Int. Seaw. Symp.*, Univ. Tokyo Press, Japan, p. 7–14.
Jensen, A. (1979), *Proc. 9th Int. Seaw. Symp.*, Santa Barbara, Science Press, Princeton, 17–34.
Jensen, A., Nebb, H. and Saeter, E.A. (1968), *Norsk-Instit. tang-og Tareforsk, Rept.* 32.
Jensen, D.S. (1919), *Soil Sci.*, **7**, 201.
John, D.N. and Asare, S.O. (1975), *Mar. Biol.*, **30**, 325–330.
Johnston, C.S. (1969), *Int. Rev. Ges. Hydrobiol.*, **54**, 473–490.
Johnston, H.W. (1965), *Tuatara*, **13**, 90.

Johnston, H.W. (1966), *Tuatara*, **14**, 30.
Johnston, H.W. (1972), *Proc. 7th Int. Seaw. Symp.*, Univ. Tokyo Press, Japan, pp. 429–435.
Johnston, R. and Percival, E.G.V. (1950), *J. Chem. Soc.*, 1994.
Johnstone, G.R. and Feeney, F.L. (1944), *Am. J. Botany*, **31**, 25.
Jones, N. S. and Kain, J.M. (1967), *Helgoland. wiss. Meeres.*, **15**, 460.
Jones, W.G.M. and Peat, S. (1942), *J. Chem. Soc.*, 225.
Jong, H.G.B. de. (1928), *Rev. Trav. Chim.*, **7**, 797.
Jong, H.G.B. de. (1929), *Kolloid Beihefte*, **29**, 454.
Jong, H.G.B. de. and Decker, W.A.L. (1932), *Biochem. Z.*, **251**, 105.
Jong, H.G.B. de. and Henneman, J.P. (1932a), *Kolloid Beihefte*, **35**, 441.
Jong, H.G.B. de. and Henneman, J.P. (1932b), *Kolloid Beihefte*, **36**, 123.
Jong, H.G.B. de., Kruyt, H.R. and Lens, J. (1932), *Kolloid Beihefte*, **36**, 149.
Jong, H.G.B. de. and Lanzing, J.C. (1932), *Kolloid Beihefte*, **35**, 89.
Jong, H.G.B. de. and Der Linde, P. van. (1934), *Rec. Trav. Chim.*, **53**, 737.

Kain, J.M. (1963), *J. Mar. Biol. Ass. U.K.*, **43**, 129.
Kain, J.M. (1964), *J. Mar. Biol. Ass. U.K.*, **44**, 415.
Kain, J.M. (1967), *Helgoland. wiss. Meeres.*, **15**, 489.
Kaliaperumal, N. and Kalimuthu, S. (1976), *Bot. Mar.*, **19**, 157–159.
Kalimuthu, S., Channubhotla, V.K., Salvarag, M., Najmuddin, M. and Panigraphy, R. (1977), *Proc. 9th Int. Seaw. Symp.*, Abst. 456.
Kalugina, A.A. and Lachko, O.A. (1966), *Ref. Zhur. Biol.*, No. 11D396.
Kalugina-Gutnik, A.A. (1968), *6th Int. Seaweed Symposium, Spain,* Abstract.
Kaminer, K.M. (1974), *Proc. all Union Res. Inst. Mar. Fish and Ocean.*, pp. 154.–163.
Kaneda, R. and Ando, H. (1972), *Proc. 7th Int. Seaw. Symp.*, Univ. Tokyo Press, Japan, pp. 553–557.
Kappanna, A.N. and Rao, V. (1963), *Indian J. Techn.*, **1**, 222.
Karlstrom, O. (1963), *Norsk Inst. tang-og Taresforsk, Rept.* 291.
Kasahara, K. and Nishibori, K. (1975), *Bull. Jap. Soc. Sci. Fish.*, **41**, 193–200.
Katada, M. (1949), *Bull. Jap. Soc. Sci. Fish.*, **15**, 359.
Katada, M. (1958), *J. Shimonoskei Coll. Fish.*, **5**, 1.
Kawaguchi, K., Yamada, S. and Miyama, S. (1953), *Bull. Jap. Soc. Sci. Fish.*, **19**, 481.
Kim, C.S. (1972), *Proc. 7th Int. Seaw. Symp.*, Univ. Tokyo Press, Japan, pp. 573–574.
Kim, D.H. (1970), *Bot. Mar.*, **8**, pp. 140–162.
Kim, D.H. and Hendriquez, P. (1979), *Proc. 9th Int. Seaw. Symp.*, Santa Barbara, Science Press, Princeton, 257–262.
King, J.G. (1924), Fuel Res. Bḑ. Tech. Paper 9.
King, J.G. (1925), *Analyst*, **50**, 371.
Kingman, A.R. and Senn, T.L. (1977), *J. Phycol.*, **13**, Suppl. Abst. 199, p. 36.
Kirby, R.H. (1953), *Seaweeds in Commerce,* H.M. S.O., London.
Kireeva, M.S. (1964), *Ref. Z. Biol.*, (1964), **33** (15).
Kireeva, M.S. (1965), *Okeanologiya*, **5**, 14.
Kirigeva, T.S. and Schapora, T.F. (1939), *Trans. Inst. Mar. Fish Oceanog. U.S.S.R.* **7**, 5
Kizevetter, I.V. (1936a), *Bull. Far East Branch Acad. Sci., U.S.S.R.*, **20**, 57.
Kizevetter, I.V. (1936b), *Bull. Far East Branch Acad. Sci., U.S.S.R.*, **21**, 85.
Kizevetter, I.V. (1937a), *Bull. Far East Branch Acad. Sci., U.S.S.R.*, **23**, 53.
Kizevetter, I.V. (1937b), *Bull. Pac. Sci. Inst. Fish. Ocean.*, **13**, 1.
Kizevetter, I.V. (1938), *Bull. Far East Branch Acad. Sci., U.S.S.R.*, **31**, 49.

Kizevetter, I.V. (1941), *J. Appl. Chem. U.S.S.R.*, **14**, 250.
Klostermann, (1921), *Z. Hyg. Infeckt.*, **94**, 262.
Klugh, A.B. (1918), *Canadian Fisherman*, **5**, 1024.
Kohn, K. and Kohn, M. (1942), *Inst. Res. Indust. Raw Mat. Jerus. Bull.* I.
Koizumi, T. and Kakuwaw, (1940), *Sci. Rep. Tohoku Imp. Univ. Jap.* IV. **15**, 105.
Kojima, Y., Fukushima, Y. and Kono, M. (1960), *J. Shimonosekei Coll. Fish.*, **9**, 43.
Kojima, Y., Tawawa, S. and Yamada, Y. (1960), *J. Shimonosekei Coll. Fish.*, **10**, 43.
Kok, B. and Oorschot, J.L.P. van (1954), *Act. Bot. Neerland*, **3**, 533.
Konig, T. and Bettels, J. (1904), *Z. Unters Nahrgs. genussmitte.* **10**, 457.
Korentzvit, A. (1934), *Khim. Farm. Prom.*, **4**, 36.
Korentzvit, A. (1935), *Khim. Farm. Prom.*, **3**, 153.
Korentzvit, A. (1935), *J. Appl. Chem. U.S.S.R.*, **8**, 912.
Korentzvit, A. (1938), *J. Appl. Chem. U.S.S.R.*, **11**, 351.
Koryakin, A.I. (1939), *Lab. Prakt., U.S.S.R.*, **1**, 8.
Kowarski, I.G. and Puderau, R. (1936), *Rev. Path. Comp. Hyg. Gen.*, **35**, 913.
Kraemer, H. (1899), *Am. J. Pharm.*, **71**, 479.
Krefting, A. (1896), *Soc. Chem. Ind. J.*, **15**, 726.
Krefting, A. (1898), *Soc. Chem. Ind. J.*, **17**, 794, 846.
Krefting, A. (1900), *Soc. Chem. Ind. J.*, **19**, 361.
Krim-Ko Company (1943), Mimeog. Separate, Scituate. Mass., U.S.A.
Kringstad, H. and Lunde, G. (1938), *Kolloid Z.*, **83**, 202.
Krishnamurthy, V., Venugopal, R., Thiagaraj, J.G. and Shah, H.N. (1967), In: *Proc. Seminar on Sea, Salt and Plants*, (Krishnamurthy, V. ed.), Central Salt and Marine Chemicals Res. Inst. pp. 315–320.
Kruyt, K. and Bungenbende Jong, H.G. (1928), *Kolloid Beihefte*, **28**, 1.
Kurogi, M. (1953), *Bull. Tohoku Fish. Res. Lab.*, 2.
Kurogi, M. (1959), *Bull. Tohoku Fish. Res. Labb.*, **15**, 33.
Kurogi, M. (1961), *Bull. Tohoku Fish. Res. Lab.*, **18**.
Kurogi, M. (1963), Fishing News International July/September.
Kurogi, M. and Akiyama, K. (1957), *Bull. Tohoku Fish. Res. Lab.*, **10**.
Kurogi, M. and Akiyama, K. (1962), *Bull. Tohoku Fish. Res. Lab.*, **20**, 121, 127, 138.
Kurogi, M. and Akiyama, K. (1965), *Bull. Tohoku Fish. Res. Lab.*, **25**, 171.
Kurogi, M. and Akiyama, K. (1966), *Bull. Tohoku Fish. Res. Lab.*, **26**, 77.
Kurogi, M. and Hirano, K. (1956), *Bull. Tohoku Fish. Res. Lab.*, **8**, 27, 45.
Kurogi, M. and Sato, S. (1967), *Bull. Tohoku Fish. Res. Lab.*, **27**, 111.
Kurogi, M., Sato, S. and Yoshida, T. (1967), *Bull. Tohoku Fish. Res. Lab.*, **27**, 131.
Kurogi, M. and Yoshida, T. (1966), *Bull. Tohoku Fish. Res. Lab.*, **26**, 91.
Kusakabe, J. (1967), *Bull. Jap. Soc. Sci. Fish.*, **33**, 984.
Kuzneitzov, V.V. (1946), *C.K. Acad. Sci. U.S.S.R.*, **54**, 533.
Kylin, H. (1913), *Z. Physiol. Chem.*, **83**, 171.
Kylin, H. (1915), *Z. Physiol. Chem.*, **94**, 337.
Kylin, H. (1929), *Z. Physiol. Chem.*, **186**, 50.
Kylin, H. (1931), *Z. Physiol. Chem.*, **203**, 58.

Lami, H. (1949), *Proc. 7th Pan-Pacific Science Congress*, Auckland, N.Z., **5**, 80.
Lapin, P.M. and Lipatov, S.M. (1939), *Colloid J., U.S.S.R.*, **5**, 690.

Lapique, L. (1920), *Compt. Rend. Soc. Biol.*, **83**, 1610.
Lapique, L. and Brocq-Rousseau (1920), *C. R. Acad. Sci. Paris*, **170**, 1600.
Lapointe, B.E., Williams, L.D., Goldman, J.C. and Ryther, J.H. (1976), *Aquaculture*, **8**, 9–2
Laroste De Diaz, E.N. (1957), *Rev. Invest. Agric.*, **11**, 163.
Larsen, B. and Haug, A. (1972), *Proc. 7th Int. Seaw. Symp.*, Univ. Tokyo Press, Japan, pp. 491–495.
Larsen, B., Haug, A. and Painter, T.J. (1966), *Proc. 5th Int. Seaw. Symp.*, p. 287, Pergamon Press, Oxford.
Lawson, G.W. (1954), *Am. J. Botany*, **41**, 212.
Le Clerc, H. (1940), *Presse. Med.*, **48**, 700.
Le Cornu, C.P. (1859), *J.R. Agric. Soc.*, **2**, 40.
Lee, C.F. and Stoloff, L.S. (1946), Spec. Sci. Rept. 37, Fish and Wild Life Service, U.S. Dept. Int.
Lee, S.W. (1967), Preliminary studies on the ecology and physiology of *Pterocladia pinnata* (Huds) Papenfuss. M. Sc Thesis, Auckland University, New Zealand.
Lendner, A. (1926), *Pharm. Acta. Helv.*, **1**, 183.
Leonard, J. and Compère, P. (1967), *Bull. Jard. bot. nat. Belg.*, **37**, Suppl.
Lepik, E. (1926), *Mitt. Phytophathol. Vesuchssta. Univers. Tartu*, **1**, 1.
Levring, T., Hoppe, H.A. and Schmid, O. (1969), *Marine Algae*, Cram, De Gruyter and Co., Hamburg.
Lian-Ching-Li (1934), *Peking Nat. Hist. Bull.*, **8** (4) pp. 363–374.
Liaw, J.P. and Chiang, Y.M. (1977), *J. Phycol.*, **13**, Suppl. Abst., 231, p. 41.
Liesegang, R.E. (1919), *Farben-Ztg.*, **24**, 971.
Ligthelm, S.P., von Holdt, M.M. and Schumann, H.I. (1953), *Proc. 1st Int. Seaw. Symp.*, Edinburgh, p. 51.
Lillig, R. (1928), *Pharm. Ztg.*, **79**, 632, 644, 658.
Lindberg, B. (1953), *Acta Chem. Scand.*, **7**, 1119.
Lindner, (1920), *Z. techn. Biol.*, **10**, 193.
Lipatov, S.M. and Morozov, A.A. (1935a), *Kolloid Z.*, **71**, 317.
Lipatov, S.M. and Morozov, A.A. (1935b), *Kolloid Z.*, **72**, 325.
Lipskii, V.J. (1932), *C.R. Acad. Sci., U.S.S.R.*, **3**, 60.
Little, E.C.S. (1948), M. Sc. Thesis, University, Auckland, New Zealand.
Little, E.C.S. (1949), *Proc. 7th Pacific Sci. Cong.*, **5**, 83.
Lockwood, H.C. and Hayes, R.S. (1931), *Soc. Chem. Ind. J.*, **50**, 145.
Loeper, B. and Verpy, G. (1916), *C.R. Acad. Soc. Biol.*, **79**, 660.
Lohrisch, H. (1908), *Z. Exp. Path. Pharm.*, **5**, 478.
London, E.S. (1897), *Centrb. Bakt. Parasitenk Abt. I.*, **21**, 686.
Loose, G. (1966), *Bot. Marina*, **9**, Suppl. 7–14.
Lucas, H.J. and Steward, W.T. (1940), *J. Am. Chem. Soc.*, **62**, 1070.
Lund, S. and Bjerre-Petersen, L. (1953), *Proc. 1st Int. Seaw. Symp.*, Edinburgh, p. 85.
Lund, S. and Bjerre-Petersen, L. (1964), *Proc. 4th Int. Seaw. Symp.*, Biarritz, p. 410. Pergamon Press, Oxford.
Lunde, G. (1937a), *Tecknisk. Ukeblad.*, **84**, 192.
Lunde, G. (1937b), *Angewande. Chem.*, **50**, 731.
Lunde, G., Heen, E. and Oy, E. (1937), *Z. Physiol. Chem.*, **247**, 189.
Lunde, G., Heen, E. and Oy, E. (1938), *Kolloid Z.*, **83**, 196.
Lunde, G. and Lie, J. (1938), *Z. Physiol. Chem.*, **254**, 227.

Lunde, G. and Lunde, S. (1938), *Rep. Norweg. Fish. Mar. Invest.*, **5**.
Lunde, G., Lunde, S. and Jakobsen, A. (1938), *Fiskeridivektor Skrifter. Ser. Havundenk.*, **5**, 1.
Lundestad, J. (1929), *Centrabl. Bakt. Abt. II*, **75**, 32.
Luxton, D.M. (1977), *Aspects of the Biology and Utilisation of Pterocladia and Gracilaria*, PhD Thesis, Auckland University, pp. 237.
Lyle, L. (1938), *J. Botany Lond.*, **76**, 193.

McCandless, L.E. and Craigie, J.S. (1974), *Bot. Mar.*, **17**, pp. 125–129.
McCandless, E.L., Craigie, J.S. and Hansen, J.E. (1975), *Canad. I. Bot.*, **53**, pp. 2315–2318.
McCandless, E.L., Craigie, J.S. and Walter, J.A. (1973), *Planta*, **112**, pp. 201–212.
McCandless, E.L., Evelegh, M., Volmer, C. and Di Ninno, V., (1979), *Proc. 9th Int. Seaw. Symp., Santa Barbara*, Science Press, Princeton.
MacDougal, D.T. and Clarke, B.L. (1925), *Science N.Y.*, **62**, 126.
MacDougal, D.T. and Spoehr, H.A. (1918), *Proc. Soc. Exp. Biol. Med.*, **16**, 33.
MacDougal, D.T. and Spoehr, H.A. (1920), *Botan. Gaz*, **70**, 268.
McDowell, R.H. (1966), *Proc. 5th Int. Seaw. Symp.*, p. 379, Pergamon Press, Oxford.
MacFarlane, C.I. (1952), *Can. J. Botany*, **30**, 78.
MacFarlane, C.I. (1956), *Proc. 2nd Int. Seaw. Symp. Trondheim*, pp. 186–191, Pergamon Press, Oxford.
MacFarlane, C.I. (1964), *Proc. 4th Int. Seaw. Symp. Biarritz*, p. 414, Pergamon Press, Oxford.
MacFarlane, C.I. (1966), *Proc. 5th Int. Seaw. Symp. Halifax*, p. 169, Pergamon Press, Oxford.
McIlwain, H. (1938), *Br. J. Exp. Path.*, **19**, 411.
MacKinnon, H.D. (1930), *Food Ind.*, **2**, 123.
McLendon, J.F. (1933), *J. Biol. Chem.*, **102**, 91.
Macmorine, H.G. (1942a), *Can. Publ. Hlth. J.*, **33**, 461.
Macmorine, H.G. (1942b), *Water Sewage*, **80**, 26.
McNeely, W.H. and Kovacs, P. (1975), ACS Sympos. Ser. 15, Amer. Chem. Soc., pp. 269–281.
McNeely, W.H. and Pettitt, D.J. (1973), Algin In: *Industrial Gums*, (Whistler, R.L. and Bemiller, I.N., eds.), 2nd edn., Academic Press, New York, pp. 49–81.
MacPherson, M.G. and Young, E.G. (1949), *Can. J. Res.*, **27**, 73.
MacPherson, M.G. and Young, E.G. (1952), *Can. J. Botany*, **30**, 67.
Maass, H. (1959), *Alginsäure und Alginate, Strassenbau. Chemie und Technik Verlagsgesellschaft m.b.H.*, Heidelberg.
Maass, H. (1962), *Bot. Marina*, 3, Suppl. 86.
Madge, H.A. (1936), *Ann. Botany*, **50**, 677.
Madgwick, J. Haug, A. and Larsen, B. (1978), *Bot. Mar.*, **21**, pp. 1–3.
Madlener, J.C. (1977), *The Sea Vegetable Book*, Potter, New York.
Magidson, O.J. (1929), *Bull. Pac. Sci. Fish. Res. Stat.*, **3**.
Makaroff, A. (1946), *Food Ind.*, **18**, 1545.
Malhado, P. (1931), *Ann. Soc. Pharm. Chim. Sao Paulo*, **2**, 89.
Malin. (1921), *Commerce Repts.*, **202**, 1036.

Mallman, W.L. and Breed, R.S. (1941), *Am. J. Publ. Hlth.*, **31**, 341.
Mangenot, G. (1883), These d'Agregation, Paris.
Mangenot, G. (1928), *Bull. Soc. Bot. Fr.*, **75**, 519.
Mangon, M.H. (1859), *C.R. Acad. Sci. Paris*, **49**, 322.
Mann, K.H. (1972a), *Mar. Biol.*, **12**, 1–10.
Mann, K.H. (1972b), *Mar. Biol.*, **14**, 199–209.
Mann, K.H. (1973), *Science*, **182**, 975–981.
Mantner, H.G. (1954), *Econ. Bot.*, **8**, 174.
Manures (1937), *Min. Agric. Fish. Bull.*, **36**.
Marchand, E. (1879), *Bull. Soc. Bot. Fr.*, **26**, 287.
Marini-Bettolo, G.B. (1948), *Ann. Chim. Applicata*, **38**, 294.
Marquardt, J.C. (1930), *Food Ind.*, **2**, 76.
Marsh, J.T. and Wood, F.C. (1942), *An Introduction to the Chemistry of Cellulose*, London.
Marshall, S.M. Newton, L. and Orr, A.P. (1949), *A Study of Certain British Seaweeds and their Utilisation in the Preparation of Agar*, London.
Martinez, N., Noemi, G., Rodriguez, L.V. and Casillas, C. (1962), *J. Pharmaceut. Sci.*, **52**, 498.
Masera, E. (1932), *Boll. Sci. Ital. Soc. Internag. Microbiol*, **4**, 126.
Matens, H., Regenstein, J.M. and Baker, R.C. (1977), *Econ. Bot.*, **31** (1), 24–27.
Mathieson, A.C. and Burns, R.L. (1971), *J. Exp. Mar. Biol. Ecol.*, **7**, 197–206.
Mathieson, A.C. and Burns, R.L. (1975), *J. Exp. Mar. Biol. Ecol.*, **17**, 137–156.
Mathieson, A.C. and Prince, J.S. (1973), In: *Chondrus crispus* (Harvey, M.J. and McLachlan, J. eds.), pp. 53–79, Nova Scotia Institute Science.
Matignon, C. (1914), *Rev. gen. Sci.*, **25**.
Matsudaira, C. and Wasaki, H. (1953), *Tohoku J. Agric. Res.*, **3**, 277.
Matsui, H. (1916), *J. Coll. Agric. Imp. Univ. Tokyo*, **5**, 391, 413.
Matsui, T. (1962), *J. Shimonoseki Coll. Fish.*, **11**, 587.
May, V. (1945), *J. Council Sci. Ind. Res. Australia*, **18**, 62.
Mazur, A. and Clarke, H.T. (1938), *J. Biol. Chem.*, **123**, 729.
Mehta, B.R. and Parekh, R.G. (1978), *Bot. Mar.*, **21**, 251–252.
Meier, F.E. (1935), *Ann. Rept. Smiths. Inst.*, 409.
Mendel, L.B. and Swartz, M.D. (1910), *Am. J. Med. Sci.*, **139**, 422.
Merz, A.R. (1914), *J. Ind. Engng. Chem.*, **6**, 19.
Merz, A.R. and Lindemuth, I.R. (1913), *J. Ind. Engng. Chem.*, **5**, 729.
Micara, F.A.E. (1945), *Duke Univ. Mar. Sta. Bull.*, **3**, 40.
Michanek, G. (1975), FAO Fisheries Tech., Paper 138, Rome.
Michanek, G. (1978), *Bot. Mar.*, **21**, 469–475.
Miller, C.D. (1933), *Hawaii Agric. Expt. Sta. Bull.*, **68**.
Miller, D. (1938), *N.Z.J. Forestry*, **4**, 170.
Miller, R.E. and Rose, S.B. (1939), *J. Bacteriol.*, **38**, 539.
Mitchell, (1748), *Trans. R. Soc. Lond.*, **45**, 541.
Mitchell, H.S. (1922), *Am. J. Physiol.*, **62**, 557.
Mirsubishi T. and Hayashi, K. (1972), *Proc. 7th Seaw. Symp.*, Univ. Tokyo Press, Japan, 464–468.
Mitsukuri, S. and Toraishi, S. (1928), *J. Chem. Soc. Japan*, **49**, 244.
Miura, A. (1975), In: *Advance of Phycology in Japan*, (Tokida, J. and Hirose, H. eds.), 273–304, W. Junk Hague.

Miura, I. (1927), *J. Coll. Agric. Imp. Univ. Tokyo,* **9**, 101.
Miwa, T. (1930), *J. Chem. Soc. Japan,* **51**, 738.
Miwa, T. (1932a), *Sci. Repts. Tok. Bun. Daig.,* **2**, 23.
Miwa, T. (1932b), *Botan. Mag., Tokyo,* **46**, 261.
Miyabe, K. (1902), *Publ. Fish. Bur. Hokkaido Gov.,* **3**.
Miyake, S. and Hayashi, K. (1939), *J. Soc. Trop. Agr. Tohoku Imp. Univ.,* **11**, 200.
Moffat, (1915), *Trans. Highl. Agric. Soc. Scot.*
Mohamed, A.F. and Halim, Y. (1952), *Am. J. Botany,* **39**, 689.
Moirano, A.L. (1977), *Sulphated seaweed Polysaccharides,* in *Food Colloids,* (Graham, H.D. ed.), Avi Publ. Co., Connect. 347–381.
Mollion, J. (1972), *Proc. 7th Int. Seaw. Symp.,* Univ. Tokyo Press, Japan, p. 26.
Mollion, J. (1973), *Bot. Mar.,* **16**, 221–225.
Mollion, J. (1979), *Proc. 9th Int. Seaw. Symp., Santa Barbara,* Science Press, Princeton, 233–240.
Mollion, J. (1980), *Bot. Mar.,* **23**, 197–199.
Monie, W.D. (1952), *Water and Sewage Works,* **99**, 96.
Montagne, C. (1846), *Rev. Bot.,* **2**, 363.
Moore, L.B. (1941), *Dept. Sci. Ind. Res. N.Z. Bull.,* **85**, 355.
Moore, L.B. (1943), *Chron. Bot.,* **7**, 406.
Moore, L.B. (1944), *N. Z. J. Sci. Tech.,* **25**, 183.
Morel, Th. (1950), *Chron. Nat.,* **106**, 37.
Moreland, C.F. (1937), *Am. J. Botany,* **24**, 592.
Morozov, A.A. (1935), *Colloid J., U.S.S.R.,* **1**, 37.
Morrison, R. (1973), *Irish Moss and its Potential in New Brunswick,* Dept. Fish and Environment, Res. and Develop. Branch, Fredericton, N.B., pp. 65.
Morse, J.L. (1910), *J. Am. Med. Assoc.,* **55**, 934.
Moss, B. (1948), *Ann. Botany N.S.,* **12**, 267.
Moss, B. and Naylor, M. (1954), *Trans. Roy. Soc. N.Z.,* **81**, 473.
Moss, J.R. (1977), *J. Phycol.,* **13**, Suppl. Abst., 265, p. 47.
Mowat, J.A. (1964), *Proc. 4th Int. Seaw. Symp. Biarritz,* p. 352, Pergamon Press, Oxford.
Mueller, and Rees, D.A. (1968), *Drugs from the Sea,* Symposium, p. 271. (Frudenthal, H.D. ed.), Mar. Tech. Soc., Washington D.C.
Mumford, T.F. (1977), In: *Marine Plant Biomass of the Pacific Northwest.* (Krauss, R.W. ed.), pp. 139–161, Oregon State University Press.
Mumford, T.F. (1979), *Proc. 9th Int. Seaw. Symp., Santa Barbara,* Science Press, Princeton, 515–524.
Munda, I. (1964), *Bot. Mar.,* **7**, 76.
Munda, I. (1970), *Nova Hedwigia,* **19**, 535–550.
Munda, I. (1972), *Bot. Mar.,* **15**, 1–45.
Muther, A. and Tollens, B. (1924), *Berlin Bericht.,* **37**, 1, 298.
Myers, J. (1964), Life Sciences and Space Research II, *4th Int. Space Sci Symp.,* p. 323, (1963).
Myers, J. and Graham, J.R. (1961), *Plant Physiol.,* **36**, 342.
Myklestad, S. and Haug, A. (1966), *Proc. 5th Int. Seaw. Symp. Halifax,* p. 297, Pergamon Press, Oxford.

Nakoa, Y., Onohara, T., Matsubara, T., Fujita, Y. and Zenitani, B. (1972), *Bull. Jap. Soc. Sci. Fish.*, **38** (6), 561.
Nebb, H. (1967), *Norsk Landbruk* nr., **16**, 3–7.
Nebb, H. and Jensen, A. (1966), *Proc. 5th Int. Seaw. Symp. Halifax*, p. 387, Pergamon Press, Oxford.
Needler, A.W.H. (1944), *Atlantic Biol. Stat. Circ. G.*, **3**.
Neish, I.C. (1972), *A. Technological Development Programme for Dulse Cultivation on Great Manan Island*, N.B. Ocean Science Ass. Tld.
Neish, I.C. (1973), *The distribution of kelp and other commercially useful marine algae in Charlotte County, New Brunswick*, N.B. Dept. Fish and Environment.
Neish, I.C. (1976), *J. Fish. Res. Bd. Canad.*, **33**, 1007–1014.
Neish, I.C. and Fox, C.H. (1971), Atlantic Regional Lab., Halifax, Rept. 12, 35 pp. NCR 12034, Canada.
Neish, I.C., Fox, C.H. and Shacklock, P.C. (1971), Atlantic Regional Lab., Halifax, Rept. 14, 25 pp, NCR 12253, Canada.
Nelson, W.L. and Cretcher, L.H. (1929), *J. Am. Chem. Soc.*, **51**, 1914.
Nelson, W.L. and Cretcher, L.H. (1930), *J. Am. Chem. Soc.*, **51**, 2130.
Neuberg, C. and Ohle, H. (1921), *Biochem. Z.*, **125**, 311.
Neuberg, C. and Schweitzer, C.H. (1937), *Monatchr. Chem.*, **71**, 46.
Newcomb, E.L. and Smythe, C.E. (1921), *J. Am. Pharm. Assoc.*, **X**, 524.
Newton, L. (1945), *Endeavour*, **4**, 69.
Newton, L. (1948), *Proc. Linn. Soc. Lond.*, **159**, 84.
Nichols, A.A. (1933), *Centrbl. Bakt.* Abt. II., **88**, 177.
Nicol, H. (1931), *Nature, Lond.*, **128**, 1041.
Nicotri, M.E. (1977), *Aquaculture*, **12**, 127–136.
Nielsen, K. (1977), *J. Phycol.*, **13**, Suppl. Abst., 276, p. 48.
Nielsen, K., Nielsen, P.S. and Twide (1977), *9th Int. Seaw. Symp.*, Copenhagen Section Factory Publ.
Nilson, H.W. and Schaller, J.W. (1941), *Food Res.*, **6**, 461.
Nishigeni, K.E. and Semesi, A.K. (1977), *Bot. Mar.*, **20**, 239–247.
Nishimura, T. (1903), *J. Imp. Fish. Bur.*, **12**.
Noda, H. (1971), *Bull. Jap. Soc. Sci. Fish.*, **37**, 30–39.
Noda, H., Horiguchi, Y. and Araki, S. (1975), *Bull. Jap. Soc. Sci. Fish.*, **41**, 1299–1304.
Noda, M. and Kitami, T. (1964), *J. Fac. Sci. Niigata Univ. Ser. II*, **4**, 25.
Norman, A.G. (1937), *The Biochemistry of Cellulose, Polyuronides, Lignin, etc.*, Clarendon Press, Oxford.
Norris, R., Simeon, M.K. and Williams, K.B. (1937), *J. Nut.*, **13**, 425.
North, W.H. (1961), *Nature, Lond.*, **190**, 124.
North, W.H. (1971), Kelp programme Ann. Rept. 1970–71, W. Keck Mar. Biol Lab., Cal. Inst. Technology.
North, W.H. (1972), Kelp programme Ann. Rept. 1971–72, W. Keck Mar. Biol. Lab., Cal. Inst. Technology.
North, W.H. (1973), Kelp programme Ann. Rept. 1972–73, W. Keck Mar. Biol. Lab., Cal. Inst. Technology.
North, W.H. (1974), Kelp programme Ann. Rept. 1973–74, W. Keck Mar. Biol. Lab., Cal. Inst. Technology.

North, W.J. and Wheeler, P.A. (1979), *Proc. 9th Int. Seaw. Symp., Santa Barbara,* Science Press, Princeton, 67–78.
Northrup, Z. (1919), *Abstr. Bact.,* **3**, 7.
Novy, E.G. and De Kruif, P.H. (1917), *J. Infect. Dis.,* **20**, 629.
Noyes, H.A. (1918), *Chem. Analyst.* **25**, 12.

Obereck De Meijer, Van (1891), *Centrb. Bakt. Parasitenk. Abt. I.,* **9**, 163.
Oden. (1917), *Intern.-Zeit. Phys. Chem. Biol.,* **3**, 83.
Ohta, F. and Tanaka, T. (1964), *Mem. Fac. Fish. Kagoshima Univ.,* **13**, 38.
Oishi, E. and Kunisake, N. (1970), *Bull. Jap. Soc. Sci. Fish.,* **36**, 1187–1285.
Okamura, K. (1909–32), *Icones of Japanese Algae,* Tokyo.
Okamura, K. (1925), *J. Imp. Fish. Inst.,* **21**, 10.
Okamura, K. (1932), *Proc. 5th Pacif. Sci. Cong.,* **4**, 3153.
Okamura, K. (1934), *J. Imp. Fish. Inst.,* **29**, 47.
Okazaki, A. (1971), *Seaweeds and their Uses in Japan,* pp. 1–165, Tokyo Univ. Press, Tokyo.
Okuda, Y. and Nakayamu, D. (1916), *J. Agric. Imp. Univ. Tokyo,* **5**, 341.
O'Neill, A.N. (1955), *J. Am. Chem. Soc.,* **77**, 2837.
O'Neill, A.N. and Stewart, D.K.R. (1956), *Can. J. Chem.,* **34**, 1700.
Onokhin, I.P. (1938), *Vodoroslego. Nauch. Issledovatel. Inst. Vodorsli. Belogo. Morya,* 228.
Oohusa, T. (1971), *Occas. Pap. Yamamoto Nosi Res. Lab.,* 1971.
Opotzkii, V.F. and Bortnik, L.A., *Ukr-Khem. Zh. 10. Wiss. Tech.,* **1**, 331.
Ortega, M. (1977a), *Sargazo gigante,* Comm. 12. Inst. Invest. Sobre Recursos Bioticos, Mexico.
Ortega, M.A. (1976), *El Agar,* Comm. 9th Inst. Invest Sobre Recursos Bioticos ac., Mexico.
Ortega, M.M. (1972), *Rev. lat-amer. Microbiol.,* **14** (2), 85–97.
Ortega, M. (1977b), *Pelo de Cochi,* Comm. 15. Inst. Invest Sobre Recursos Bioticos, Mexico.
Ortiz, A.G. (1968), *La Industria de Derivados de Algas en Espana,* Santiago de Compostela.
Oshima, (1905), U.S. Dept. Agric. Off. of Exp. Sta. Bull. 159.
Owen, G. (1849), *J.R. Agric. Soc.,* **10**, 142.
Oza, R.M. (1978), *Bot. Mar.,* **21**, 165–167.

Painter, E. (1887), *Proc. Am. Pharm. Assoc.,* **35**, 678.
Painter, T.J. (1966), *Proc. 5th Int. Seaw. Symp. Halifax,* p. 305, Pergamon Press, Oxford.
Palminha, F.P. (1971), *Bot. Junta. Nac. Fom. Pescas,* **18**, (7), 25–36.
Parenzan, P. (1970), *La Cladophora prolifera Kütz. dal Golfo di Taranto e Possibilità di una sua Valorizzaxoine Economica* in *Possibilità di Utilizzazione Industriale delle Alghe in Italia,* Vol. 5, Dagli incontri, tecnici, CNR Lab. di Technol. Pesca ed Ente Fiera Internaz. della Pexa, Ancona, 1970, pp. 14–21.
Parke, M. (1948), *J. Mar. Biol. Ass.,* **27**, 706.
Parker, H.S. (1973), *Aquaculture,* **3**, 425–439.
Parr, A.E. (1939), *Bull. Bingham Oceanog. Collect.,* **6**, 1–94.
Parsons, M.J. Pickmere, S.E. and Bailley, R.W. (1977), *N.Z. Journ. Bot.,* **15**, 589–595.

Partridge, S.M. and Morgan, W.T.J. (1942), *Br. J. Exp. Path.*, **23**, 84.
Paul, T. (1901), *Centrb. Bakt. Abt.* II., **29**, 270.
Paul, T.M., Edward, D.W. and Skoryna, S.C. (1964), *Med. Assoc. J.*, **91**, 553–557.
Pauli, W. and Palmrich, L. (1937), *Kolloid. Z.*, **79**, 174.
Pauli, W. and Steinback, (1941), *Helv. Chim. Acta.*, **24**, 317.
Pavlov, P.N. and Borshim, M.I. (1941), *Chimia*, 147.
Pavlov, P.N. and Engel'stein, M.A. (1936), *Colloid J. U.S.S.R.*, **2**, 821.
Payen, M. (1859), *C.R. Acad. Sci. Paris*, **49**, 521.
Pehorey, J. (1937), *Rev. Gener. Matieres. plast.*, **13**, 270.
Pekelharing, C.A. (1921), *Archs. Intern. Physiol.*, **18**, 495.
Penman, A. and Sanderson, G.R. (1972), *Carbohydrate Res.*, **25**, 273–282.
Pennington, W. (1942), *Nature, Lond.*, **148**, 314.
Pentegow, B.P. (1929), *Bull. Pacif. Sci. Fish. Res. Sta. Vladivos.*, **3**, 1.
Pentegow, B.P. (1930), *Chem. Wirtsch.*, **1**, 134.
Percival, E. (1964), *Proc. 4th Int. Seaw. Symp. Biarritz*, p. 18. Pergamon Press, London.
Percival, E. (1969), *Oceanogr. Mar. Biol. Ann. Rev.*, **6**, 137.
Percival, E.G.V. (1939), *Pharm. J.*, **142**, 189.
Percival, E.G.V. (1944), *Nature, Lond.*, **154**, 673.
Percival, E.G.V. and Buchannan, J. (1940), *Nature, Lond.*, **145**, 1020.
Percival, E.G.V. and Forbes, I.A. (1938), *Nature, Lond.*, **142**, 1076.
Percival, E.G.V., Munro, J. and Somerville, J.C. (1937), *Nature, Lond.*, **139**, 512.
Percival, E.G.V. and Ross, A.G. (1948), *J. Soc. Chem. Ind.*, **67**, 420.
Percival, E.G.V. and Ross, A.G. (1950), *J. Chem. Soc.*, 717.
Percival, E.G.V. and Ross, A.G. (1951), *J. Chem. Soc.*, 720.
Percival, E.G.V. and Sim, W.S. (1936), *Nature, Lond.*, **137**, 997.
Percival, E.G.V. and Somerville, J.C., (1937), *J. Chem. Soc.*, 1615.
Percival, E.G.V., Somerville, J.C. and Forbes, I.A. (1938), *Nature, Lond.*, **142**, 797.
Percival, E.G.V. and Thomson, T.G.H. (1942), *J. Chem. Soc.*, 750.
Perez, R. (1970), *Rev. Trav. Inst. Tech. Marit.*, **34**, 351–362.
Perrot, E. and Gatin, C.K. (1912), *Ann. Instit. Oceanog.*, **3**, Fasc 1., Monaco.
Pethybridge, G.H. (1915a), *J. Dept. Agric. Tech. Ind. Ireland*, **15**, 546.
Pethybridge, G.H. (1915b), *Int. Agric. Tech. Rundschau*, **6**, 1129.
Petrov, K.M. (1963), *Ref. Zh. Biol.*, No. 6V 30.
Pickmere, S.E., Parsons, M.J. and Bailey, R.W. (1973), *Phytochem.*, **12**, 2441–2444.
Pickmere, S.E., Parsons, M.J. and Bailey, R.W. (1975), *N.Z. J. Sci.*, **18**, 585–590.
Pierre, I. (1853), *Mem. Soc. Linn. Norm.*, **9**.
Pijer, A. and Kraan, G.J. (1921), *Med. J.S. Africa*, **18**, 221.
Pillai, K.S. and Varier, N.S. (1953), *Proc. 1st Int. Seaw. Symp.*, pp. 53, 101, Edinburgh.
Pillay, R.S. (1977), *J. Phycol.*, **13**, Suppl. Abst. 307, 308, p. 54.
Pirie, N.W. (1936a), *Br. J. Exp. Path.*, **17**, 269.
Pirie, N.W. (1936b), *Biochem. J.*, **30**, 369.
Pleskatsevich, P. (1938), *Vosoroslerogo Nauch. Issledovatel Inst. Vodorsli. Belogo, Morya.* pp. 221–234.
Polunin, N. (1942a), *Nature, Lond.*, **148**, 143, 374.
Polunin, N. (1942b), *Chron. Bot.*, **7**, 133.

Porodko, T.M. (1909), *J.R. Microsc. Soc.,* **19**, 256.
Porumbaru, A. (1880), *C.R. Acad. Sci. Paris,* **90**, 1081.
Post, E. Von (1939), *Planta,* **28**, 743.
Povolny, M. (1966), *Rostlinna Vyroba,* **12**, 355–340.
Povolny, M. (1969a), *Proc. 6th Int. Seaw. Symp.,* La Marina Merchante, Madrid, Spain, p. 703–722.
Povolny, M. (1969b), *Rostlinna Vyroba,* **15**, 545–554.
Povolny, M. (1971), *Rostlinna Vyroba,* **17**, 877–888.
Povolny, M. (1972), *Rostlinna Vyroba,* **18**, 703–710.
Powell, J.H. and Meeuse, B.J.D. (1964), *Econ. Botany,* **18**, 164.
Pownall, P.C. (1964), *Fish. Newsletter,* **23**, 11, 13, 15.
Prat, S. (1927), *Am. J. Botany,* **14**, 167.
Prideaux, E.B.R. and Howitt, F.O. (1932), *Trans. Faraday Soc.,* **28**, 79.
Prince, E.E. (1917), *Canadian Fisherman,* **4**, 48.
Prince, J.R. (1974), *Aquaculture,* **4**, 69–79.
Prince, J.S. (1971), *An Ecological Study of the Marine Red Alga, Chondrus crispus, in the Waters off Plymouth, Massachusetts,* PhD Thesis, Cornell Univ.
Pringle, J.D. (1976), *The Marine Plant Industry–Commercially Important Species and Resource Management* in *Proc. Bras dor Lakes Aquacult. Conf.,* (McKay, G. and K. ed.), Sydney, Nova Scotia.
Pringle, J.D. (1979), *Proc. 9th Int. Seaw. Symp.,* Santa Barbara, Science Press, Princeton, 225–232.
Pringle, J.D. and Semple, R.E. (1976), *A preliminary Assessment of the Ecological Impact of an Experimental Chondrus (Irish Moss) Harvester off Coastal Prince Edward Island.* Tech. Rept. Ser. No. Mar. T-76-1, Dept. Environment, Halifax, pp. 28.
Pringsheim, H. and Pringsheim, E. (1910), *Centrb. Bakt. Abt.* II., **26**, 227.
Printz, H. (1959), *Norske Videnskap.-Akad. Oslo I. Mat-Nature,* **3**, 4–15.
Pshenin, L.N. (1959), *Mikrobil. Transl.,* **28**, 886.

Quastel, J.H. and Webley, D.M. (1947), *J. Agric. Sci.,* **37**, 257.

Rao, M.U. (1968), *Proc. Symp. Living Resources of Seas around India.* Central Fish Inst.
Rao, M.U. (1969), *Proc. 6th Int. Seaw. Symp.,* La Marina Merchante, Madrid, Spain, pp. 579–584.
Rao, M.U. (1970), *Bull. Cent. Mar. Fish. Res. Inst.,* Mandapan Camp.
Rao, M.U. and Kalimuthu, S. (1972), *Bot. Mar.,* **15**, 56–59.
Rao, P.V.S., Rao, K.R. and Subbaramiah, J. (1977), *Seaweed Res. Util.,* **2**, 82–86.
Rapson, A.M., Moore, L.B. and Elliot, I.L. (1943), *N.Z. J. Sci. Tech.,* **23**, 149–170.
Rauch, G. (1943), *Arch. Hyg. Bakt.,* **130**, 57.
Reed, M. (1907), *Ann. Rept. Hawaii Agric. Exp. Sta.,* 1906.
Reedman, E.J. and Buckley, L., *Can. J. Res.,* **21**, 348.
Rees, D.A. (1963), *J. Chem. Soc.,* 1821.
Reinbold, Th. (1896), *Schr. Naturw. verf. f. Schl.-Holst.,* **9**, 145.
Reindemeester, W. (1908), *Z. Wiss. Mikroscop.,* **25**, 42.
Reinke, O. (1918), *Chem. Ztg.,* **42**, 230.

Ricard, M.P. (1931), *Bull. Soc. Chim. Biol.,* **13**, 417.
Richards, H.M. (1905), *Science, N.Y.,* **21**, 895.
Richter, E. (1887), *Berlin Klin. Wochenschr.,* 600.
Ricohermosa, M.A. and Deveau, L.E. (1977), *Commercial propagation of Eucheuma spp. Clones in the South China Sea. A Discussion of Trends in Cultivation Technology and Commercial Production Pattern,* Mar. Coll. Maine, U.S.A.
Ricohermosa, M.A. and Deveau, L.E. (1979), *Proc. 9th Int. Seaw. Symp.,* Santa Barbara, Science Press, Princeton, 525—532.
Ridley, J.E. (1967), *Brit. Phycol. Bull.,* **3**, 410.
Rigg, G.B. (1912a), *Plant World,* **15**, 83.
Rigg, G.B. (1912b), *Ecological and Economic Notes on Puget Sound Kelps in Preliminary Report on the Fertiliser Resources on the United States,* (Cameron, F.K. ed.), Wash. D.C. Doc. 190 62nd Congress. 2nd Session.
Rinck, E. and Brondardel, J. (1949), *C.R. Acad. Sci., Paris,* **228**, 263.
Robbins, W.J. (1939), *Am. J. Botany,* **26**, 772.
Robertson, D.J. (1911), *Highland Industries.*
Robertson, G.R. (1930), *Ind. Engng. Chem.,* **22**, 1074.
Rodriguez, O. (1953), *Proc. 1st Int. Seaw. Symp.,* Edinburgh, pp. 75—76.
Roe, A.F. (1941), *J. Bacteriol.,* **41**, 48.
Roe, A.F. and Thaller, H.I. (1942), *Science, N.Y.,* **96**, 43.
Roels, O.A., Haines, K.C. and Sunderlin, J.B. (1975), *Proc. 10th European Symp. Mar. Biol.,* Universa Press, Belgium. Vol. 1. pp. 381—390.
Rogachev, V.I. (1935), *Colloid J., U.S.S.R.,* **1**, 79.
Roman, W. (1930), *Z. Angew. Chem.,* **44**.
Rosam, K. (1904), *Centrb. Bakt. Abt.* II, **12**, 464.
Rose, R.C. (1937), Ph. D. Thesis, University of London.
Rose, R.C. (1949), *Proc. 7th Pan-Pacific Science Congress,* Auckland, N.Z., **5**, 66.
Rossi, G. and Marescotti, A. (1933), *Gazz. Chim. Ital.,* **63**, 121.
Rossi, G. and Marescotti, A. (1936), *Gazz. Chim. Ital.,* **66**, 223.
Rost, E. (1915), *Mitt. Dtsch. Seefisherei,* **31**, 160.
Rost, E. (1917), *Mitt. Dtsch. Seefisherei,* **33**, 28, 237.
Round, F.E. (1965), *The Biology of the Algae,* St. Martins.
Rudiger, M. (1922), *Z. Untersuch. Lebensmitt.,* **64**, 77.
Russell, E.J. (1910), *J. Bd. Agric.,* **17**, 458.
Russell-Wells, B. (1922), *Biochem. J.,* **16**, 578.
Russell-Wells, B. (1929), *Nature, Lond.,* **124**, 654.
Ryther, J.H., De Boer, J.A. and Lapointe, B.E., *Proc. 9th Int. Seaw. Symp.,* Santa Barbara, Science Press, Princeton, 1—16.
Ryther, J.H., Dunstan, W.M., Tenore and Huguenin, J.E. (1972), *Bioscience,* **22**, 144—152.
Ryther, J.H., Williams, L.D., Kneale, D.C., Lapointe, B.E., Loftus, J.B. and Stenberg, W.B. (1976), Ann. Rept. Harbor Branch Foundation pp. 39, Florida Power Light Co. and U.S. ERDA E (11-!)-2948.

Saeter, E.A. and Jensen, A. (1957), *Norsk. Inst. for Tang-og Tareforsk. Rept. 17.*
Saiki, T. (1906), *J. Biol. Chem.,* **2**, 261.
Saint-Yves, A. (1879), *De l'Utilité Algues Marines,* Paris.

Saito, A. and Idler, D.R. (1966), *Can. J. Biochem.*, **44**, 1195.
Saito, T. and Ueda, T. (1953), *Mem. Fac. Fish.* (Kagoshima University), **3**, 141–148.
Saito, Y. (1956a), *Bull. Jap. Soc. Sci. Fish*, **22**, 21.
Saito, Y. (1956b), *Bull. Jap. Sci. Fish.*, **22**, 229, 235.
Saito, Y. (1975), *Undaria* in *Advances in Phycology in Japan*, (Tokida, J. and Hirose, H. eds.), pp. 304–320, Junk.
Saito, Y., Iso, N., Mizuro, H., Fujii, S. and Suzuki, Y. (1977), *Bull. Jap. Soc. Sci. Fish.*, **43**, 1299–1306.
Salle, H. et cie. (1912), *Ann. de la Drogue et de ses Dérives.*
Saller (1916), *Prometheus*, 726.
Samec, M. and Isajevic, V. (1921), *C.R. Acad. Sci. Paris*, **173**, 1474.
Samec, M. and Isajevic, V. (1922), *Kolloid Beihefte*, **16**, 285.
Sanbonsuga, Y. and Neushul, M. (1977), *J. Phycol.*, **13**, Suppl. Abst. 341, p. 59.
Sanbonsuga, Y. and Torii, S. (1973), *Bull. Hokk. Reg. Fish. Res. Lab.*, **39**, 61–68.
Sanbonsuga, Y. and Torii, S. (1974), *Bull. Hokk. Rep. Fish. Res. Lab.*, **40**, 48–59.
Santos, G.A. and Doty, M.S. (1979), *Proc. 9th Int. Seaw. Symp., Santa Barbara*, Science Press, Princeton, 361–368.
Sargent, M.C. and Lantrip, L.W. (1952), *Am. J. Botany*, **39**, 99.
Sarochan, V.F. (1962), *Ref. Zh. Biol.*, (1962) No. 19 V36.
Sato, S. (1971), *Bull. Jap. Soc. Sci. Fish.*, **37**, 326–332.
Saunders, R.G. and Lindsay, G.J. (1979), *Proc. 9th Int. Seaw. Symp., Santa Barbara*, Science Press, Princeton, 249–256.
Sauvageau, C. (1918), *Rev. Gen. Sci., Paris*, **29**.
Sauvageau, C. (1920), *Utilisation des Algues Marines*, Paris.
Sauvageau, C. (1921), *Bull. Stat. Biol. d'Arcachon*, **18**, 53.
Sawasaki, M., Torii, S. and Nakamura, K. (1965), *Sci. Rep. Hokk. Fish. Res. Stat.*, **3**, 51–56.
Scagel, R.F. (1948), *Rept. B.B. Depart. Fish.*, **1**, 70.
Schachat, R.E. and Glicksman, M. (1959), *Econ. Botany*, **13**, 365.
Schaefer, M.B. (1964), *An investigation of the effects of discharged wastes on kelp*, State Water Quality Control Bd., Sacramento.
Scheffer, V.B. (1943), *Fishery Market News*, **5**, 1.
Schimizu, Y. (1971), *Bull. Jap. Soc. Sci. Fish.*, **37**, 540.
Schmid, O.J. (1962), *Bot. Marina*, **3**, Suppl., 67.
Schmid, O.J. and Hoppe, H.A. (1962), *Bot. Marina*, **3**, Suppl., 101.
Schmidt, C. (1844), *Ann. Chem. Pharm.*, **51**, 29.
Schmidt, E. and Vocke, F. (1926), *Ber. Deutsch. Chem. Ges.*, **59**, 1585.
Schnetter, R. (1966), *Bot. Mar.*, **9**, 1–4.
Schnetter, R. and Schnetter, M.L. (1967), *Mitt. Inst. Colombo-Ateman Invest. Gent.*, **1**, 45–52.
Schoffel, E. and Link, K.P. (1933), *J. Biol. Chem.*, **100**, 397.
Schotteliens, M. (1877), *Centrb. Bakt. Parasitenk, Abt. I.*, **2**, 1042.
Schulzen, H. (1962), *Bot. Marina*, **3**, Suppl., 75.
Schwenke, H. (1965), *Kiel. Meeresf.*, **21**, 144.
Schwimmer, M. and D. (1955), *The Role of Algae and Plankton in Medicine*, Grune and Stratton, N.Y.

Scott, W.R. (1914), *Bd. Agric. Scot. Rept. on Home Ind. Highl. Islands*, p. 118.
Scottish Seaweed Research Association, *Ann. Repts.*, 1945–57.
Scottish Inst. Seaweed Research Association, *Ann. Repts.*, 1952–68.
Scruti, F. (1906), *Gazz. Chim. Ital.*, **36**, II, 619.
Segers-Laureys, A. (1913), *Rec. Inst. Bot. Leo. Errera*, **9**.
Segi, T. (1966), On the Species of *Gelidium* from Japan and its vicinity. Department of Marine, University of Mie Prefecture, Japan.
Selby, H.H. and Selby, T.A. (1959), *Agar* in *Industrial Gums*, Acad. Press. N.Y.
Semenenko, V.E. and Vladimirova, M.G. (1961), *Fiziol. Rast*, **8**, 743.
Senn, T.L. and Kingman, A.R. (1977), *J. Phycol.*, **13**, Suppl. Abst. 356, p. 62.
Senn, T.L. and Skelton, J. (1969), *Proc. 6th Int. Seaw. Symp.*, La Marina Merchante, Madrid, Spain, pp. 723–730.
Serft, (1906), *Pharmaz. Praxis*, **5**.
Sergev, H. (1916), *Pharm. Zentralhalle*, **57**, 407.
Sernov, S.R. (1909), *Int. Rev. Hydrob. 1910*, **3**, 226.
Shacklock, P.F., Robson, D., Forsyth, I. and Neish, I.C. (1973), Atlantic Regional Lab., Halifax, Rept. 18., 22 pp. NRC 13113, Canada.
Shacklock, P.F., Robson, D. and Simpson, F.J. (1975), Atlantic Regional Lab., Halifax, Rept., 21, 27 pp. NRC 14735, Canada.
Shang, Y.C. (1976), *Aquaculture*, **8**, 1–7.
Sharp, S.S. (1939), *J. Econ. Entomol.*, **32**, 394.
Shaw, T.I. (1962), *Halogens*, in Lewin, *Physiology and Biochemistry of Algae*, Acad. Press, 247–251. N.Y.
Sheehy, E.J., Brophy, J., Dillon, J. and O'Muineachin, P. (1942), *Econ. Proc. R. Dublin Soc.*, **3**, 150.
Shiraiwa, Y., Abe, K., Sasaki, S.F., Ikawa, T. and Nisizawa, K. (1975), *Bot. Mar.*, **18**, 97–104.
Shitanaka, M. and Suto, S. (1954), *Bull. Jap. Soc. Sci. Fish.*, **20**, 487.
Shmeler, V. (1938), *Spirto-Vodachnaya Prom.*, **15**, 19.
Shrum, G.M. (1953), *Proc. 1st Int. Seaw. Symp.*, p. 107, Edinburgh.
Shulman, M.S. (1940), *Colloid J. U.S.S.R.*, **6**, 747.
Shurtleff, W.R. and Aoyagi, A. (1979), The book of Sea Vegetables, Autumn Press, Brookline, Mass. U.S.A.
Sieburth, J. and Jensen, A. (1967), *Appl. Microbiol.*, **15**, 830.
Silverthosne, W. and Sorenson, P.E. (1971), *Prepr. Mar. Tech. Soc. Ann. Conf.*, **7**, 523–533.
Simmons, P.L. (1883), *The Commercial Product of the Sea*, 3rd Ed.
Simonetti, G., Giaccone, G. and Pignatti, S. (1970), *Helgol. wiss. Meeresunters*. **20**, 89–96.
Simpson, F.J., Neish, A.C., Shacklock, P.E. and Robson, D.R. (1978), *Bot. Mar.*, **21**, 229–236.
Simpson, F.J., Shacklock, P., Robson, D. and Neish, A.C. (1979), *Proc. 9th Int. Seaw. Symp., Santa Barbara*, Science Press, Princeton, 509–514.
Singh, R.N. (1942), *Ind. J. Agric. Sci.*, **12**, 743.
Sinova, E.S. (1928), *Bull. Pac. Oc. Sci. Fish. Res. Stat. Vlad.*, **1**, 77.
Sinova, E.S. (1929), *Trav. Inst. Rech. Indust. Commite. Exer. d'Archangel.*, **6**, 1.
Sinova, E.S. *Trav. Stat. Biol. Sebast.*, **4**, 1.

Skelton, B.J. and Senn, T.L. (1969), *Proc. 6th Int. Seaw. Symp.*, La Marina Merchante, Madrid, Spain, p. 731.
Skoryna, S.C., Paul, T.M. and Edward, D.W. (1964), *Canad. Med. Ass. J.*, **91**, 285–288.
Skoryna, S.C., Paul, T.M. and Edward, D.W., (1965), *Canad. Med. Ass. J.*, **93**, 404–407.
Skottsberg, C. (1921), *Svensk. Vet. Akad. Handl.*, **61**, No. 11.
Smidsrød, O. and Haug, A. (1968), *Acta. Chem. Scand.*, **22**, 797.
Smith, D.B., Cook, W.H. and Neal, J.L. (1954), *Arch. Biochem. Biophys.*, **53**, 192.
Smith, D.B., O'Neill, A.N. and Perlin, A.S. (1955), *Can. J. Chem.*, **33**, 1352.
Smith, D.B. and Young, E.G. (1955), *J. Biol. Chem.*, **217**, 845.
Smith, F. and Montgomery, R. (1959), *The Chemistry of Plant Gums and Mucilages*, Reinhold, N.Y.
Smith, H.M. (1894), *Bull. U.S. Fish. Comm.*, 1893.
Smith, H.M. (1905a), *Bull. U.S. Bur. Fish.*, **24**, 133.
Smith, H.M. (1905b), *Nat. Geog.*, **16**, 201.
Smith, T. (1931), *Science, N.Y.*, **74**, 21.
Smith, W. (1885), *J. Soc. Chem. Ind.*, **4**, 518.
Soerjodinote, (1969), Occas. Pap. IPFC 69/6.
Sorensen, N.A. and Kristensen, K. (1950), U.S. Patent 2, 516, 350.
South, G.R. (1979), *Proc. 9th Int. Seaw. Symp., Santa Barbara*, Science Press, Princeton, 133–142.
Span, A. (1969), *Proc. 6th Int. Seaw. Symp., Madrid*, p. 383–437.
Spaulding, M.F. (1940), *Soil Sci. Soc. Amer. Proc.*, **5**, 259.
Speakman, J.B. (1945), *Nature, Lond.*, **155**, 655.
Speakman, J.B. and Chamberlain, N.H. (1944), *J. Soc. Dyers Colourists*, **60**, 264.
Spencer, G.S. (1920), *J. Ind. Engng. Chem.*, **12**, 682, 786.
Spoon, W. (1951), *K. Inst. Tropen. Ber. Afd. Trop. Prod.*, **234**, 1.
Stancioff, D.J. (1969), *6th Seaw. Symp., Spain*, Abstract.
Stancioff, D.J. and Stanley, N.F. (1969), *Proc. 6th Int. Seaw. Symp.*, La Marina Merchante, Madrid, Spain, pp. 595–609.
Standt, A.J. (1888), *Am. J. Pharm.*, **60**, 170.
Stanford, E.C.C. (1862a), *Chem. News*, **5**, 167.
Stanford, E.C.C. (1862b), *J. Soc. Arts*, **10**, 185.
Stanford, E.C.C. (1876), *Chem. News*, **34**.
Stanford, E.C.C. (1883a), *J. Soc. Chem. Ind.*, **3**, 297.
Stanford, E.C.C. (1883b), *Chem. News*, **47**, 254, 267.
Stanford, E.C.C. (1883c), *Pharm. J. Ser III.*, **13**, 1019.
Stanford, E.C.C. (1884), *J. Soc. Asts.*, **32**, 717.
Stanford, E.C.C. (1885), *J. Soc. Chem. Ind.*, **4**, 594.
Stanford, E.C.C. (1886), *J. Soc. Chem. Ind.*, **5**, 218.
Stanford, E.C.C. (1899), *J. Soc. Chem. Ind.*, **18**, 398.
Stanier, R.Y. (1941), *J. Bacteriol.*, **42**, 427.
Starr, M.P. (1941), *Science N.Y.*, **93**, 333.
Steiner, A.B. and McNeeley, W.H. (1954), *Advances in Chemistry*, Series II. p. 68.
Stenhouse, J. (1844), *Ann. Chem.*, **51**, 349.
Stephens, E.L. (1949), *Trans. Roy. Soc. S. Africa*, **32**, 105.

Stephenson, W.A. (1974), *Seaweed in Agriculture and Horticulture,* 3rd ed., Rateaver, Pauma Valley, Cal., pp. 241.
Stephenson, W.M. (1966), *Proc. 5th Int. Seaw. Symp. Halifax,* p. 405, Pergamon Press, Oxford.
Stewart, C.M., Higgins, H.G. and Austin, S. (1961), *Nature, Lond.,* **192**, 1208.
Stewart, G.R. (1915), *J. Agric. Res.,* **4**, 39.
Stiles, W. (1919), *Biochem. J.,* **14**, 58.
Stirn, J. (1968), *Rev. Int. Oceanogr. Med.,* **9**, 99–106.
Stokes, J.H. (1916), *J. Infect. Dis.,* **18**, 415.
Stokes, J.H. (1917), *J. Amer. Med. Soc.,* **68**, 1092.
Stoloff, L. (1959), *Carrageenan* in *Industrial Gums,* Acad. Press. N.Y.
Stoloff, L. (1962), *Econ. Botany,* **16**, 86.
Stoloff, L.S. (1943), *Fishery Market News,* **5**, 1
Street, J.P. (1917), *Modern Hosp.,* **9**, 398.
Sulit, J.I., Salcedo, L.G. and Panganibam, P.C. (1956), *Indo-Pacific Fish. Counc. Proc.,* **6**, 213, 280.
Sundene, O. (1961), *Nytt. Mag. Bot.,* **9**, 5.
Sundene, O. (1964), *Nytt. Mag. Bot.,* **11**, 83.
Suneson, S. (1932), *Z. Physiol. Chem.,* **213**, 270.
Suto, S. (1949), *Bull. Jap. Soc. Sci. Fish.,* **15**, 226.
Suto, S. (1950a), *Bull. Jap. Soc. Sci. Fish.,* **15**, 649, 671.
Suto, S. (1950b), *Bull. Jap. Soc. Sci. Fish.,* **16**, 137.
Suto, S. (1951), *Bull. Jap. Soc. Sci. Fish.,* **17**, 9, 13.
Suto, S. (1954), *Bull. Jap. Soc. Sci. Fish.,* **20**, 494.
Suto, S., Maruyama, T. and Umebayashi, O. (1954), *Bull. Jap. Soc. Sci. Fish.,* **20**, 490.
Suzuki, N., Nishikawa, I. and Aoki, S. (1931), *Nipp. Tiksan. Gkw. Ho.,* **4**, 227, 263.
Svendsen, Per. (1968), *6th Int. Seaw. Symp., Spain,* Abstract.
Swan, J.G. (1894), *Bull. U.S. Fish. Comm.,* **13**.
Swartz, M.D. (1911), *Trans. Conn. Acad. Sci.,* **16**.

Taboury, F. and Bernuchon, J. (1937), *Bull. Soc. Chem.,* **4**, 1857.
Tagawa, S. and Kojima, Y. (1972), *Proc. 7th Int. Seaw. Symp.,* University Tokyo Press, Japan, 447–450.
Tagawa, S., Kojima, Y., Yamada, Y. and Kono, M. (1960), *J. Shimonosekei Coll. Fish.,* **10**, 35.
Takagi, M.L. (1975), *Seaweeds as Medicine,* in *Advance of Phycology in Japan,* (Tokida, J. and Hirose, H. eds.), pp. 321–326. Junk.
Takahashi, E. (1914), *J. Coll. Agric. Japan,* **6**, 109.
Takahashi, E. (1920), *J. Coll. Ag. Hokk. Imp. Univ. Japan,* **8**, 183.
Takahashi, E. and Shiragama, K. (1931), *J. Agric. Chem. Soc. Japan,* **7**, 45.
Takahashi, E. and Shiragama, K. (1932), *J. Agric. Chem. Soc. Japan,* **8**, 8, 659, 1259.
Takahashi, E. and Shiragama, K. (1934), *J. Coll. Agric. Hokk. Imp. Univ. Japan,* **35**, 101.
Takamatsu, M. (1938), *Saito Ho-on Kai Mus. Bull.,* **14**, Bot. 5.
Takao, V. (1916), *J. Pharm. Soc. Japan,* **5**, 1061.

Takeuchi, T., Shitanaka, M., Fukuhara, A. and Yamasaki, H. (1956), *Bull. Jap. Soc. Fish.*, **22**, 16.
Tang, P.S. and Whang, P. Ch. (1935), *Chinese J. Physiol.*, **9**, 285.
Tanner, H. (1848), *J. R. Agric. Soc.*, **9**, 469.
Tanner, H.G. (1922), *J. Ind. Engng. Chem.*, **14**, 441.
Tassily, E. and Leroide, J. (1911), *Bull. Soc. Chim.*, **4** ser. **9**, 63.
Taylor, A.R.A. (1972), *Proc. 7th Int. Seaw. Symp.*, Univ. Tokyo Press, Japan, 263–267.
Taylor, I.E.P. and Wilkinson, A.J. (1977), *Phycologia*, **16**, 37–42.
Tendeloo, H.J.C. (1941), *Rec. Trav. Chim.*, **60**, 347.
Thiercelin, (1880), *Bull. Soc. Chim. Fr.*, **33**, 559.
Thompson, L. (1922), *J. Lab. Chin. Med.*, **7**, 758.
Thone, F. (1940), *Science, N.Y.*, **91**, Mar. 8.
Thorn, J., Young, M.W. and Skeed, E. (1937–38), Rep. Sea Fisheries Invest. Comm. N.Z. Gov. Printer.
Tilden, J.E. (1935), *The Algae and Their Life Relations.* Minneap.
Togasawa, Y. (1954), *Bull. Jap. Soc. Sci. Fish*, **20**, 193.
Togasawa, Y. and Miue, T. (1954), *Bull. Jap. Soc. Sci. Fish*, **20**, 189.
Tokareva, J.P. (1936), *Konservanaya Prom.*, **3**, 36.
Tokida, J. (1954), *J. Fac. Fish. Hokkaido Univ.*, **2** (1).
Tokida, J. and Kaneko, T. (1963), *Bull. Jap. Soc. Phyc.*, **11**, 24.
Tokida, J. and Kaneko, T. (1964), *Jap. J. Phycol.*, **12**, 24–30.
Tokuda, H. (1977), *Bull. Jap. Soc. Sci. Fish.*, **43**, 587–594.
Tolstikova, N.E. (1977), *Proc. All-Union Res. Inst. Mar. Fish and Oceanog.*, **124**, 31–36.
Tondo, M.F.D. (1931), *Bull. Sta. Biol. d'Arcachon.*, **27**, 175.
Torii, S. and Kawashima, S. (1979), *Proc. 9th Int. Seaw. Symp.*, Santa Barbara, Science Press, Princeton, 473–478.
Towle, D.W. and Pearse, J.S. (1973), *Limnol. and Oceanog.*, **18**, 155–159.
Townsend, C.T. and Zuch, F.L. (1943), *J. Bacteriol.*, **46**, 269.
Transactions. (1962), *Trans. of the all-Union Conf. of Workers in the Algae Industry of the U.S.S.R.*, Vol. I. Archangel Pub. Ho.
Tressler, D.K. (1923), *Marine Products of Commerce*, New York.
Treumann, J. (1933), *Bull. Far Eastern Branch Acad. Sci., U.S.S.R.*, **114**.
Trunova, O.N. and Crintal, A.R. (1977), *Proc. All-Union Res. Inst. Mar. Fish. and Oceanog.*, **124**, 61–64.
Trzcinski, P. and Czapte, K. (1966), *Proc. 5th Int. Seaw. Symp. Halifax*, p. 443, Pergamon Press, Oxford.
Tschirsch, A. (1912), *Handbuch der Pharmakognosie*, Vol. 2, Leipzig.
Tseng, C.K. (1933), *Lingnan Sci. J.*, **12**, 14.
Tseng, C.K. (1944a), *Scientific Monthly*, **58**, 24.
Tseng, C.K. (1944b), *Scientific Monthly*, **59**, 37.
Tseng, C.K. (1944c), *California Monthly*, May.
Tseng, C.K. (1945a), *Science, N.Y.*, **101**, 597.
Tseng, C.K. (1945b), *Food Industries*, **17**, 10 et seq.
Tseng, C.K. (1945c), *Chem. Met. Eng.* June.
Tseng, C.K. (1946a), *J.N.Y. Bot. Gard.*, **47**, 1.
Tseng, C.K. (1946b), J. Alexander, *Colloid Chemistry, Theoretical and Applied*, **6**, 629.

Tseng, C.K. (1947), *Econ. Bot.*, **1**, 69–97.
Tshudy, R.H. and Sargent, M.C. (1943), *Science, N.Y.*, **97**, 89.
Tsuboi, S. (1918), *J. Chem. Ind. Japan*, **21**, 648.
Tsuchiya, M. and Hong, K. (1966), *Proc. 5th Int. Seaw. Symp. Halifax*, p. 315, Pergamon Press, Oxford.
Tsuchiya, M. and Suzuki, Y. (1953), *Bull. Jap. Soc. Sci. Fish.*, **20**, 1092.
Turrentine, I.W. (1907), *Pharm. Cent. Deuts.*, **48**, 505.
Turrentine, I.W. and Shoaff, P.S. (1919), *J. Engng. Chem.*, **11**, 864.
Turrentine, I.W. and Shoaff, P.S. (1921), *J. Engng. Chem.*, **13**, 605.
Turrentine, I.W. and Shoaff, P.S. (1923), *J. Engng. Chem.*, **15**, 159.
Turrentine, I.W. and Tanner, H.G. (1922), *J. Engng. Chem.*, **14**, 19.

Ueda, S. (1929), *J. Imp. Fish. Inst. Tokyo*, **24**, 139.
Ueda, S. (1937), *Bull. Japan Soc. Sci. Fish.*, **6**, 91.
Ueda, S. and Katada, M. (1949), *Bull. Jap. Soc. Sci. Fish.*, **15**, 354.
U.S. Dept. Agriculture (1899), *Farm. Bull.*, 105.
U.S. Dept. Agriculture (1913), *Bur. Soils Circ.*, 76.
Usov, A.I. (1977), *Proc. All-Union Res. Inst. Mar. Fish. and Oceanog.*, **124**, 65–70.

Val, M. del and Montequi, D. (1951), *Bol. Inst. Espanol. Oceanog.*, **40**, 1.
Varma, R.P. and Rao, K.K. (1964), *Ind. J. Fish.*, **9**, 205–211.
Vasil'ev, V.V. and Guenis, A.L. (1936), *Chem. Ind.*, **36**, 1136.
Villon, A.M. (1893), *Chem. News*, **68**, 311.
Vincent, P.L. (1960), *Chem. Ind.*, London, 1109.
Vincent, P.L., Goring, D.A.I. and Young, E.G. (1955), *J. Appl. Chem.*, **5**, 374.
Vincent, V. (1924), *Les Algues Marines et leurs Emplois Agricoles, Alimentaires, Industriels*, Paris.
Vinogradova, A.P. (1953), *The Elementary Chemical Composition of Marine Organisms*, Sears Foundn. Mar. Res., Yale Univ.
Vinogradov, K.A. (1962), *Odesk. Univ. Sci. Geol. l'geog. Nauk*, **152**, 179.
Vinogradov, K.A. (1964), *Zh. Biol.*, No. 11D, 281.
Voigtlander, F. (1899a), *J. Chem. Soc., Lond.*, **56**, 817.
Voigtlander (1899b), *Z. Phys. Chem.*, **3**, 316.
Vozzhinskaya, V.B. (1966), *Inst. Okeanol.*, **81**, 153–175.

Waaland, J.R. (1975), *Phytochem.*, **14**, 1359–1362.
Waaland, J.R. (1976), *J. Exptl. Mar. Bio. Ecol.*, **23**, 45–53.
Waaland, J.R. (1977), in *Marine Plant Biomass of the Pacific Northwest*. (Krauss, R.W. ed.), pp. 117–137, Oregon State Univ. Press.
Waaland, J.R. (1979), *Proc. 9th Int. Seaw. Symp., Santa Barbara*, Science Press, Princeton, 241–248.
Waele, H. De (1929), *Ann. Physiol. Chim. Biol.*, **5**, 877.
Waksman, S.A. and Allen, M.C. (1934), *J. Am. Chem. Soc.*, **56**, 2701.
Waksman, S.A. and Bavendamm, W. (1931), *J. Bacteriol.*, **22**, 91.
Waksman, S.A., Carey, C.L. and Allen, M.C. (1934), *J. Bacteriol.*, **28**, 213.
Walker, A.W. (1943), *Food Res.*, **8**, 435.
Walker, A.W. and Day, A.A. (1943), *J. Bacteriol.*, **45**, 20.

Walker, E. (1941), *Nature, Lond.*, **147**, 808.
Walker, F.T. (1947), *J. Ecol.*, **35**, 166.
Walker, F.T. (1948), *Proc. Linn Soc. Lond.*, **159**, 90.
Walker, F.T. (1950), *J. Ecol.*, **38**, 139.
Walker, F.T. (1952), *J. Ecol.*, **40**, 74.
Walker, F.T. (1953), *Proc. 1st Int. Seaw. Symp.*, Edinburgh, p. 91–92.
Walker, F.T. (1954), *J. Cons. Explor. Mer.*, **20**, 160–166.
Walker, F.T. (1958), *Acta Adriatica*, **8**, 1–8.
Walker, F.T. and Richardson, W.D. (1957), *J. Ecol.*, **45**, 225.
Watanbe, T. and Kato, S. (1972), *Bull. Jap. Soc. Sci. Fish.*, **38**, 431.
Watkinson, J.G. and Smith, R. (1972), *New Zealand Fisheries*, Wton. 91 pp.
Warren, L.E. (1925), *J. Am. Med. Assoc.*, **84**, 1682.
Wassermann, A. (1946), *Nature, Lond.*, **158**, 271.
Weber, U. and Gerhard, H. (1938), *Deutsch. Apotheker-Ztg.*, 91, 92.
Weigand, T.S. (1894), *Am. J. Pharm.*, **66**, 596.
Weimarn, P.P. Van (1910), *J. Russ. Phys. Chem. Soc.*, **42**, 653.
Weinberger, L.W., Stephan, D.G. and Middleton, F.M. (1960), *Ann. N.Y. Acad. Sci.*, **136**, 131–154.
West, G.S. and Fritsch, F.E. (1912), *The British Fresh Water Algae*, Cambridge.
Whistler, R.L. and Kirby, K.W. (1959), *Z. Physiol. Chem. Hoppe-Seyler's*, **314**, 46.
Whittaker, H.A. (1911), *J. Am. Publ. Hlth. Assoc.*, **1**, 632.
Whyte, N.C. and Englar, J.R. (1979), *Proc. 9th Int. Seaw. Symp.*, Santa Barbara, Science Press, Princeton, 437–444.
Whyte, N.C., Englar, J.R. and Kung, M. (1977), *J. Phycol.*, **13**, Suppl. Abst. 425, p. 73.
Wilcox, H.A. and Leese, R.M. (1976), *Hydrocarbon Processing*, Apr. p. 86–89.
Wilcox, W.A. (1887), *U.S. Fish Comm. Fish. Ind. U.S.*, Sect. 2.
Williams, R.H. (1944), *Quart. J. Fla. Acad. Sci.*, **8**, 161.
Wilson, E. (1943), *Nature Magazine*, **36**, 127.
Wing, W.T. (1942), *Pharm. J.*, **149**, 103.
Wirth, H.E. and Rigg, G.B. (1937), *Am. J. Botany*, **24**, 68.
Wise, J.J. and Silvestri, A.J. (1976), *Oil and Gas J.*, Nov. 22, p. 1140–1142.
Woessner, J.W., Sorenson, P. and Coon, D. (1972), *Proc. 7th Int. Seaw. Symp.*, Univ. Tokyo Press, Japan, p. 450.
Wohnus, J.F. (1942), *Calif. Fish Game*, **28**, 199.
Woodward, F.N. (1966), *Proc. 5th Int. Seaw. Symp, Halifax*, p. 55, Pergamon Press, Oxford.
Wort, D.J. (1955), *Can. J. Botany*, **33**, 323.
Wutoh, J.G. (1977), *The Procurement, Isolation, Identification and Toxicology of Biologically Active Substances Derived from Marine Algae and Stingray*. Status Rept., Center for Envir. and Est. Stud., Univ. Maryland, p. 45.

Yamada, N. (1976), *J. Fish. Res. Bd. Canada*, **33**, 1024–1030.
Yamasaki, H. (1954), *Bull. Jap. Soc. Sci. Fish.*, **20**, 442–447.
Yamazoto, K.Y. (1976), *Abst. Intern. Symp. Ecol. and Manag. Trop. Shallow Water Communities*, Indonesia Inst. Sciences.
Yanagawa, T. (1929), *Rept. Imp. Indus. Res. Inst. Osaka, Japan*, **10**, 6.
Yanagawa, T. (1936), *Rept. Imp. Indus. Res. Inst. Osaka, Japan*, **17**.

Yanagawa, T. (1937a), *Rept. Imp. Indus. Res. Inst. Osaka, Japan,* **18**, 1, 29.
Yanagawa, T. (1937b), *Bull. Japan. Soc. Scient. Fish.,* **6**, 185.
Yanagawa, T. and Nishida, Y. (1930), *Rept. Imp. Indus. Res. Inst., Osaka, Japan,* **11**, 14.
Yanagawa, T. and Nishida, Y. (1932), *Rept. Imp. Indus. Res. Inst., Osaka, Japan,* **12**, 16.
Yanagawa, T. and Nishida, Y. (1933), *Rept. Imp. Indus. Res. Inst., Osaka, Japan,* **14**, 1.
Yanagawa, T. and Yositaka, Y. (1939), *Rept. Imp. Indus. Res. Inst., Osaka, Japan,* **20**, 1.
Yaphe, W. (1957), *Can. J. Microbiol.,* **3**, 987.
Yaphe, W. (1959), *Can. J. Bot.,* **37**, 751.
Yaphe, W. (1966), *Proc. 5th Int. Seaw. Symp. Halifax,* p. 333, Pergamon Press, Oxford.
Yaphe, W. (1973), *Chemistry of Carrageenan* in *Chondrus crispus.* (Harvey, M.J. and McLachlan, J., eds.), pp. 103−108.
Yaphe, W. and Duckworth, M. (1972), *Proc. 7th Int. Seaw. Symp.,* Univ. Tokyo Press, Japan, pp. 15−22, Tokyo, Also *Carb. Res.,* **16**, 189, 435, 446; **18**, 1−9.
Yarham, E.R. (1944), *Country Life,* **95**, 814.
Yendo, K. (1902), *Postelsia,* **1**.
Yendo, K. (1914), *Econ. Proc. R. Dublin Soc.,* **2**, 105.
Yokote, T. (1899), *Centrb. Bakt. Parasitenk. Abt.* I., **25**, 379.
Yoshida, R. (1966), *Bull. Tohoku Fish. Res. Lab.,* **26**, 109.
Yoshida, R., Sakuri, Y. and Kurogi, M. (1964), *Bull. Tohoku. Fish. Res. Lab.,* **24**, 88.
Yoshida, R. (1972), *Bull. Tohoku. Fish. Res. Lab.,* **32**, 89.
Yoshimura, A., Tuda, H., Sakari, M., Harada, T. and Oishi, K. (1976), *Bull. Jap. Soc. Sci. Fish.,* **42**, 661−664.
Young, K., Hong, K.C., Duckworth, M. and Yaphe, W. (1972), *Proc. 7th Int. Seaw. Symp.,* Univ. Tokyo Press, Japan, 469−472.
Young, E.G. and Rice, F.A.H. (1945), *J. Biol. Chem.,* **156**, 781.

Zhelezkov, P.S. (1938), *Colloid. U.S.S.R.,* **4**, 423.
Zhelezkov, P.S. (1939a), *Konservnaya Prom.,* **10**, 30.
Zhelezkov, P.S. (1939b), *Colloid J. U.S.S.R.,* **5**, 409.
Zhelezkov, P.S. (1939c), *Colloid J. U.S.S.R.,* **5**, 733.
Zhelezkov, P.S. (1940), *Colloid J. U.S.S.R.,* **6**, 403.
Zobrist, L. and Gruber, M. (1936), *Kolloid Z.,* **77**, 333.
Zuntz, N. and Beckmann, E. (1916), *Mitt. d'eusch. Seefisher. Ver.,* **32**, 144.
Zunz, E. and Gelat, M. (1916), *J. exp. Med.,* **24**, 247.
Zupink, L. (1895), *Centrb. Bakt. Parasitenk. Abt.* I., **18**, 202.

Author Index

Abbott, 70
Abe, 201, 237
Abdel-Fattah, 179
Akiyama, 106
Aleem, 273
Allen, 221, 244
Allsobrook, 143
Alvarez, 138, 139, 241
Anderson, N.S., 117, 119, 126, 127
Anderson, W., 133
Ando, 109
Andrew, 213
Anglo, 179
Anno, 232
Anon, 16, 45, 67, 69
Aoyagi, 72, 86, 159
Apinis, 229
Arai, 184
Araki, 183, 184, 185, 188
Arasaki, 83
Astbury, 194
Aston, 42
Atlas, 247
Audoin, 259
Augier, 34, 35, 47, 50
Azare, 179

Baardseth, 196, 257, 261
Bailey, 124
Baily, 18
Baker, 234, 236, 237
Bardach, 241, 266, 269
Barilotti, 165
Barry, D., 183
Barry, V.C., 183, 208, 226
Beckmann, 37, 48
Bellan-Santini, 261
Benitez, 138

Berad-Thierrault, 201
Berg, 234
Berger, 220
Bergeron, 39, 40
Bettels, 88
Bhakuni, 133, 220, 236
Bird, 196
Bjorndal, 94
Black, 33, 36, 37, 45, 49, 50, 128, 196 *et seq.*, 201, 228, 229, 231, 239
Blunden, 50, 51, 52
Bodard, 182
Bold, 2
Boney, 2
Booth, 49
Bose, 179
Boua, 50
Breton-Provencher, 201, 202, 230
Broser, 133, 236
Brunei, 173
Bunger, 236
Burger, 133
Burkholder, 236
Burlew, 18
Burns, 119, 120, 252, 256
Butler, 121, 183
Bunyankina, 61

Caassini-Lokeur, 173
Caces-Borja, 271
Cameron, 270
Campa, 31
Caraes, 231
Cardinal, 201, 230
Cauer, 22
Causey, 168
Cavaliere, 262

313

Cavi, 234
Chamberlain, 194, 217, 219, 232
Chapman, A.R., 203, 229, 232
Chapman, D.J., 2
Chapman, V.J., 2, 19, 28, 62, 237
Chase, 170
Chaveaux, 233
Chen, 244
Cheney, 114, 138
Cheng, 241, 266
Chester, 229
Chiang, 104
Chidambaram, 43
Christensen, 254
Christian, 182
Claudio, 234
Clement, 69
Clingmam, 184
Colin, 94
Colinvaux, 256
Compère, 16, 69
Connell, 226
Cook, 122, 195, 228
Coon, 98
Cooper, 164
Cornhill, 239
Cotton, 41
Cottrell, 205, 206, 212, 213
Courtois, 19
Craigie, 121, 124, 128, 229, 232, 244, 256
Crossman, 28
Cruess, 82
Czapke, 49, 259

Dahlberg, 192
Davidson, 43, 73, 77, 86, 153, 158
Davies, 251
Davis, 261
Dawes, 138, 244
Dawson, 2
DeBoer, 189, 190, 246
Delapine, 242
De Loach, 161, 168
Del Proo, 163, 164, 273
De Paula, 204
Deschiens, 31, 215

Devaux, 138, 141, 276
Diaz-Pifferer, 31, 171, 260
Dillon, 183, 194, 196, 204, 208, 216, 226
Dizerbo, 121, 236
Dolan, 127
Dorfel, 194, 201
Doty, 135 *et seq.*, 138, 241, 246, 271
Doyle, 203
Duckworth, 188
Dunstan, 16, 245, 247, 248
Durrairatnam, 265
Durrant, 232

Earle, 260
Eddy, 16
Edelman, 50
Edelstein, 244
Edwards, 260
Ehresmann, 234
Elenkin, 69
Elliot, 42, 274
Elsner, 133, 236
Engel'shtein, 188
Esdorn, 220, 234
Evitushenko, 230

Feeny, 160, 162
Ferguson-Wood, 170, 175, 176
Ffrench, 116, 254, 256
Filho, 204
Fischer, 194, 201
Fleming, 228
Forsdike, 191
Fraser, 235
Frei, 195, 196
Friedlander, 226
Fries, 50
Fritsch, 2
Fujikawa, 109
Fujita, 102
Funaki, 153

Garcain, 234
Garcia-Pineda, 230
Gardner, 161
Gatin, 94, 234

Author Index

Gauther, 238
Giaconne, 262
Gislén, 257
Glicksman, 109, 129, 130 *et seq.*
Gloess, 78, 94, 115, 207, 208, 216, 219, 220
Goh, 54
Goldie, 237
Goldman, 16, 248
Goring, 213
Gortner, 183
Grasdalen, 195
Grenager, 196
Griffith, 68
Grimmett, 42
Grintal, 237
Gros, 241
Grua, 265
Guegen, 94
Guerrero, 44
Guiseppe, 234
Guven, 142

Haas, 121, 183, 196, 238
Habekost, 235
Haines, 246, 251
Halim, 179
Hallsson, 30
Halstead, 235
Hamail, 103
Hand, 33
Hanic, 117
Hart, 192
Harvey, 119, 121, 122, 124, 128
Hasegawa, 74, 78, 79, 241
Hatch, 235
Haug, 195, 200, 201, 213
Hawkins, 229
Hayashi, 161
Heilbron, 122
Hendrik, 46, 47
Henriques, 169
Hill, J., 98
Hill, T.G., 183
Hirano, 106
Hirase, 143, 144, 147, 183
Hirst, 194, 195, 226, 228

Hitchins, 191
Hjerten, 188
Hoag, 190
Hoagland, 46, 203
Hoffmann, C., 27, 71, 78, 108, 145, 148, 161, 230, 233
Hoffmann, W.F., 183
Hoie, 33, 37
Holmes, 182
Hong, 82, 188
Hoppe, 43, 78, 81, 82, 83, 86, 89, 96, 98, 100, 141, 143, 148, 175, 188, 197 *et seq.*, 213, 226, 229, 236, 259
Horiuschi, 159
Hosada, 88
Hotta, 103
Howe, 194
Hoygaard, 96
Hoyle, 169
Hua, 18, 239
Huegenen, 250
Humm, 161, 167, 170, 190, 191, 254
Hussain, 50

Idler, 96, 122, 254
Imada, 103, 104
Isaac, 203
Ishibashi, 88
Ishio, 101
Iwasaki, 103, 104, 108

Jackson, 59
Jensen, 32, 33, 36, 37, 38, 49, 202, 204, 228, 230
John, 179
Johnston, 48, 63, 71, 72, 88, 91, 92, 261
Johnstone, 160, 162, 164
Jones, 182, 183, 194

Kalimuthu, 203, 230, 231
Kaneda, 109, 237
Kappanna, 181
Kardakova-Prejeutzoffa, 31
Karlstrom, 96
Kasahara, 109
Katada, 153

Kato, 101
Kawaguchi, 234
Kawashima, 79
Kelly, 18
Khoklov, 240
Kim, 37, 169, 274
Kingman, 50, 51, 52
Kirby, 31, 43, 68, 115, 121, 130, 176, 177, 179, 197 *et seq.*, 217, 233, 237
Kireeva, 257
Kitami, 83
Kizevetter, 167, 172
Knoepffler-Peguy, 241
Kojima, 153, 160
Kok, 18
Konig, 88
Kovacs, 205, 206, 212, 213, 235
Krefting, 194
Kristenson, 230, 254
Krisnamurthy, 265
Kunisaki, 90
Kurogi, 99, 103, 104, 106
Kusakabe, 82
Kylin, 228

Lami, 115
Lantrip, 6
La Pointe, 246
Larsen, 194, 195, 200
Lawson, 179
Lee, 176, 190
Leese, 56, 59
Leikind, 191
Leonard, 229
Léonard, 16, 69
Levring, 83, 86, 89, 96, 148, 175, 188, 213, 226, 230, 236, 259
Li, 18, 43
Liaw, 104
Lie, 96
Lillig, 220, 234, 235
Lindberg, 230
Lindsay, 171
Little, 42
Lohrisch, 93, 94
Loose, 96

Losa, 31
Lucas, 194
Lunde, 45, 96, 194, 197 *et seq.*, 201, 229, 230, 232
Luxton, 111, 124, 176, 177, 184, 186

Maas, 208, 215
Macaranas, 138
McCandless, 121, 122, 124, 127, 244
McDowell, 221
McFarlane, 31, 67, 117, 256
McGittrick, 183
McGuiness, 196
McLachlan, 119, 121, 122, 124, 128
McLarney, 266
McLendon, 95
McLeod, 213
McNeely, 216, 221, 235
McPherson, 45, 229, 230
Madlnener, 62, 68, 86, 88, 90, 95
Madgwick, 200
Maeki, 103
Mangon, 39
Mann, 256
Manners, 228
Mantner, 232
Marchand, 154
Marquardt, 192
Marsh, 194, 217
Marshall, 115, 173
Martin, 228
Martinez, 230
Mathieson, 114, 119, 138, 244, 252, 256
Matsudaira, 103, 108
Matsui, 90, 182
Matsuoka, 165
Meier, 233
Mehta, 230, 231
Menzel, 248
Michanek, 62, 71, 87, 88, 103, 109, 151, 168, 171, 173, 204, 207, 253, 254, 257, 259, 262 *et seq,*, 277
Mitchell, 49
Mitsubishi, 161
Miue, 196

Author Index

Miura, 98 *et seq.*, 108, 241
Miyake, 73, 81
Mohammed, 179
Moirano, 122, 124, 129, 134
Molteno, 264
Mollion, 122, 142, 261
Monteque, 229, 230
Montgomery, 109
Moon, 138, 251
Moore, 70, 176, 233, 274
Morel, 190
Morrisson, 254
Moss, 111, 113, 114, 135, 150, 159, 161, 171, 173, 193, 196, 204, 254, 260, 264, 266, 267, 271, 277, 278
Mowat, 50
Mumford, 241
Munda, 256, 262
Murphy, 234, 236, 237
Myers, 239
Mykelstad, 200

Nakamura, 106
Nakao, 101
Naylor, 231
Neal, 122
Nebb, 37
Neish, 241 *et seq.*, 256
Neuschel, 203
Newton, 115, 173
Nicotri, 244, 246
Nielsen, 112, 174, 197 *et seq.*
Nishibori, 109
Noda, 83, 108, 109
North, 6, 56, 271, 273
Nunn, 184

Ohta, 153
Oishi, 90
Okamura, 145
Okazaki, 152, 241, 267 *et seq.*
O'Neill, 184
Oorhusa, 267
Oorschott, 18
Oshima, 90, 94
Orr, 115, 173

Ortega, 69, 121, 162, 204, 210
Oza, 181

Painter, 200
Palminha, 260
Panganiban, 179
Parenzan, 262
Parekh, 230, 231
Parker, 241, 246
Parsons, 124, 125, 244
Parr, 261
Pavlov, 188
Pearce, 273
Peat, 182, 183
Pehorey, 216
Penman, 196, 201
Pentegow, 20, 90
Percival, 94, 126, 134, 183, 187, 195, 226, 232
Perrot, 94, 234
Perez, 171, 259, 260
Pickmere, 124, 244
Pierre, 40
Pignatti, 262
Pillay, 188, 190
Post, 88
Povolny, 52, 54, 55
Powell, 227
Pownall, 205
Preston, 195, 196
Prince, 119, 244, 246, 251
Pringle, 117, 119, 255

Quege, 204

Rao, 181, 203, 230, 236, 265
Rapson, 12, 42, 274
Rasmussen, 96
Rees, 126, 127
Reisner, 18
Ricard, 201, 229
Ricohermosa, 138, 141, 276
Rigg, 270
Robbins, 190
Roebs, 251
Rodriquez, 261
Ross, 226, 232

Ryther, 245 et seq., 252, 266

Saeter, 33, 37
Saiki, 93
Saito, 81, 82, 83, 103, 104, 122, 212, 241
Salcedo, 179
Sanbonsuga, 78, 79, 203
Sandvik, 37
Sanderson, 196, 201
Sannay, 33
Santimone, 34, 35, 47
Santos, 135 et seq.
Sargent, 6, 165
Sato, 104, 106, 109
Saunders, 171
Sauvageau, 31, 62, 182
Sawasaki, 106
Seagel, 270
Schachat, 109, 134
Schiralisa, 200
Schimizu, 109
Schmid, 43, 81, 83, 86, 89, 96, 98, 141, 148, 175, 188, 197 et seq., 213, 226, 229, 236
Schoffel, 194
Schulzen, 213
Schwimmer, 193
Selby, 182
Semple, 119
Senn, 50 et seq.
Shacklock, 244
Shang, 241
Sheehy, 38
Shephard, 191
Shirigama, 86, 183
Shitanaka, 102
Shurtleff, 62, 72, 78, 79, 82 et seq., 108, 159
Sieburth, 32
Silva, 133, 220, 236
Silverthorne, 165
Silvestri, 55
Simonetti, 262
Simpson, 117
Skelton, 53

Skottsburg, 13
Skoryna, 220
Smidsrd, 195
Smith, D.B., 195
Smith, F., 109, 122
Smith, H.M., 2, 72, 86
Smith, R., 274
Soerjodinote, 271
Somerville, 183
Sorenson, 98, 230
South, 203
Span, 262
Speakman, 194, 217 et seq., 232
Spoon, 179
Stancioff, 126, 127, 129, 133
Stendardo, 234
Stanford, 19, 121, 208
Stanley, 16, 126, 127, 138, 248, 251
Steiner, 216, 221
Stenhouse, 229
Stephen, 184
Stephenson, 31, 33, 34, 35, 37, 43, 50 et seq.
Steward, 194
Stewart, 184, 228, 231, 232
Stoloff, 111, 183, 190
Subrahmangan, 2
Sulit, 179
Sunderlin, 251
Suto, 99, 102, 106, 153
Suzuki, 109
Swartz, 93, 129, 130, 192

Tanaka, 153
Tagawa, 160
Takagi, 220, 234, 236
Takahashi, 86, 183
Takamatsu, 145, 153
Takenchi, 104
Tang, 95
Taylor, 50, 119
Tenore, 16, 245, 247, 248
Thompson, 18
Togasawa, 196
Tokido, 86
Tokudo, 102
Tolstikova, 207

Author Index

Tondo, 87, 159
Torii, 78, 79, 106
Towle, 273
Tressler, 161
Trunova, 237
Tseng, 43, 110, 117, 141, 148, 151, 160, 161, 168, 170, 179, 188, 194, 210, 218, 237, 273
Tshudy, 165
Tsuchiya, 109, 188
Turner, 229, 237

Ueda, 153
Unny, 43
Usov, 111

Val, 229, 230
Van der Hoeck, 260
Varma, 265
Varier, 199, 231
Vincent, 48, 195, 213
Vinogradova, 20
Vozzhinskaya, 270

Waaland, 244, 245
Waksman, 221
Walker, 259
Walter, 124
Watanabe, 101, 143, 144, 147
Watkinson, 274
Weinberger, 245
Whang, 95

Whyte, 210
Wilcox, 56, 59
Wilkinson, 50
Williamson, 70
Wilson, 68
Winton, 238
Wise, 55
Wiseman, 96
Woessner, 98
Wold, 195
Wolff, 190
Wong, 128
Wood, 194, 217
Woods, 55
Woodward, 36
Wort, 45
Wynn, 2

Yamada, 141, 151, 154, 159, 171, 173, 175, 179, 181
Yamasaki, 99, 102, 106
Yaphe, 126, 149, 188, 190
Yarham, 173, 233, 239
Yoshido, 98, 99, 103, 106
Yoshimura, 88
Young, 45, 149, 213, 229, 230

Zamazoto, 236
Zenitani, 102

Plant Index

Acanthopeltis, 111, 155, 267
    japonica, 86, 148, 149, 153, 185
Acanthophora spicifera, 87
Aeodes orbitosa, 111, 143
    ulvoidea, 143
Agardhiella, 88, 124, 260
    tenera, 124
Agarum, 254, 256
    fimbriatum, 227
Ahnfeltia, 111, 172, 173, 180, 181, 256, 257
    concinna, 70, 94, 145
    gigartinoides, 64
    plicata, 148, 149, 160, 167, 171, 270
Alaria, 24, 26, 30, 228, 254, 256, 267, 269
    crassifolia, 64, 91
    esculenta, 4, 8, 18, 33, 63, 64, 96, 97, 197, 201
    fistulosa, 8, 25, 31, 42, 227
    grandifolia, 254
    valida, 96
Alginobacter, 221
Alginomonas, 221
Alsidium, 234
    helminthocorton, 234
Amansia, 236
Analipus japonicus, 64
Anatheca montagnei, 122, 124
Argopecten, 248
Arthrothamnus, 268
    bifidus, 73, 89, 93, 236
    kurilensis, 73
Ascophyllum, 3, 18, 22, 24, 31 et seq., 37, 38, 44, 46, 48 et seq., 196, 197, 200 et seq., 207, 214, 228 et seq., 250, 254, 256 et seq.
    mackaii, 64, 250
    nodosum, 2, 4, 34, 97, 255
Asparagopsis sanfordiana, 70
    taxiformis, 64
Asterionella japonica, 238
Azotobacter vinelandii, 194, 196, 201

Bacillus megatherium, 236
Bacterium terrestralginicum, 55, 221
Bangia fusco-purpurea, 64
Beckerella, 267
Blossevillea, 42
Bostrychia, 88
Botrytis, 54
Brongniartella mucronata, 172
Bryothamnion seaforthii, 172
    triquetrum, 172

Campylaeophora, 89, 155, 267
    hypneoides, 64, 86, 183 et seq.
Carpopeltis affinis, 86
    flabellata, 86
Carpophyllum, 42
Catenella nipae, 88
Caulerpa, 237
    clarifera, 70, 89
    laetvirens, 87
    peltata, 87
    prolifera, 263
    racemosa, 64, 87, 88
    racemosa var. clavifera, 87
Centroceras clavulatum, 237
Ceramium, 155, 267
    boydeni, 184, 185
Chaetomorpha, 16, 18, 87, 238
    crassa, 64, 88
    linum, 246
Chlamydomonas, 96

Chlorella, 16, 239
Chnoospora pacifica, 88
Chondria, 234
   armata, 234
   littoralis, 236
   oppositiclada, 234
Chondrus, 5, 94, 114 *et seq.*, 120 *et seq.*, 145, 182, 236, 238, 241, 244, 250, 256, 260, 267
   armatus, 86
   crispus, 4, 8, 63, 64, 96, 111 *et seq.*, 124, 126 *et seq.*, 173, 246, 252, 254, 255, 259
   elatus, 72, 86, 145
   ocellatus, 86, 96, 112
Chorda, 238
   filum, 18, 64
Chroococcus turgidus, 69
Chaetoceras lauderi, 238
Chrysymenia wrightii, 236
Cladophora, 18, 88
   prolifera, 262
Codium, 14
   divaricatum, 85
   fragile, 64, 85
   intricatum, 95
   muelleri, 70
   tomentosum, 87, 96
Corallina officinalis, 87
Corallopsis minor, 234
   salicornia, 87
Costaria, 9
   costata, 227, 236
Crassostrea, 248
Cutleria multifida, 236
Cymathere, 9, 268
   triplicata, 227
Cystophora, 42
Cystoseira, 241, 261, 263
   barbata, 262
   corniculata, 262
   crinata, 261
   fimbriata, 262
   mediterranea, 261
   spicata, 262
   spinosa, 262
   stricta, 261

Delessaria sanguinea, 236
Desmarestia, 198
   herbacea, 227
   ligulata, 227
   munda, 227
Diadema setosum, 138
Dictyopteris justii, 236
   plagiogramma, 70
   polypodioides, 68, *235*
   zonarioides, 237
Dictyota, 87
   dichotoma, 236
Digenes, 234, 260, 263, 267, 269
   simplex, 95, 172, 234
Dilsea, 111
Dumontia incrassata, 111
Dunaliella, 16
Durvillea, 42, 196, 200, 205, 237, 277
   antarctica, 13, 56, 63, 71, 198, 203, 231, 239, 253, 265, 274
   harveyi, 13
   utilis, 63
   potatorum, 71, 203
   willana, 198, 203, 231

Ecklonia, 12, 42, 204, 230, 267, 269
   buccinalis, 56, 203
   brevipes, 250
   cava, 11, 19, 64, 72, 204
   maxima, 16, 22, 43, 69, 198, 203, 264
   radiata, 42, 196, 198, 202, 227, 231, 232, 237
   stolonifera, 83
Eckloniopsis, 267
   radicosa, 239
Ectocarpus, 80
Egregia, 198
   laevigata, 9, 11, 235
   menziesii, 9
Eisenia, 22, 267, 269
   arborea, 163
   bicyclis, 11, 19, 64, 83, 84, 201, 204, 212, 239
Enantiocladia duperryi, 172

Endarachne, 267
Endocladia, 111
  muricata, 167
Enteromorpha, 43, 72, 83, 90, 94, 96, 236, 267, 269
  clathrata, 64, 246
  compressa, 64, 89, 236
  flexuosa, 70
  intestinalis, 64, 70, 88
  prolifera, 65, 236
Eucheuma, 10, 12, 14, 65, 88, 111, 117, 120, 124, 126, 135, 140, 141, 148, 182, 241, 244, 246, 257, 263, 271, 278
  arnoldii, 136, 137
  acanthocladum, 135
  cottonii, 112, 113, 124, 136, 137, 139, 271
  denticulatum, 135
  edule, 112, 135
  gelatinosa, 179
  isiforme, 68, 135, 138
  johnstonii, 135
  muricatum f. *depauperata*, 135
  nudum, 135, 138, 251
  odontophorum, 136, 137
  okamurai, 135
  papulosa, 72, 86
  platycladum, 135, 136, 137
  procrusteanum, 135 *et seq.*
  serra, 112
  speciosum, 135, 175
  spinosum, 72, 112, 113, 124, 135, 138, 139, 175, 187, 271
  striatum, 135, 264
  uncinatum, 138, 163, 251, 273

Falkenbergia hillebrandii, 236
Fucus, 8, 50, 67, 119, 214, 227, 233, 238, 256, 258
  balticus, 34
  ceranoides, 197
  distichus, 256
  evanescens, 31, 202, 229, 231
  gardneri, 236
  platycarpus, 2
  serratus, 3, 5, 18, 32, 34, 41, 48, 97, 197, 200, 201, 228, 229, 237, 257
  spiralis, 2, 4, 48, 200, 229, 232
  vesiculosis, 2, 4, 18, 32, 34, 41, 48, 63, 65, 97, 196, 197, 200 *et seq.*, 207, 228 *et seq.*, 237, 246, 256 *et seq.*
  virsoides, 262
Furcellaria, 111, 124, 254, 256, 258
  fastigiata, 112, 124, 133, 255, 257

Gelidiella, 111, 171, 180, 263
  acerosa, 149, 172, 181, 188, 236, 265
Gelidiopsis rigida, 87, 150
Gelidium, 2, 10, 11, 14, 70, 95, 111, 142, 148, 154 *et seq.*, 160, 165, 168, 170 *et seq.*, 180, 182, 188, 189, 256, 260, 263, 266, 267, 269, 274
  amansii, 11, 12, 65, 72, 95, 149, 153, 155, 180, 183 *et seq.*, 264
  arborescens, 150, 161
  australe, 161, 189
  cartilagineum, 150, 161 *et seq.*, 175, 184, 189, 264, 273
  corneum, 161, 172, 173, 182
    var. *sesquipedale*, 149
  densum, 161
  divaricatum, 72, 149
  japonicum, 150, 153, 155, 185
  latifolium, 173
  liatulum, 150
  lingulatum, 150
  micropterum, 181
  nudifrons, 150, 161
  pacificum, 150, 153
  pristoides, 150, 175, 264
  pulchellum, 173
  pulchrum, 161, 189
  pusillum, 153
  pyramidale, 161
  sesquipedale, 161, 173, 261

Gelidium (continued)
  spinulosum, 150, 174
  subcostatum, 150, 153, 185, 187
  subfastigiatum, 150
  ungulatum, 171
  vagum, 150
Gigartina, 12, 113 *et seq.*, 125, 148, 165, 241, 244, 266, 274, 277
  acicularis, 112, 124, 128, 179, 260
  angulata, 120, 126
  atropurpurea, 126
  canaliculata, 112, 121, 163, 167
  chamissoi, 112
  decipiens, 126, 238
  exasperata, 244 *et seq.*
  mamillosa, 97, 111
  papillata, 65, 96
  pistillata, 112, 124, 128
  radula, 112
  serrata, 167
  skottsbergii, 112
  stellata, 5, 63, 111, 112, 120, 124, 173, 252, 255, 256, 260
  teedii, 86
Gloiopeltis, 146, 160, 266, 267
  coliformis, 72, 145, 179
  complanata, 145
  furcata, 11, 65, 72, 95, 145
  intricata, 145
  tenax, 72, 145
Gonyaulax catenella, 238
  monilata, 238
  tamarensis, 238
Gracilaria, 12, 14, 43, 68, 111, 142, 151, 155, 160, 168, 170 *et seq.*, 173, 176, 183, 188, 241, 244, 246, 250, 256, 260, 263, 265, 267, 278
  blodgettii, 161
  bursapastoris, 169
  caudata, 171, 172
  cervicornis, 172
  compressa, 63, 68
  confervoides, 167
  cornea, 150, 172
  coronopifolia, 70, 89, 94, 169
  corticata, 180, 181, 236

  crassa, 88
  crassissima, 172
  cylindrica, 172, 260
  damaecornis, 172, 261
  debilis, 261
  dentata, 179
  domingensis, 260
  dura, 262
  edulis, 180
  eucheumoides, 179
  ferox, 172
  foliifera, 161, 171, 172, 189, 190, 246, 252
  furcellata, 175
  henrique siana, 179, 182
  lichenoides, 87, 148, 149, 150, 179, 189, 265
  mamillaris, 172, 260
  multipartita, 150, 180
  secundata var. *pseudoflagellifera*, 177, 252, 274
  taenoides, 87
  verrucosa, 13, 65, 72, 84, 88, 149, 150, 153, 161, 167, 169, 171, 172, 175, 179, 180, 184, 185, 190, 256, 260, 261, 262, 264, 274
Gracilariopsis sjostedtii, 172
Grateloupia, 95, 145, 146
  affinis, 86
  divaricata, 86
  filicina, 12, 70, 72, 86, 145
Griffithsia, 87
Gymnogongrus, 111, 124
  flabelliformis, 86
  javanicus, 87
  pinnulatus, 86, 145
  tenuis, 172
  vermicularis, 70

Halidrys, 18, 200
  siliquosa, 198
Halisaccion glandiforme, 65
Haliseris pardalis, 94
Halymenia, 120
Hedophyllum sessile, 227
Helianthus annuus, 50

Heterochordaria, 89, 237, 267, 269
  abietina, 85, 91, 236
Hesperophycus, 8
Himanthalia, 39, 97, 197, 200, 232
  lorea, 5, 18
Hijikia fusiforme, 65, 83, 89, 90, 95, 96, 267
Hormisira banksii ecad libera, 250
Hydrolathrus cancellatus, 88
Hypnea, 14, 43, 111, 113, 124, 142, 148, 170, 173, 182, 188, 241, 260, 263
Hypnea cervicornis, 87, 112, 172
  flagelliformis, 179
  musciformis, 112, 124, 135, 142, 161, 171, 172, 179, 234, 246, 251, 252, 260, 261, 262
  nidifica, 70, 94, 124, 237
  setosa, 124
  spicifera, 142, 264

Iridaea, 111, 113, 117, 143, 145, 146, 171, 241, 244, 270, 274
  capensis, 143
  cordata, 246
  cornucopiae, 233
  edulis, 63, 68
  radula, 124
Iridophycus, 111, 143, 267
  capensis, 264
  flaccidum, 236
Ishige okamurai, 201

Jania rubens, 234

Kallymenia dentata, 86
Kjellmaniellia crassifolia, 92, 236, 268
  gyrata, 65, 73

Laminaria, 1, 4, 9, 11, 18, 19, 20, 22, 30, 33, 37, 50, 72 *et seq.*, 89, 195, 214, 236, 253, 256, 267, 268, 269, 278
  abyssalis, 205

  angustata, 22, 65, 74, 81, 92, 93,
    var. *longissima*, 65, 74, 81, 90, 93
  bongardiana, 31, 63, 67
    f. *elliptica*, 31
  bracteata, 79, 235
  brasiliensis, 205
  cichoriodes, 81, 93
  cloustoni, 6
  cuneifolia, 227
  diabolica, 74, 76, 81
  digitata, 3, 6, 22, 24, 34, 46, 48, 63, 96, 97, 195, 197, 200 *et seq.*, 226, 228 *et seq.*, 237, 257
Laminaria flexicaulis, 3, 24
  fragilis, 81, 93
  hyperborea, 3, 5, 6, 22, 32, 34, 48, 96, 97, 196, 197, 200, 201, 204, 213, 226, 227, 229 *et seq.*, 236, 257, 259
  japonica, 22, 61, 65, 74, 77, 78, 79, 81, 90, 92, 93, 95, 266, 270
    var. *ochatensis*, 65, 74, 81
  longicruris, 65, 88, 201, 203, 229, 230
  longipedalis, 76
  longipes, 78
  nigripes, 254
  pallida, 16, 43, 203, 264
  radicosa, 239
  religiosa, 22, 74, 77, 81, 92, 93, 95
  saccharina, 3, 7, 22, 34, 63, 65, 97, 196, 197, 201, 227, 229, 257, 259
  stenophylla, 3
Laurencia, 70
  obtusa, 87, 236
  pinnatifida, 63, 65, 68
  poitei, 168
Lessonia, 239, 277
  fucescens, 14
  littoralis, 227
  variegata, 42
Lithothamnion, 260
  coralloides, 40
Lyngbya, 95
  maiuscula, 236

Macrocystis, 6, 7, 8, 13, 16, 24 *et seq.*, 45, 46, 48, 56 *et seq.*, 71, 162, 195, 196, 198, 203, 205, 210, 214, 227, 235, 265, 270, 273, 277, 278
  angustifolia, 274
  integrifolia, 271
  pyrifera, 8, 10, 12, 42, 56, 163, 201, 271, 273
Mercenaria, 248
Meristhotheca japonica, 86
Mesogloia, 267
  crassa, 85
  decipiens, 85
Microcoleus chthonoplastes, 44
Monostroma, 72, 83, 96, 269
  latissimum, 65, 91
  nitidum, 236, 237
Murrayella periclados, 236
Myriocystum, 199
Mytilus, 248

Nemacystus decipeins, 66
Nemalion helminthoides, 66
  multifidum, 86
  vermiculare, 86
Nematonostoc flagelliforme, 69
Neoagardhiella, 244
  baileyii, 246
Nereocystis, 8, 24, 25, 26, 45, 46, 48, 68, 203, 210, 227, 237, 238, 239, 270
  luetkeana, 7, 63, 66, 271
Nitschia, 96
Nostoc, 66, 88
  commune, 16, 69
  commune f. *flagelliforme*, 89

Pachydiction coriaceum, 237
Pachymenia, 111, 120, 250
  carnosa, 143
Pachymeniopsis, 267
Padina australis, 87
  gymnospora, 236
Palmaria, 31, 241, 244, 250
  palmata, 5, 9, 63, 66, 68, 97, 234, 256

Pelagophycus, 8, 24, 25, 26, 46, 48, 198, 203, 239
  porra, 56
Pelvetia, 8, 31
  canaliculata, 2, 3, 200, 232
  fastigiata, 235
  wrightii, 232
Petalonia fascia, 66, 85
Phormidium tenue, 69
Phyllogigas, 71, 277
Phyllophora, 111, 112, 119, 171, 256, 257, 263
  brodiaei, 263
  nervosa, 20, 124, 141, 263
  membranifolia, 263
  rubens, 141, 167
Phyllospadix, 76
Phymatolithon calcareum, 40
Pleurophycus gardneri, 66
Porphyra, 10, 71, 85, 98, 102 *et seq.*, 236, 241, 266, 267, 269
  akasakai, 98
  amansii, 246
  angustata, 98, 99, 104
  capensis, 14
  columbina, 12, 63, 98
  dentata, 72, 98
  kuneida, 98, 99
  laciniata, 63, 98
  miniata, 66
  naiadum, 96
  nereocystis, 66, 96, 98
  okamurai, 98
  onoi, 98
  perforata, 63, 66, 68
  pseudo-linearis, 98, 99
  seriata, 98
  suborbiculata, 66, 72
  tenera, 66, 89, 95, 96, 98, 106, 109
  umbilicalis, 5, 13, 63, 66, 97, 106
  variegata, 99
  yezoensis, 66, 98, 99, 104
Postelsia, 227
  palmaeformis, 66
Prymnesium parvum, 238
Prorocentrum micans, 238

Pseudomonas atlanticum, 149
Pterocladia, 10, 12, 111, 148, 149, 153, 167, 171, 181, 263, 267
   capillacea, 150, 176, 260, 274
   densa, 150
   lucida, 150, 176, 178, 274
   pinnata, 150, 153, 161, 173, 176, 179, 263
   tenuis, 150, 184, 185
Pterygophora, 227
Pythium, 102

Rhizoclonium rivulare, 234
Rhodoglossum, 111
   pulchrum, 233
Rhodomela subfusca, 96
Rhodymenia, 13, 68, 94, 239
   pertusa, 9, 96

Sacchoriza polychides, 4, 7, 18
Salmonella, 237
   gallinarum, 37
Sarcodia ceylanica, 86
   montagneana, 87
Sarcophycus potatorum, 71
Sargassum, 11, 19, 31, 42, 43, 87, 88, 207, 217, 244, 263 *et seq.*
   bacciferum, 235
   cinctum, 230
   echinocarpum, 70
   enerve, 66, 85, 239
   fulvellum, 66
   fusiforme, 72
   horneri, 204
   linifolium, 235
   longifolium, 199
   muticum, 9, 236
   natans, 37, 204, 230, 250, 261
   serratifolium, 72
   siliquosum, 88
   swartzii, 230
   tenerrimum, 199, 230
   thunbergii, 236
   vulgare, 230, 234
   wightii, 199, 203, 230
Scytosiphi lomentaria, 66
Spirulina maxima, 69
   platensis, 16, 69
Staphylococcus aureus, 236
Stilophora rhizoides, 235
Stoechospermum marginatum, 199, 230
Suhria, 14, 149, 175, 181
   vittata, 69, 150, 174, 264

Thalassiophyllum, 9
Tichocarpus crinitus, 111, 124
Turbinaria, 87
   conoides, 203, 230
   decurrens, 199
   ornata, 203, 230
Turnerella mertensiana, 233

Ulva, 1, 14, 31, 39, 43, 68, 83, 87, 94, 234, 267, 269
   fasciata, 70, 89
   lactuca, 62, 63, 66, 71, 72, 89, 95, 96, 97, 266
   linza, 66, 88, 89, 246
   pertusa, 66, 72, 236, 266
Undaria, 72, 83, 88, 96, 236, 241, 267, 269, 278
   distans, 81
   peterseniana, 81
   pinnatifida, 11, 66, 72, 80, 81, 82, 89, 91
   undarioides, 81

Vibrio agar-liquifasciens, 191

Wildemannia perforata, 96
Wrangelia, 236

Yatabella, 111
Yucca, 238

Subject Index

Acetone, 28
Adriatic, 262
Aeodan, 111
Aethiops vegetabilis, 233
Africa, 179, 193
Agar, 12, 13, 84, 93, 110, 111, 129 et seq., 141, 142, 143, 148 et seq., 233, 236, 238, 260, 274, 277, 278
Agarabiose, 187, 188
Agar-Mex, 161, 162, 164
Agaroid, 182
Agarol, 193
Agaropectin, 148, 183, 184, 187, 188
Agarophyte, 148 et seq., 152, 160, 171, 172, 184, 246, 253, 260, 261, 264, 265
Agarose, 109, 175, 183 et seq., 193
Alaska, 8, 25, 42, 63, 65, 67, 237, 238, 270
Algasol, 234
Algea Produkter, 34, 35, 38, 52
Algeria, 240
Algifert, 52
Algin, 19, 20, 55, 94, 110, 194 et seq., 238, 277, 278
Algin Corp, 208
Alginate(s), 115, 129 et seq., 200, 204, 207, 233, 257, 260, 261, 264, 274
Alginate Industries, 18, 19, 32, 204, 205
Alginic acid, 43, 194 et seq., 210, 212 et seq.
Alginure, 43, 44
Algit, 36, 38
Aluminium alginate, 210, 215
Amanori, 85, 88, 95, 96, 98, 99

America, 5, 24, 25, 42, 45, 62, 67, 111, 115, 160, 167, 171, 192
American Agar Co., 161, 162
American Gas Ass., 55
Ammonium alginate, 216
Ammonium sulphate, 21
Amoy, 72, 266
Anaesthetics, 237
Animal meal, 31 et seq.
Antarctic, 277
Antibiotic, 133, 238
Anti-coagulant, 133, 229, 236
Antigua, 14, 62, 68
Anti-helminthic, 237
Antilles, 260
Ao-nori, 83, 269
Aquaculture, 18, 267
Arame, 83, 239
Archangel, 172
Arctic Sea, 253, 256
Argentine, 180, 263, 269
Ariake Sea, 102
Asakusa-nori, 83, 90, 93, 98, 107
Ascophyllan, 200, 215
Asia, 88, 241
Atherosclerosis, 236
Atlantic, 256, 277
Atlantic Mariculture Ltd., 241
Auckland, 250, 252
Australia, 13, 43, 56, 61, 71, 135, 151, 168, 175, 180, 191, 240, 274
Auvergne, 240
Auxin, 50
Azores, 173, 260

Bacillariophyceae, 2

Bahrein, 43
Baja California, 10, 121, 161, 162, 165, 181, 210, 271
Baltic, 135, 257, 259
Banyuls-sur-mer, 261
Barbados, 14, 43, 62, 68
Basedow's disease, 233
Bay of Fundy, 68, 254, 256
Beer, 131, 133, 143, 157, 193, 216, 220
Belgium, 43, 259
Bengal Isinglass, 149
Behring Sea, 31, 87
Bermuda, 43
Beryllium alginate, 218, 219
Biomass, 117, 118, 120, 205, 241, 249, 253, 254, 262, 270, 273
Black Sea, 20, 167, 257, 263, 277
Bladder wrack, 2
Bohemia, 240
Bonda Co., 31
Bootlace weed, 238
Botany Bay, 274
Brazil, 43, 142, 171, 180, 204, 205, 263
British Columbia, 8, 45, 171, 180, 204
British Guiana, 43
Brittany, 31, 40, 115, 260
Bronchitis, 234
Bryoza, 20
Bulgaria, 263
Bull kelp, 7, 8, 13, 71, 237, 239, 274
Burma, 88
Busse, 270
Button weed, 39

Cadmium alginate, 219
Calcium alginate, 196, 209, 210, 216 *et seq.*
Calcium carbonate, 40
California, 8, 24, 27, 45, 56, 57, 59, 61, 68, 98, 136, 137, 138, 160, 161, 240, 242, 271, 273
Calvados, 259
Canada, 31, 43, 66, 68, 114, 115, 117, 119, 202, 230, 270

Canadian N.R.C., 244
Canary Islands, 261
Cancer disease, 101
Canning (industry), 191
Cape Agulhas, 264
Cape Comorin, 265
Cape Flattery, 271
Cape Horn, 14
Cape Juby, 261
Cape Town, 16, 264
Carotene, 32, 109
Carragheen, 12, 69, 111, 260
Carrageenan, 14, 109 *et seq.*, 148, 149, 179, 182, 213, 233, 236, 251, 254, 271, 274, 277, 278
Caribbean, 10, 14, 261
Casablanca, 261
Cattle, 30, 31
Ceará, 263
Cedros Is., 271
Ceiling board, 216
Cellophane, 216
Central Africa, 16
Central Pacific, 271
Ceylon, 43, 149
Ceylon Moss, 265
Chad, 69
Channel Is., 41, 115
Charcoal, 20, 21, 154
Char process, 19, 21
Chase Organics, 52
Cheese, 131, 192, 216
Chefoo, 79, 160
Chesapeake Bay, 256
Chiba, 19
Chile, 14, 43, 71, 117, 151, 169, 180, 205, 207, 212, 267, 274, 277
Chiloe, 71
China, 16, 43, 62, 64 *et seq.*, 71, 78, 79, 86, 135, 143, 147, 151, 160, 180, 190, 235, 239, 240, 265, 266
Chinese Moss, 149
Chlorophyceae, 1, 2, 62, 70
Cholesterol, 122, 146, 220

Chondriol, 234
Chondriosterol, 236
Christmas Is., 138
Chromium alginate, 218, 219
Clemson University, 51, 53
Clones, 244
Cobalt alginate, 219
Cochin China, 86
Coimbatore, 179
Commander Is., 31, 67
Consumption, 234
Cook Strait, 274
Copper alginate, 219
Cornwall, 41
Coronado Kelp Co., 26
Corsican Moss, 234
Cosmetics, 250
Cough syrup, 130
Cream stabilisers, 130
Cryptonemiales, 111
Cuba, 31, 171, 234, 261
Curacao, 43
Custard, 192
Cyanophyceae, 2, 69
Cytokinin, 50, 51

Dakayama, 150
Darien, 79
Davis Gelatine Co., 176
Decoctum chondi, 133
Denmark, 31, 133, 135, 256, 257
Dentocoll, 193
Diatoms, 16, 107, 244, 248
Dinophyceae, 2
Djibouti, 138
Dog food, 129
Domoic acid, 234, 235
Driftweed, 39, 41, 42, 43
Dulse, 31, 68, 94, 95, 96, 234, 237, 239, 254, 256
Dysentery, 133
Dynamite, 240

Egypt, 263
Eire, 31, 65, 66, 68, 98, 115, 256
Elk kelp, 8
Emphysema, 234

England, 20, 41, 63, 204
English Channel, 56
Epiphytes, 20, 244
Ergosterol, 236
Ethanol, 55, 56
Eucheuman, 112, 135, 148
Europe, 2, 18, 22, 31, 62, 85, 87, 192

Falkland Is., 14, 43
Fanning Is., 139
Farms, 31, 55 *et seq.*, 138, 139, 271
Federal German Republic, 259
Fertiliser, 25, 245, 262, 266
Ficoguanoide, 44
Films, 192, 239
Finistère, 115
Finland, 30, 43, 259
Florida, 10, 14, 138, 161, 168, 250, 252, 260
Food chain, 241
Formosa, 151
Foveau Strait, 274
France, 20, 31, 39, 40, 52, 56, 64 *et seq.*, 69, 114, 115, 117, 133, 135, 173, 181, 204, 207, 231, 240, 256, 260
Fregastel, 115
Fucaceae, 226
Fucin, 196
Fucoidan (Fucoidin), 208, 215, 232
Fucoids, 201
Fucose, 200, 232, 233
Fucosan, 233
Fucoxanthin, 33
Fungal culture, 190
Funoran, 111, 143 *et seq.*, 187
Furcellaran, 111, 112, 129 *et seq.*, 277

Galactan, 93, 94, 121, 124, 148, 232
Galactose, 182 *et seq.*, 232, 233
Gelan, 111
Gel-Mex, 162, 164
Gelose, 130, 182
Gelidiales, 111
Genu Products Ltd., 119
Georgia Strait, 270
Germany, 24

Ghana, 43
Gibberellin, 50, 51
Gigartinales, 111
Ginnanso, 233
Glasgow, 19
Glazing, 215
Gloahec-Herter process, 208, 210, 211
Goémon, 31, 39, 111
Goitre, 71, 95, 233, 235, 236
Gold Coast, 179
Goro Bay, 262
Grand Manan, 256
Great Britain, 18, 19, 98, 121, 130, 133, 173, 274
Greece, 43, 234, 262
Greenland, 64, 96, 254, 256
Green's process, 208
Grenada, 62
Growth hormones, 49, 50
Guam, 70, 88, 138
Gulf of California, 10
Gulf of Kutch, 265
Gulf of Mannar, 265
Gulf of Riga, 257
Gulf of Taranto, 262
Guluronic acid, 195, 213

Hainan, 43
Hakodate, 72, 78
Halifax, 242, 244
Haptophyceae, 2
Hawaii, 10, 64, 69, 70, 87, 169, 237, 274
Hebrides, 18, 41
Heligoland, 234, 259
Helminal, 234
Heparin, 229, 236
Hercules Powder Co., 26, 27, 28, 203
Herpes, 234
Hibi, 100
Higo, 86
Hijiki, 83
Hiroshima, 99
Hiumaa Is., 257
Hokkaido, 19, 73, 106, 145, 150, 269
Hoku-riko, 86
Holland, 157

Hondawara, 239
Hondo, 145
Honduras, 43
Hongkong, 31, 72, 160, 234, 266
Honshu, 106, 153
Horses, 38
Hoshi-nori, 90, 98, 106 *et seq.*
Hypnean, 111, 142, 179

Ice cream, 130, 131, 192, 216
Iceland, 30, 43, 63 *et seq.*, 69, 231, 256
Icings, 192, 222
Ifni, 261
Ile de Re, 39
India, 43, 179, 180, 188, 207, 230, 235
Indonesia, 11, 64, 65, 87, 135, 179, 180
Insecticides, 220
Iodine, 18 *et seq.*, 26, 28, 33, 46, 70, 75, 95, 108, 220, 233
Irban Strait, 257
Ireland, 18, 38, 41, 63, 111, 114, 117, 173
Iridophycan, 143, 172, 236
Irish Moss, 4, 86, 94, 111, 115, 116, 121, 122, 129, 133, 171, 238, 254, 259
Isle of Man, 41
Ise Bay, 83
Israel, 18
Italy, 181, 262
Izmen Bay, 270
Ize, 86
Izu, 103, 153

Jamaica, 14, 16, 43, 62, 68
Japan, 9, 10, 11, 18, 19, 22, 42, 62 *et seq.*, 68, 70, 72, 79, 80, 86, 87, 95, 98, 109, 121, 135, 143, 146, 147, 149, 150, 153 *et seq.*, 160, 172, 180, 181, 193, 204, 207, 267 *et seq.*
Japan Agar Distribution Co., 158
Japan Seaweed Industry Co., 15

Japanese gelatine, 151
Japanese isinglass, 151
Jelly, 191, 192, 220
Jersey, 41

Kagoshima, 269
Kainic acid, 234, 235
Kali syndicate, 24
Kamchatka, 238, 239
Kanagawa, 19
Kansu, 69
Kanten, 149, 153, 157, 158
Kara Sea, 253
Karna Vita Co., 176
Kathiawar, 265
Kattegatt, 134
Kelco Co., 204, 208, 276
Kelp(s), 6, 9, 11, 12, 18, 19, 22, 25, 26, 27, 45, 48, 55, 61, 67, 74, 76, 207, 209, 233, 253, 264, 266, 271, 273
Kelp Industries Pty., 205
Kenya, 43
Kerguelen, 265
Kieselguhr, 240
Kinetin, 50
Kinukusa, 153
Kombu, 72 et sq., 88, 90, 93, 94, 95, 204, 233, 277
Korea, 64 et seq., 88, 109, 180, 266, 267
Kozu, 86
Kurile Is., 270
Kuronori, 98
Kyoto, 150
Kyowa Hakko Kogyo Co., 104, 150

Labrador, 254
Lacquers, 232
La Jolla, 273
Lake Texcoco, 69
Laminaran, 38, 55, 94, 110, 210, 226 et seq.
Laminariaceae, 4, 9, 226
Laminine, 236, 237
Laos, 88
Laver, 68, 69, 96, 98 et seq., 266

Leather, 130, 192, 193, 216, 232
Limu, 70
Linoleum, 193, 215, 216
Litex Co., 133
Lixiviation process, 19, 208
Loch Feochan, 31
Long Island, 3
Lorned Manufacturing Co., 27
Los Angeles, 31
Luta, 266

Macao, 72, 266
Macassar, 135
Madagascar, 264
Madras, 265
Maerl, 40, 260
Magdalena Is., 121
Magellan, Straits of, 13
Main, 117, 119
Makusa, 153
Malaya, 64, 233
Malayan Archipelago, 135
Malawi, 43
Mandapan, 265
Mangaia, 70
Mannite, 3
Mannitol, 55, 210, 226, 229 et seq.
Mannuronic acid, 194 et seq., 213
Manure, 39, 41 et seq., 49, 55, 154, 254, 257
Mariculture, 16, 62, 241 et seq.
Marine colloids, 115, 117, 119, 130, 139
Marseilles, 261
Massachussetts, 118, 119
Mastitis, 38
Matushima, 103
Mauritius, 43
Maxicrop, 43, 44, 52
Mediterranean, 63, 261, 277
Mekong River, 88
Mercury alginate, 216
Methane, 60
Methanol, 55
Mexico, 69, 121, 151, 161, 162, 204, 260, 261, 274
Mie, 269

Subject Index

Mikawa Bay, 83, 103
Milk, 130, 131, 134, 192, 212, 216
Mink, 31
Miru, 85
Miye, 150
Moheji, 106
Monterey, 271, 273
Morocco, 173, 181, 204, 261
Mount Lompoc, 240
Murmansk Sea, 256
Murphy Enterprises, 31

Nagano, 150
Netherlands, 259
Newfoundland, 117, 154
New Brunswick, 67, 117, 118, 231, 241, 254, 256
New England, 120, 256
New Hampshire, 118, 120, 256
New South Wales, 175, 203
New Zealand, 12, 31, 42, 43, 54, 56, 61, 71, 98, 120, 124, 149, 162, 176, 180, 191, 203, 233, 237, 239, 274
Niacin, 96, 109
Nicaragua, 43
Nicotine spray, 193, 220
Nigeria, 43
Ningpo, 160
Nitrogen, 22, 37, 39, 42 *et seq.*, 88, 90 *et seq.*, 106, 190, 245
Nitromannite, 232
Noirmontier, 115
Nori, 98 *et seq.*, 246, 269, 277
Normandy, 31
North America, 9, 62, 68, 142
North Atlantic, 2
North Carolina, 142, 168, 175, 190, 260
North Ronaldsay, 30
North Sea, 256, 259
Northumberland Strait, 254
Norway, 30 *et seq.*, 52, 67, 202, 204, 207, 229, 256, 257
Nova Scotia, 31, 67, 68, 117, 118, 231, 254

Oarweeds, 3, 4, 6
Odessa, 141
Ogonori, 84
Oman, 264
Omori, 99
Onigusa, 150, 153
Ontario, 31
Oppositol, 236
Orkney Islands, 18, 30, 238
Osaka, 72, 146, 150

Pacific Kelp Mulch Co., 26, 42
Pacific Grove, 46
Pacific Ocean, 277
Paint(s), 143, 193, 215, 232
Pakistan, 18
Palk Bay, 265
Pamban, 265
Pantothenic acid, 36, 96
Paper, 192
Paraiba, 263
Patagonia, 264
Pelt, 239
Penicillin, 215
Pentosans, 93, 94
Pepper dulse, 68
Peru, 13, 277
Phaeophyceae, 1, 10, 70
Phaeophytin, 237
Pharmaceuticals, 130, 134, 191, 220, 232, 236
Phenols, 201
Philippines, 64 *et seq.*, 68, 88, 109, 120, 135, 138, 139, 179, 180, 246, 271
Phosphates, 39, 44, 46, 47
Phycolloids, 110, 111, 148, 234
Phyllophoran, 112, 141, 172
Phytoplankton, 1, 248, 250
Pierrefitte Auby, 119
Pigs, 31, 32, 37, 38
Plastics, 215
Point Conception, 273
Point Loma, 272
Point Sur, 25
Polishes, 130, 215

Subject Index

Porphyran, 187
Portugal, 117, 162, 173, 181, 256, 260
Potash, 21, 25, 26, 28, 42, 45, 46, 47
Poultry, 31, 33
Praha, 52
Pratas Is., 234
Prince Edward Is., 117, 118, 119, 254
Priomorye, 61
Protein, 88, 89, 90, 93, 108, 109, 120
Puerto Rico, 171, 260
Puget Sound, 270

Qualidia, 261
Queen Charlotte Is., 270

Raft culture, 266, 269
Rangoon, 88
Rayon, 217
Rhuematism, 233
Rhodesia, 43
Rhode Island, 245
Rhodophyceae, 1, 16, 20, 67, 70
Rhodophyta, 120, 235
Riboflavin, 96, 108
Rio Grande do Norte, 263
Roscoff, 39
Russia, 19, 20, 61, 149, 167, 171, 239

Saaremaa Is., 257
Sado, Isle of, 83
St. Kitts, 62
St. Lawrence, 254
Sakhalin, 160, 172, 270
Salad, 64 *et seq.*
Samoa, 70, 138, 274
San Diego, 28, 46, 60, 273
San-in, 86
Santiago, 71
Sargasso Sea, 261
Saxitoxin, 238
Scabies, 237
Schizuoka, 19, 150
Scilly Isles, 41
Scituate, 115
Scotia Marine Products, 204

Scotland, 18, 19, 30, 31, 41, 63, 68, 256, 258
Scottish Seaweed Res. Ass., 22, 33
Scrofula, 235
Seagro, 43, 44
Sea lettuce, 39, 62, 96
Sea of Azov, 20
Sea of Okhotzk, 270
Seatron, 68
Seattle, 68
Senegal, 142, 261
Sewage, 250, 251, 252
Shampoos, 220
Shaving soap, 133, 193
Sheep, 30, 32, 33, 37, 38
Sherbets, 131, 192, 216
Shetland Is., 238
Shima, 86
Shizuoka, 269
Shoe stain, 193
Sicily, 262
Silk(s), 191, 193, 217
Singapore, 43
Sinkiang, 69
Size, 114, 146, 238
Sizing, 143, 191, 192, 215, 224
Skye, 237
Soap, 193, 220
Soda, 18, 21
Sodium alginate, 212, 213, 218, 219, 221
Soups, 64 *et seq.*, 78
South Africa, 14, 16, 43, 56, 69, 143, 149, 151, 162, 174, 181, 191, 203, 264, 269
South America, 13, 56, 66, 71, 162, 236, 239
South Carolina, 51, 161
Spain, 44, 115, 173, 181, 256, 260
Sprains, 233
Squalene, 237
Sri Lanka, 265
Stauffer Chemicals, 117, 119
Stewart Is., 120
Stipites Laminariae, 234
Sudan, 264

Sumidagawa R., 99
Sugar wrack, 3
Sydney, 203
Syphilis, 236
Syrup, 222
Sweden, 257

Tahiti, 70
Taiwan, 64 *et seq.*, 153, 180, 234, 252, 267
Tanzania, 135, 264
Tasmania, 13, 203, 205, 274
Tenassarim, 88
Tengusa, 149, 153, 183
Thailand, 43, 88
Tobago, 64
Tokyo Bay, 98, 99
Tokyo, 72, 83
Toothpaste, 130, 134
Tremeti Is., 262
Trincomalee, 265
Trinidad, 14, 43, 62, 68
Tsing tao, 79
Tunisia, 263
Turkey, 234
Tuticorin, 265
Tyrrehenian Sea, 262

Ulcers, 133, 134, 236
United Kingdom, 43, 65, 66, 135, 259
USA, 33, 43, 56, 62 *et seq.*, 114, 115, 117, 121, 149, 191, 193, 203, 220
USSR, 65, 68, 207, 256, 270

Vancouver, 68
Vancouver Is., 25, 270
Venezuela, 142, 171
Venice, 262
Vermifuge, 234
Vietnam, 65, 88
Virginia, 240
Vitamin, 33, 37, 38, 95, 96, 97, 108, 109
Vladimir, 172
Vladistock, 20, 172, 270
Voiles, 191
Vulcanite fibre, 215

Wakame, 81, 82, 269, 277
Wales, 98
Wedgeport, 31
West Indies, 56, 62, 63, 66, 68, 193
White Sea, 20, 257
Wines, 193
Wood's Hole, 242, 248, 250, 252

Xylose, 232

Yamaguchi, 19
Yarmouth, 31
Yegonori, 183
Yugoslavia, 262

Zambia, 43
Zernov's Phyllophora Sea, 263
Zelex, 220
Zonarol, 237